C. H. Calisher and D. E. Griffin (eds.)

Emergence and Control of Zoonotic Viral Encephalitides

Archives of Virology
Supplement 18

SpringerWienNewYork

Prof. Charles H. Calisher
Arthropod-borne and Infectious Diseases Laboratory, Department of Microbiology, Immunology and Pathology, College of Veterinary Medicine and Biomedical Sciences, Colorado State University, Fort Collins, U.S.A.

Prof. Diane E. Griffin
Department of Molecular Microbiology and Immunology, Johns Hopkins Bloomberg School of Public Health, Baltimore, U.S.A.

Springer-Verlag Wien New York is a part of Springer Science + Business Media
springeronline.com

Typesetting: Thomson Press (India) Ltd., Chennai

Printed on acid-free and chlorine-free bleached paper
SPIN: 10969762

With 43 (partly coloured) Figures

CIP data applied for

ISBN 3-211-20454-7 (soft cover) Springer-Verlag Wien New York
ISBN 3-211-20455-5 (hard cover) Springer-Verlag Wien New York

Preface

From 6–8 April 2003, more than 50 researchers and administrators from more than a dozen countries attended a symposium entitled "Emergence and control of zoonotic viral encephalitis". Held in a convivial setting at Les Pensières, Veyrier du Lac, near Annecy in the French Alps, this was the fourth in a series of meetings on the emergence and control of infectious diseases organized by the Mérieux Foundation.

This Special Issue of Archives of Virology presents the papers delivered at this symposium. The general objectives of the meeting were to review the biology of viral encephalitis, the virulence and genetic evolution of encephalitis viruses, and the factors involved in emergence of these diseases. Emergence or reemergence of viruses may be due to virus evolution, to the impact and influence of human populations on previously undisturbed ecosystems, or to better recognition. Clearly, if we are to prevent or ameliorate treatment of these diseases we must understand the basic mechanisms by which these viruses emerge or reemerge and cause illnesses. Methods for detection of infections caused by neurotropic viruses and for detection of the viruses or their genome sequences are available and improving and methods for detection of antibody have improved; all are clinically- and epidemiologically-relevant.

Examples of relatively recently recognized viruses causing encephalitis in humans, livestock or wildlife include (family *Paramyxoviridae*, genus *Henipavirus*) Hendra and Nipah viruses (henipaviruses), both of which are neurotropic, and (family *Rhabdoviridae*, genus *Lyssavirus*) Australian bat lyssavirus, also a neurotrope. All three viruses, and others related to them, have been shown to have fruit bats (*Pteropus* spp.) as their natural hosts. Progress is being made in understanding transcription regulation and cell fusion by henipaviruses. In addition, basic epidemiologic procedures and classical prevention strategies have been put into place to prevent infections caused by these viruses.

Since 1988, there has been a worldwide effort to eradicate the non-zoonotic but neurotropic poliomyelitis viruses. The number of cases has been reduced by 99% and the natural occurrence of these viruses now is limited to 7 countries. The system established to conduct surveillance and response may provide a model for use in tracking and controlling viruses causing encephalitis. Long-term studies of ecologic parameters, seasonality, and changing virus and vector prevalences are being used to determine risk factors in various arbovirus infections, including Japanese encephalitis virus in Thailand, and are being applied for prevention and control.

In Russia, where West Nile virus has long been recognized without causing significant disease problems, recent detection of various virus genotypes suggest a mélange of genotypes circulating in various areas and moved between areas by birds. Generation and maintenance of continuous genetic variation may lead to partial protection and escape mutants, which could provide a "pump" that generates more variants and "new" viruses. When adapted to naïve populations of birds, horses, and humans, in the presence of competent arthropod vectors, epidemics may arise and new opportunities for these viruses and virus variants may occur, perhaps such as the incursion of West Nile virus into the New World in 1999. Evidence presented suggests that there has been little genomic variation in New World West Nile virus since its 1999 recognition there; this situation could change quickly, should the virus adapt to a new vector or host.

It is important to continue to make the classical epidemiologic observations that have characterized disease investigations in the previous half-century. However, to understand the overall effects of virus outbreaks, what are needed are denominators. Numerous presentations demonstrated that we are beginning to understand the molecular mechanisms leading to pathogenetic events and further studies may provide information useful for the development of antiviral compounds and candidate vaccines.

Attendees were provided with an overview of various transmission cycles of arboviruses, which are concomitantly diverse in regards to their hosts and vectors. Viral neuroinvasiveness appears to depend on the uniqueness of phylogenetically diverse hosts, their ages, genetic predispositions, and immune statuses, and virus origin, passage level, dose, and other factors – a complex situation to investigate and comprehend. Critical factors impacting neuroinvasiveness and neuronotropism must be coupled to cause encephalitis. Viral mutations may affect the ability of the virus to replicate in cells, altering viral virulence, but it may be specific genomic and polyprotein sequence changes that affect high viremias and replication in neurons that are central to emergence. The extent of the roles of various proteins in virus infections, neuronal involvement, and apoptosis are being recognized. Now we are beginning to understand complex signaling mechanisms, antibody-producing cell types, cytokines, and the cellular responses and the pathways leading to both disease and protection from disease.

Considerable progress has been made in understanding the relationships between genetic and functional diversities, neuronal receptors, transport, and cellular protein-virus interactions, an understanding critical to further insights to neurotropism, pathogenesis, pathogenetic mechanisms, and immunogenicity.

Phylogenetic trees were used to describe the evolution of encephalitic flaviviruses, geographic exclusion, virus persistence, and flaviviral recombination as a mechanism of flaviviral evolution. In addition, data were presented that illustrated persistence of and immune modulation by alphaviruses, which, in concert, allow the virus to replicate while preventing the host from responding to its benefit.

Control must be based on rapid recognition of early cases and subsequent immunization of people or animals at-risk or immunization of people or animals with the potential to become at-risk, such as travelers, laboratory personnel, and attending clinicians. Attendees were told of diverse methods being used to develop vaccines. Representatives from W.H.O. explained that organization's plans for responses to disease emergence and for preventing zoonotic diseases from reaching human populations.

In summary, it is clear that new paradigms for field studies of zoonotic diseases are necessary and that these approaches must include longitudinal and in-depth investigations of agent, host, habitat, and environment if we are to predict risk and respond in an appropriate manner. At this time zoonotic disease control comprises prevention and public education and not much more. It is encouraging that progress is being made in rapid diagnosis, production of sophisticated vaccines, and in understanding the molecular mechanisms by which zoonotic viruses persist and cause disease.

We are indebted to many people for making this meeting and Special Issue possible, including the members of the Scientific Committee who helped us devise the scientific program (Jean-Christophe Audonnet, Vincent Deubel, Franz Heinz, John Mackenzie, and Jean-François Saluzzo) and all the experts who participated in the meeting. We are grateful to personnel of the Merieux Foundation, especially Marissa Vicari for her assistance in the scientific program and for collecting manuscripts, and also Marlene Crozet and Jeannine Camu, who assisted with general logistics and in making the attendees comfortable. We thank Aventis Pasteur for its support.

Charles H. Calisher
Diane Griffin
Betty Dodet

Contents

Listed in Current Contents

Arch Virol (2004) [Suppl] 18: 1–11

Archives of Virology

Zoonotic viruses of wildlife: hither from yon

J. E. Childs*

Viral and Rickettsial Zoonoses Branch, National Center for Infectious Diseases, Centers for Disease Control and Prevention, Atlanta, Georgia, U.S.A.

Summary. The emergence of zoonotic viruses maintained by wildlife reservoir hosts is poorly understood. Recent discoveries of Hendra (HENV) and Nipah (NIPV) viruses in Australasia and the emergence of epidemic West Nile virus (WNV) in the United States have added urgency to the study of cross-species transmission.

The processes by which zoonotic viruses are transmitted and infect other species are examined as four transitions. Two of these, inter-species contact and cross-species virus transmission (spillover), are essential and sufficient to cause epidemic emergence. Sustained transmission and virus adaptation within the spillover host are transitions not required for virus emergence, but determine the magnitude and scope of subsequent disease outbreaks. Ecologic, anthropogenic, and evolutionary factors modify the probability that viruses complete or move through transitions. As surveillance for wildlife diseases is rare and often outbreak-driven, targeted studies are required to elucidate the means by which important zoonotic viruses are maintained and spillover occurs.

Introduction

Human encephalitides caused by zoonotic viruses are unfortunate, but incidental occurrences. With rare exceptions, humans do not contribute to the maintenance of viruses circulating among wildlife populations; viruses may be repeatedly introduced, but most only briefly colonize the human host.

An appreciation that specific human-animal interactions – a bite received from a deranged dog – preceded the onset of certain diseases – a progressive delirium followed by death – was chronicled in ancient texts from Mesopotamia and China; these unmistakable descriptions of rabies encephalitis indicate some humans were quite familiar with zoonotic viruses by the second millennium B.C. [1]. A renewed

*Current address: Department of Biology and Program in Population Biology and Evolutionary Ecology, Emory University, 1510 Clifton Road, Atlanta, Georgia 30345, USA

interest in zoonotic viruses maintained by wildlife was kindled by near-coincident discoveries of novel viruses, HENV, NIPV, and a bat-associated lyssavirus [22], causing fatal human encephalitis in Australasia, and by the forceful reminder provided by WNV in the United States [20] of how rapidly exotic vector-borne viruses can spread when introduced into receptive new environments. Interest has led to assessment. What is our core understanding of how zoonotic viruses are maintained in wildlife? How and when do these viruses cross species barriers? Where and why do certain viruses emerge on a global scale? Assessment has led to new public health activities and research, but glaring deficiencies in our global capacity to detect outbreaks and to diagnose zoonotic diseases have been exposed. Efforts to improve detection and diagnosis of human and domestic animal diseases are important, but alone they will contribute little to the broader objectives to elucidate natural maintenance cycles and to limit future cross-species transmission of zoonotic viruses. Although there are effective vaccines that protect against zoonotic encephalitides (e.g. rabies and Japanese encephalitis [JE]), developing additional new vaccines and identifying human populations to vaccinate remain daunting challenges. Ultimately, in the effort to reduce the risk of human exposure, most prevention and control programs for zoonotic diseases are predicated upon knowledge of wildlife reservoirs and arthropod vectors.

Transitions: wildlife infection to human zoonosis

Emergence of zoonotic viruses of wildlife is a complex and poorly understood process. Several loosely related phenomena are typically grouped under the rubric of "emerging disease": new diseases caused by new pathogens, increased disease incidence of pre-existing pathogens, or diagnosis of extant disease caused by a pre-existing but previously undetected pathogen [16]. Herein, the focus is on two transitions essential for successful cross-species introduction of a zoonotic virus and two other transitions influencing the magnitude and scope (e.g., sporadic, epidemic, or pandemic) of emergence. In an analogous manner, invasion biologists differentiate the transitions required of exotic species successfully overcoming dispersal barriers such as transport, introduction, establishment, and invasion

Fig. 1. The emergence of a zoonotic virus maintained by a reservoir host (H_R; top gray box) requires transmission (solid black arrow) to a secondary host population (H_S; lower gray box), which may exist as local, partially isolated communities or demes (graded dark cone within the H_S population box). The process of emergence is partitioned into four transitions (left vertical scale); two, contact and spillover (cross-species transmission) are required for all emerging viruses, and reiterations of virus through these transitions (indicated and marked as the long solid arrow) is sufficient for emergence (e.g., WNV and rabies virus). Transitions to sustained intra- H_S transmission (broken solid line leading to emergence), and virus spread and possible adaptation to H_S, establish limits to the scope and magnitude of virus emergence (sporadic, epidemic, pandemic). Broken lines leading outside of population boxes indicate failure to complete the transition or local extinction. Examples of factors that modify the probability of a virus completing a transition are listed to the right

[17]. An important conceptual distinction is that successive completion of each transition (i.e. a strict transition hierarchy) is not required for a zoonotic virus to emerge.

The two essential transitions for virus emergence are: 1) contact between infectious products (e.g., secretions, tissues, vectors, fomites, etc.) generated by a reservoir host (H_R) and susceptible individuals of the secondary host (H_S) and 2) cross-species transmission of virus from H_R to H_S, defined by introduction, replication, and release of virus from the H_S (Fig. 1). Recurrence of these

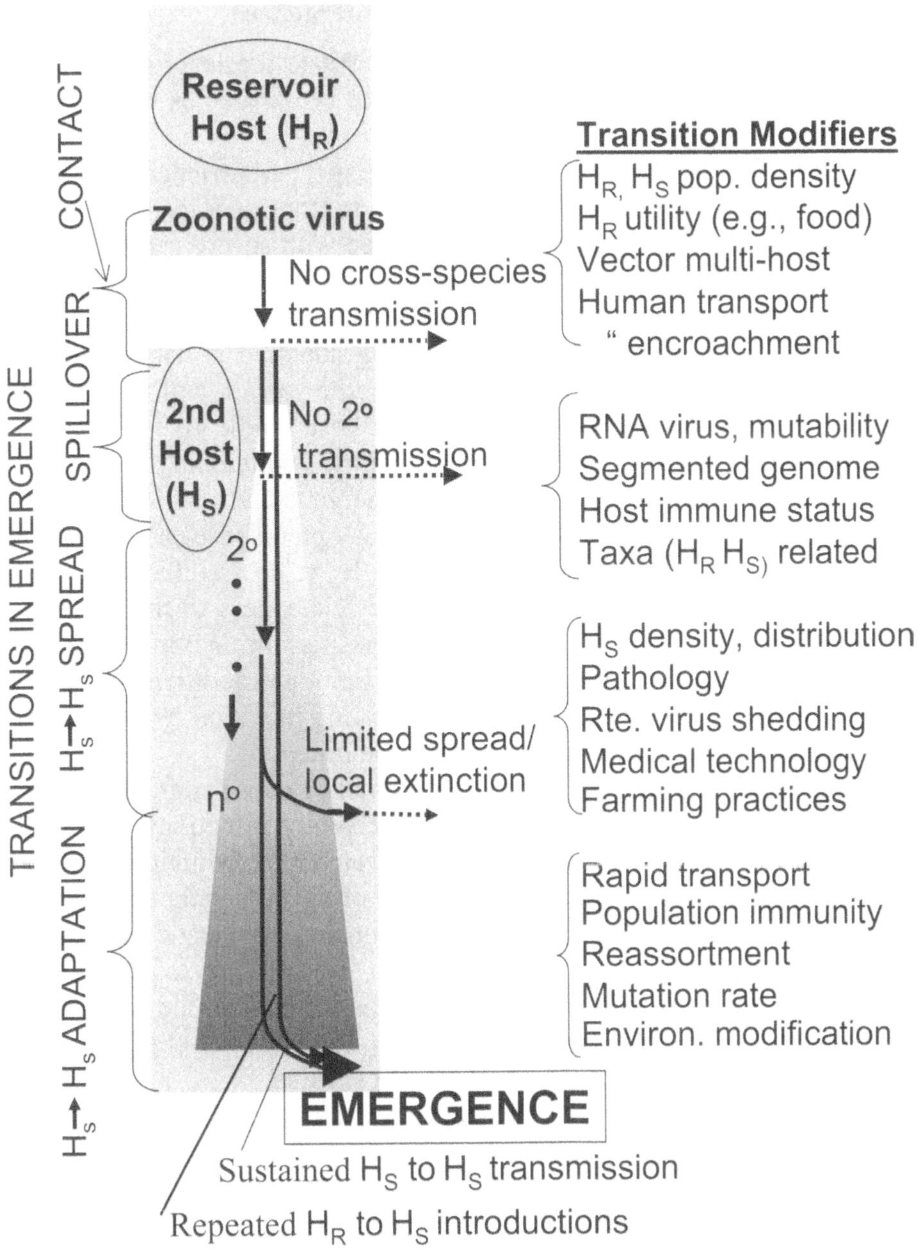

two transitions is sufficient to cause epidemic emergence (e.g. WNV and rabies virus).

The two transitions establishing the scope and magnitude of emergence are 3) sustained intra-H_S transmission of virus and: 4) subsequent spread of virus among the H_S with or without adaptations indicating H_S-specific evolution. These two transitions are the likely prerequisites of pandemic emergence [25], and the processes by which viruses are introduced and by which sustained intra-H_S transmission occur require differentiation [26].

Increasing the odds for emergence

Aspects of the virus, host, and environment modulate the level of opportunity afforded to hosts and pathogens to mix, produce cross-species infections, and generate disease (Fig. 1).

The profile of a virus successfully crossing the species barrier and establishing sustained intra-H_S transmission might include 1) an RNA genome providing high mutability (N.B. the novel coronavirus associated with severe acute respiratory disease (SARS) provides a notable exception [19]), 2) a genome segmentation permissive to reassortment upon superinfection, 3) a genetic predisposition to recombine, 4) a high replication rate providing adaptive potential, 5) a high transmission rate achieved by direct or vector-borne spread, and 6) virus maintenance among multiple-R_Hs (multiple reservoir hosts) indicative of high innate invasiveness [4, 6]. These characteristics reflect the evolutionary history of the R_H and the virus. However, in the process of emergence, virus life-history traits derived in an evolutionary context now remote may predispose success in a H_S.

Characters modifying the susceptibility of a H_S to cross-infection at the individual or population level include; taxonomic relatedness to the H_R; genetic background (e.g. MHC) and constitutive factors permitting virus attachment, uptake, replication, and release from H_S cells, immune modifiers such as age, stress, and nutritional status, and previous or existing infection by antigenically-related viruses or by pathogens affecting immune functions.

Most zoonotic viruses crossing the species barrier cause sporadic disease or self-limited local outbreaks in a H_S. When secondary infections produced by spillover events fail to replace themselves by repeated reintroduction or sustained intra-H_S transmission (i.e., the reproduction rate of the pathogen, R_0, is less than 1), epidemics are self-limited. Each case of human or domestic animal rabies typically results from a new introduction of virus from an infected, terminally ill animal; as rabies removes individuals from the H_R population, the corresponding risk of rabies spillover to H_S may decline [2, 34]. In temperate climes epidemics of West Nile fever decline with the onset of cold weather.

Once cross-species transmission has occurred, anthropogenic and ecological factors, such as rapid transport, agricultural-farming practices, medical technologies, and habitat encroachment, modification and degradation become the classes of transition modifiers most frequently cited as leading to emergence [8, 26, 30]. However, other factors also are crucial. The tissue tropism and pathogenesis of infections in the H_R and H_S influences the likelihood of virus spillover and subsequent intra-H_S transmission. Differences in the pathology and pathogenesis of NIPV in pigs (H_{S1}; first secondary host) and HENV in horses greatly influenced the transition to sustained intra-H_{S1} transmission of NIPV, as described below. When assessing the pandemic threat posed by an emerging virus, the potential to cause respiratory infection with airborne spread is uniquely important [25].

Ecological and environmental factors act at each transition, modifying the likelihood of species contact, influencing the potential for cross-species virus transmission, and enhancing or retarding disease spread. Seasonal migrations of birds introduce WNV [23] and alter the population densities of H_Rs. Warmer temperatures may shorten the prerequisite extrinsic incubation period of arboviruses in vectors. Local rainfall or snowmelt influence vector breeding success and global climatic factors, such as the El Niño Southern Oscillation [21], affect regional weather patterns modifying species interactions at different trophic levels [10]. Droughts may force aggregation of wildlife reservoir hosts and vectors, enhancing opportunities for virus transmission [31].

The emergence of NIPV in Malaysia illustrates the interplay of essential transitions and modifying factors. Anthropogenic factors have been credited as the primary elements in NIPV emergence; deforestation and pig farming practices established high-density aggregates of pigs interspersed within habitat frequented by flying foxes (*Pteropus* spp.). Humans disseminated NIPV-infected pigs between local communities and to international markets [8]. These factors greatly increased potential contact between bats (H_R) and pigs (H_S), between infected and susceptible pigs, and between infected pigs (H_{S1}) and humans (H_{S2}; second secondary host). However, successful NIPV transitions to spillover and sustained intra-H_{S1} transmission were the antecedents magnified by subsequent human activities.

Although the molecular basis is yet to be established, the rapid transitions posted by NIPV presumably were mediated by genetic endowments enhancing virus capacity to cross species-barriers. A larger genome, dissimilar from other members of the virus family *Paramyxoviridae*, warranted placement of HENV and NIPV within a new genus, *Henipavirus*. The ability of these viruses to colonize new hosts, as attested to by NIPV transmission to five mammalian Orders (Chiroptera, Carnivora, Perissodactyla, Artiodactyla, and Primata), also appears unique [33]. In pigs, NIPV infection can cause a severe, sometimes fatal, respiratory disease and also meningoencephalitis [14]. The initial tissue tropism of NIPV in pigs was probably endothelial tissue, with subsequent dissemination or further adaptation of the virus to replicate in neurons and respiratory epithelium [14]. Respiratory secretions and possibly expired air from infected pigs contained NIPV [14]; airborne dissemination of infectious material was enhanced through the violent and distinctive cough that developed in some infected pigs. The respiratory route of NIPV transmission facilitated sustained H_{S1} to H_{S1} transmission but also established pigs as a secondary reservoir or amplifying host mediating H_{S1} to H_{S2} transmission of NIPV to humans and other animals [12].

Surveillance for infection and disease among wildlife: still a pipedream?

Every animal diagnosed as diseased has been so only after multiple interactions with humans. Locating diseased or dead wildlife is notoriously difficult and, except in rare circumstances (e.g., rabies and recent dead bird surveillance for WNV), rarely coordinated as a national surveillance activity. Most efforts to investigate wildlife H_Rs and prevalences of viral infection are highly focused, short-lived, and outbreak driven; these efforts rarely build the infrastructure for sustained activity. Surveillance and laboratory capacity for monitoring and diagnosing wildlife disease (or infection) is inadequate and, at best, poorly integrated with other surveillance activities (see [9]). The hurdles and prohibitive costs of instituting any formal national wildlife disease surveillance effort indicate that targeted research initiatives remain the only viable option to investigate the complexities of zoonotic virus maintenance in wildlife. Long-term studies are essential for studying zoonotic viruses of wildlife because the dynamics of infection, transmission, and spillover are influenced by the population dynamics of the hosts, vectors, and pathogens and by habitat, environmental factors and more [24, 34]. Because of the demands on resources, not every zoonotic virus will receive attention. Although it is a difficult task, identifying priorities for collaborative study and securing national and international support for sustainable efforts should be the goal of focused studies of zoonotic diseases and disease emergence.

Priorities

Elucidating maintenance cycles, establishing the incidence And prevalence of viral infections among wildlife reservoir hosts, documenting and explaining temporal fluctuations of infection, and assessing the comparative pathogenesis of infection and disease among relevant species, are vitally important for our understanding, and potential forecasting, of the risk and consequences of zoonotic virus spillover to humans. Obtaining information of sufficient detail and quality to satisfactorily address any one of these areas is difficult but certain zoonotic viruses or classes of virus require these efforts. I suggest and then briefly illustrate three categories of zoonotic virus deserving immediate attention; 1) zoonotic viruses of a type with an established history of pandemic emergence; 2) zoonotic viruses with a strong potential for pandemic emergence, most notably those transmitted by the respiratory route, and; 3) vector-borne zoonotic viruses with highly complex maintenance cycles, for which detailed investigations of a single system should yield results of broad significance and applicability to other systems.

Emergence is an ongoing process. Minimally, simian immunodeficiency viruses have been introduced and have established sustained inter-human transmission on at least eight independent occasions [11, 28]. It is nearly certain that other introductions of simian immunodeficiency viruses to human communities have gone locally extinct, failed to spread, or remain undetected. The extant human immunodeficiency viruses represent genetic lineages derived from only two of the >30 simian immunodeficiency viruses circulating among non-human primates in Africa [28]. We know little about the prevalence of simian immunodeficiency viruses among different H_Rs, contact rates among reservoir and secondary hosts, or the ability of different simian immunodeficiency viruses to infect and replicate in human cells. Even if access to isolated human populations in West Africa were achievable, detection of novel viruses circulating in human communities would prove difficult. In addition, simian immunodeficiency viruses of the same or even different subtype, isolated from different primate species, can recombine *in vivo* and in test tube [13]. Recombination between simian immunodeficiency viruses within primate H_Rs, or possibly between human immunodeficiency virus and a simian immunodeficiency virus in co-infected humans, could result in genetic recombinants with novel phenotypes. Given our experiences with the human immunodeficiency viruses, any effort that prevents, limits, or even delays the emergence of another human-adapted simian immunodeficiency virus would be a crowning public health achievement and should be a goal.

Other zoonotic viruses of high priority for study are those with the potential to cause pandemic human or animal disease. Consider NIPV (or the coronavirus associated with SARS) with unusual and documented capabilities to cross species boundaries. NIPV causes respiratory infection in different secondary hosts and can be transmitted by airborne routes. Although human-to-human transmission of NIPV was not documented, the pathogenesis of disease in pigs and humans appears quite similar [14]; recovery of NIPV from respiratory secretions of infected humans suggests that airborne human-to-human transmission is epidemiologically plausible [3]. Little is known about the natural history and behavioral ecology of the pteropid bats implicated as reservoir hosts for NIPV and HENV [8] or the range of amplifying hosts susceptible to spillover. It would be premature to state that pteropid bats are the sole reservoir hosts for HENV or NIPV. A broad effort to investigate the natural history of these zoonotic viruses is not only justified but essential. As additional species of pteropid bats and other mammals will be found to host new paramyxoviruses, studies with complementary objectives in Asia and Africa are warranted. Recruiting specialists from disciplines with limited precedence for interaction (e.g., virologists and mammalogists) will be vital to these efforts.

Identification of a single or principal reservoir host for many zoonotic pathogens is difficult or impossible [12]. Consider WNV with >160 species of birds and 37 species of mosquitoes found infected in the United States since 1999 (J. Roehrig, CDC, personal communication, 2002). Developing the matrix of outcomes between potential vertebrate H_Rs and vectors implicated in WNV maintenance is neither feasible nor necessary. Establishing a construct for WNV transmission that uses knowledge or estimates of the relative capacities of species to serve as competent H_Rs (e.g., develop high viremia, sufficient population density, attractiveness for vectors, etc.) and vectors (e.g., disseminate and transmit WNV, blood meal-host preferences, etc.) may identify broad gaps in our knowledge of arbovirus maintenance. As an example, increases in the population size of a single species or changes in the diversity of avian species could lead to contradictory effects on epidemic spillover of WNV [27]. Epidemic enhancement could result if changes increase the density of WNV-infected blood meals available to competent mosquito vectors. However, not all avian species develop a WNV viremia sufficient to infect mosquito vectors [18]. If the avian fauna changes and the proportion of birds attractive to mosquitoes but variably resistant to developing West Nile virus viremia increases, then the potential for epidemics may diminish [6], even as overall bird density increases. As most quantitative models of host-pathogen interactions and population dynamics involve a single H_R and a single pathogen, data accrued from WNV studies could help extend theory and provide data for model assessments.

Forecasting epizootics and extending data

Sentinel animal-based surveillance systems, such as pigs for JEV or NIPV and sentinel bird flocks for St. Louis encephalitis virus and Western equine encephalitis virus, can serve as important indicators of enzootic virus activity. However, for sentinel species to provide an early warning of impending virus spillover to humans requires distinguishing the signal of increased viral transmission (to "epizootic" levels) from the expected enzootic background; long term sampling of sentinels to collect the necessary time series data is uncommon [29, 31].

Epizootics precipitated directly by environmental triggers (e.g., rainfall influences mosquito breeding influences epidemic arboviral encephalitides) or indirectly via complex trophic cascades have been modeled to forecast virus spillover to humans. Epidemiologic models may be useful in determining the risk for human disease, even when causal chains of events are imperfectly understood [10]. Model development complements, but does not replace, the need for prospective, long-term studies to quantify how variation in vector and H_R populations influence the risk of virus spillover [24].

Linking the risk of virus spillover to predictable but dynamic changes in numbers of infected individuals in a wildlife H_R population is an attractive approach, as external triggers are not required and counts of diseased individuals are easier to obtain than are population denominators. The temporal dynamics of epizootic rabies among raccoons (H_R) has been used to predict the risk of rabies virus spillover to domestic cats (H_S), the principal domestic animal diagnosed with rabies in the United States [34]. Dead bird surveillance for WNV infection may provide early warning of potential spillover to humans; however, the utility of these efforts as a public health tool is still being assessed [7].

Wildlife diseases and conservation

There is a growing recognition among wildlife and conservation biologists that zoonotic diseases pose an immediate and important threat [4–6]. Cross-species transmission of exotic viruses may be the final insult that drives some endangered species to extinction. Recent epidemics of phocine distemper among harbour seals in the North Sea, canine distemper and rabies among African carnivores (see [15]), and Ebola disease among lowland gorillas in the Republic of the Congo [32], illustrate the devastating impact of viral diseases on certain wildlife populations.

Quantitative methods for modeling host-pathogen interactions were largely developed by ecologists interested in predation and competition. Interest in this area of research has grown dramatically and strong advocacy exists to extend these approaches to emerging diseases of wildlife and to model multi-host-pathogen systems [15]. Resources to conduct long-term, field-based studies of wildlife reservoirs of zoonotic viruses are essential, but not assured; where overlapping interests exist, scattered resources may be leveraged by collaborations among microbiologists, wildlife ecologists, and conservation biologists [32]. The study of wildlife species can provide important information on how, when, and which types of pathogens are most likely to emerge. A more anthropocentric rationale is obvious. Wildlife can serve as sentinel species for zoonotic viruses capable of spillover to humans, and prospective studies of these existing model systems provide opportunities that cannot be duplicated at any cost. If the causes of wildlife conservation and public health are promoted by such efforts, we would be doing well by doing good.

Acknowledgments

I thank the Mérieux Foundation for the invitation to participate in the symposium on Emergence and Control of Zoonotic Encephalitis. Reviews by Drs. Charlie Calisher and Elizabeth Gordon improved this paper.

References

1. Baer GM, Neville J, Turner GS (1996) Rabbis and rabies. Laboratorios Baer, Mexico City
2. Childs JE, Curns AT, Dey ME, Real LA, Feinstein L, Bjornstad ON, Krebs JW (2000) Predicting the local dynamics of epizootic rabies among raccoons in the United States. Proc Natl Acad Sci USA 97: 13666–13671
3. Chua KB, Lam SK, Goh KJ, Hooi PS, Ksiazek TG, Kamarulzaman A, Olson J, Tan CT (2001) The presence of Nipah virus in respiratory secretions and urine of patients during an outbreak of Nipah virus encephalitis in Malaysia. J. Infect 42: 40–43
4. Cleaveland S, Laurenson MK, Taylor LH (2001) Diseases of humans and their domestic mammals: pathogen characteristics, host range and the risk of emergence. Philos Trans R Soc Lond B Biol Sci 356: 991–999
5. Daszak P, Cunningham AA, Hyatt AD (2000) Wildlife ecology – Emerging infectious diseases of wildlife – Threats to biodiversity and human health. Science 287: 443–449
6. Dobson A, Foufopoulos J (2001) Emerging infectious pathogens of wildlife. Philos Trans R Soc Lond B Biol Sci 356: 1001–1012
7. Eidson M, Kramer L, Stone W, Hagiwara Y, Schmit K (2001) Dead bird surveillance as an early warning system for West Nile virus. Emerg Infect Dis 7: 7631–7635
8. Field H, Young P, Yob JM, Mills J, Hall L, Mackenzie J (2001) The natural history of Hendra and Nipah viruses. Microbes Infect 3: 307–314
9. Friend M, McLean RG (2002) The role of native birds and other wildlife on the emergence of zoonotic diseases. In: The emergence of zoonotic diseases: understanding

the impact on animal and human health (Workshop Summary, ed). National Academy Press, Washington DC, pp 52–58
10. Glass GE, Yates TL, Fine JB, Shields TM, Kendall JB, Hope AG, Parmenter CA, Peters CJ, Ksiazek TG, Li CS, Patz JA, Mills JN (2002) Satellite imagery characterizes local animal reservoir populations of Sin Nombre virus in the southwestern United States. Proc Natl Acad Sci USA 99: 16817–16822
11. Hahn BH, Shaw GM, De Cock KM, Sharp PM (2000) AIDS as a zoonosis: scientific and public health implications. Science 287: 607–614
12. Haydon DT, Cleaveland S, Taylor LH, Laurenson MK (2002) Identifying reservoirs of infection: a conceptual and practical challenge. Emerg Infect Dis 8: 1468–1473
13. Holmes EC (2001) On the origin and evolution of the human immunodeficiency virus (HIV). Biol Rev Camb Philos Soc 76: 239–254
14. Hooper P, Zaki S, Daniels P, Middleton D (2001) Comparative pathology of the diseases caused by Hendra and Nipah viruses. Microbes Infect 3: 315–322
15. Hudson PJ, Rizzoli A, Grenfell BT, Heesterbeek H, Dobson AP (eds) (2002) The Ecology of wildlife diseases. Oxford University Press, Oxford
16. Institute of Medicine (1992) Emerging infections: microbial threats to health in the United States. National Academy Press, Washington DC
17. Kolar CS, Lodge DM (2001) Progress in invasion biology: predicting invaders. Trends Ecol Evol 16: 199–204
18. Komar N, Langevin S, Hinten S, Nemeth N, Edwards E, Hettler D, Davis B, Bowen R, Bunning M (2003) Experimental infection of North American birds with the New York 1999 strain of West Nile virus. Emerg Infect Dis 9: 311–322
19. Ksiazek TG, Erdman D, Goldsmith CS, Zaki SR, Peret T, Emery S, Tong S, Urbani C, Comer JA, Lim W, Rollin PE, Dowell SF, Ling AE, Humphrey CD, Shieh WJ, Guarner J, Paddock CD, Rota P, Fields B, DeRisi J, Yang JY, Cox N, Hughes JM, LeDuc JW, Bellini WJ, Anderson LJ (2003) A novel coronavirus associated with severe acute respiratory syndrome. N Engl J Med 348: 1953–1966
20. Lanciotti RS, Roehrig JT, Deubel V, Smith J, Parker M, Steele K, Crise B, Volpe KE, Crabtree MB, Scherret JH, Hall RA, Mackenzie JS, Cropp CB, Panigrahy B, Ostlund E, Schmitt B, Malkinson M, Banet C, Weissman J, Komar N, Savage HM, Stone W, McNamara T, Gubler DJ (1999) Origin of the West Nile virus responsible for an outbreak of encephalitis in the northeastern United States. Science 286: 2333–2337
21. Linthicum KJ, An yamba A, Tucker CJ, Kelley PW, Myers MF, Peters CJ (1999) Climate and satellite indicators to forecast Rift Valley fever epidemics in Kenya. Science 285: 397–400
22. Mackenzie JS, Chua KB, Daniels PW, Eaton BT, Field HE, Hall RA, Halpin K, Johansen CA, Kirkland PD, Lam SK, McMinn P, Nisbet DJ, Paru R, Pyke AT, Ritchie SA, Siba P, Smith DW, Smith GA, van den Hurk AF, Wang LF, Williams DT (2001) Emerging viral diseases of Southeast Asia and the Western Pacific. Emerg Infect Dis 7: 497–504
23. Malkinson M, Banet C, Weisman Y, Pokamunski S, King R, Drouet MT, Deubel V (2002) Introduction of West Nile virus in the Middle East by migrating white storks. Emerg Infect Dis 8: 392–397
24. Mills JN, Childs JE (1998) Ecological studies of rodent reservoirs: their relevance for human health. Emerg Infect Dis 4: 529–537
25. Mims CA (1995) Virology research and virulent human pandemics. Epidemiol Infect 115: 377–386
26. Morse SS (1995) Factors in the emergence of infectious diseases. Emerg Infect Dis 1: 7–15

27. Ostfeld RS, Keesing F (2000) Biodiversity and disease risk: the case of Lyme disease. Conserv Biol 14: 722–728
28. Peeters M, Courgnaud V, Abela B, Auzel P, Pourrut X, Bibollet-Ruche F, Loul S, Liegeois F, Butel C, Koulagna D, Mpoudi-Ngole E, Shaw GM, Hahn BH, Delaporte E (2002) Risk to human health from a plethora of simian immunodeficiency viruses in primate bushmeat. Emerg Infect Dis 8: 451–457
29. Reeves WC (1990) Epidemiology and control of mosquito-borne arboviruses in California. California Mosquito and Vector Control Association, Inc., Sacramento, 1943–1987
30. Schrag SJ, Wiener P (1995) Emerging infectious disease: what are the relative roles of ecology and evolution? Trends Ecol Evol 10: 319–324
31. Shaman J, Day JF, Stieglitz M (2002) Drought-induced amplification of Saint Louis encephalitis virus, Florida. Emerg Infect Dis 8: 575–580
32. Walsh PD, Abernethy KA, Bermejo M, Beyers R, De Wachter P, Akou ME, Huijbregts B, Mambounga DI, Toham AK, Kilbourn AM, Lahm SA, Latour S, Maisels F, Mbina C, Mihindou Y, Obiang SN, Effa EN, Starkey MP, Telfer P, Thibault M, Tutin CE, White LJ, Wilkie DS (2003) Catastrophic ape decline in Western equatorial Africa. Nature 422: 611–614
33. Wang L, Harcourt BH, Yu M, Tamin A, Rota PA, Bellini WJ, Eaton BT (2001) Molecular biology of Hendra and Nipah viruses. Microbes Infect 3: 279–287
34. Gordon ER, Curns AT, Krebs JW, Rupprecht CE, Real LA, Childs JE. Temporal dynamics of rabies in a wildlife host and the risk of interspecific transmission. Epidemiol Infect (in press).

Author's address: James E. Childs, Viral and Rickettsial Zoonoses Branch, National Center for Infectious Diseases, Centers for Disease Control and Prevention, MS-G13, 1600 Clifton Road, Atlanta, GA 30333, U.S.A.; e-mail: jameschilds@comcast.net

Arch Virol (2004) [Suppl] 18: 13–20

Archives of
Virology

The role of surveillance in polio eradication and identification of emerging viral encephalitis

E. De Gourville and **W. R. Dowdle**

Task Force for Child Survival and Development, Decatur, GA, U.S.A.

Summary. In 1988, when the World Health Assembly resolved to eradicate paralytic poliomyelitis, polio was endemic in 125 countries on 5 continents with an estimated 350,000 cases annually. By 2002, the number of countries was reduced to 7 and the number of cases by greater than 99%. Instrumental in this extraordinary progress is the timely detection and investigation of all cases of acute flaccid paralysis and the examination of stool samples in an accredited WHO Global Network laboratory. The Network consists of 124 National Poliovirus Laboratories, 15 Regional Reference Laboratories, and 7 Global Specialized Laboratories. Network Laboratories are held accountable to rigid performance standards for quality and timeliness in testing more than 60,000 stool samples annually. The Network is a hierarchical system in which polioviruses are isolated and identified in National Laboratories, differentiated as to wild or vaccine origin in Regional Reference Laboratories, and sequenced in Specialized Laboratories. Findings are promptly relayed to Regional and national program managers for immunization strategic planning, monitoring surveillance quality, and assessing eradication progress. Lessons from the Polio Laboratory Network demonstrate the value of a central coordinating body, an effective communication infrastructure, and full partnerships with peer epidemiology and medical sectors. Post-eradication polio surveillance will continue for many years. The Network legacy for other public health initiatives is access to an existing laboratory infrastructure and human resources with the proven ability to achieve technology transfer and quality laboratory performance even in resource-poor countries.

Introduction

Polio eradication is the largest global public health initiative ever undertaken. It began in 1985 as a Regional initiative in the Americas [4] and was adopted in 1988 as a global initiative by the World Health Assembly [5]. At the time of the WHA resolution, paralytic poliomyelitis was endemic in 125 countries on 5 continents with an estimated 350,000 cases annually [6]. By 2002, the number of countries was reduced to 7 and the number of cases by greater than 99% [7]. This remarkable progress was achieved by countries with the help of extraordinary public/private partnerships, spearheaded by the World Health Organization (WHO), the US Centers for Disease Control and Prevention (CDC), UNICEF, and Rotary International and supported by multiple funding partners. In the year 2000 alone, an estimated 10 million people, mostly volunteers, many from Rotary Clubs, immunized 550 million children in 85 countries.

The basic strategy of polio eradication consists of: 1) high routine immunization coverage of infants with oral polio vaccine (OPV), 2) supplementary OPV immunization through National Immunization Days (NIDs) and Sub-national Immunization Days (SNIDs), 3) sensitive field and laboratory surveillance for poliovirus, and 4) targeted door-to-door (mop-up) administration of OPV in areas of focal transmission [8]. In this manuscript we describe the poliovirus surveillance strategy and the Global Poliovirus Laboratory Network, including its structure and function, systems of communication, accreditation, and support, and potential for surveillance of other infectious diseases.

Poliovirus surveillance

Poliovirus surveillance provides the information for programmatic action by identifying outbreaks, assessing the effectiveness of immunization strategies, and guiding efforts toward global eradication. The benchmark for poliovirus surveillance is the detection, reporting, clinical investigation, and virologic testing of stool samples from all cases of acute flaccid paralysis (AFP). AFP includes Guillian-Barre syndrome, transverse myelitis, and paralysis associated with poliovirus as well as with non-polio enteroviruses [1]. Because AFP has limited specificity, virologic analysis of clinical specimens is required to assign a poliovirus etiology. AFP surveillance also has limited sensitivity, as less than 0.5% of poliovirus infections cause acute paralysis. However, the characteristic poliomyelitis epidemiologic pattern in a susceptible population makes AFP surveillance highly effective over time. Where AFP surveillance may not be fully functioning in endemic or recently endemic countries, supplemental surveillance activities may include sampling contacts of AFP cases, stool surveys of healthy children, or sampling sewage. Enterovirus and environmental surveillance are used in historically polio free developed countries in the absence of active AFP surveillance.

The AFP surveillance standard for all endemic countries is the detection and investigation of at least 1 non-polio AFP case per 100,000 persons <15 years of age. Cases must be investigated within 14 days of onset. Two stool samples must be collected 24–48 hours apart and transported to the laboratory within three days. Meeting these high performance standards requires an active partnership of public health workers, clinicians, epidemiologists, and virologists.

Structure and functions of the network

Key to sensitive surveillance is a global network of high quality poliovirus laboratories capable of timely detection and characterization of wild virus when and where it occurs [2]. The goal of the Network is to have a full global investigative capacity using the fewest number of laboratories to reduce operational costs, ensure good management, and maintain high performance quality and reliability. However, appointments of additional laboratories were often necessary to ensure laboratory capacity in areas with special geographic or geopolitical considerations. The majority of Laboratories are affiliated with governmental institutions to encourage national responsibility for the basic facility and staff and to assure that facilities improvement, new technologies, and staff competence provide lasting benefit to the country.

The Network consists of 124 National Laboratories, 15 Regional Reference Laboratories, and 7 Specialized Reference Laboratories (Fig. 1). It is coordinated from WHO Headquarters, Geneva, through Laboratory Coordinators in each of the six WHO Regions (Fig. 2). Each level of laboratories in this hierarchical system has specific responsibilities. Laboratories with technical capabilities and capacity to perform at the next highest level are encouraged to do so, with the provision that the flow of test materials and information is uninterrupted.

The National Poliovirus Laboratories are at the heart of the system. They work closely with clinicians, epidemiologists and immunization program managers

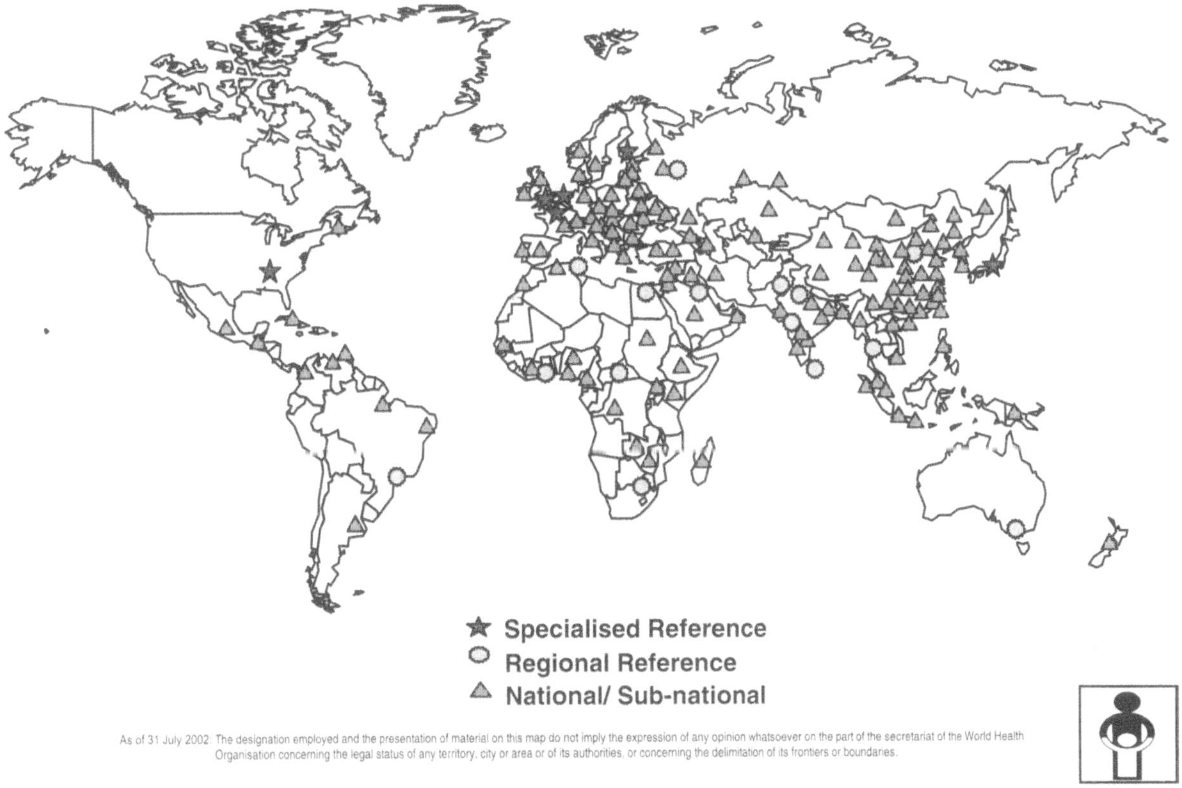

Fig. 1. Global laboratory network for polio eradication, 2002

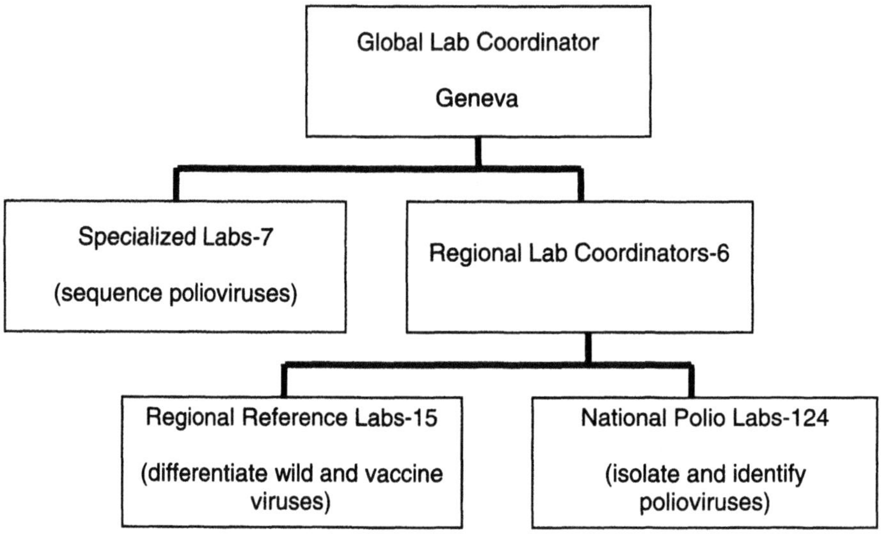

Fig. 2. Organization of polio laboratory network

to investigate cases of AFP through the examination of stool samples, using highly sensitive techniques for poliovirus isolation and serologic methods for identification. The National Laboratories are required to report 80% or more of test results to the program in 28 days or fewer after receipt of samples and forward all poliovirus isolates to the designated Regional Reference Laboratory in 7 days or fewer after identification.

The Regional Reference Laboratories are located in the Central African Republic, Ghana and South Africa in the African Region; Brazil in the Region of the Americas; Egypt, Kuwait, Pakistan, and Tunisia in the Eastern Mediterranean Region; Germany, Italy and The Russian Federation in the European Region; Sri Lanka, and Thailand in the South East Asian Region; and Australia and China in the Western Pacific Region. The Regional Reference Laboratories receive poliovirus isolates for confirmatory testing and intratypic differentiation (ITD) to determine whether poliovirus isolates are of wild or vaccine origin. They support designated National Laboratories through the distribution of reagents and reference materials, coordination of proficiency tests, and training. The latter includes hosting formal courses, accepting virologists for specialized training, or visiting countries to provide on-site instruction. Most of these Laboratories also provide basic services for their own countries and for those countries in the Region that do not have laboratories. Regional Reference Laboratories are required to meet National Laboratory performance standards as well as to report 80% or more of all ITD results in fewer than 14 days of isolate receipt and to forward 80% or more of wild viruses within 7 days to Specialized Laboratories for genome sequencing.

The Specialized Reference Laboratories are global in scope, located in Atlanta, Bilthoven, Helsinki, London, Mumbai, Paris, and Tokyo. One or more of these

Laboratories prepare the reference reagents and proficiency panels of unknown viruses on which the network-wide quality control system is based, develop training materials and offer advanced training on the molecular and antigenic characterization of polioviruses, and provide sequencing for genomic characterization of wild viruses. The Specialized Laboratories also conduct research on methods of improving the sensitivity and specificity of virus detection and some serve as National and/or Regional Laboratories.

Many of the Network Laboratories perform diagnostic testing for other virus diseases, including measles and yellow fever.

Communications

Good communication within the Laboratory Network and between the Laboratories and their respective National and Regional programs is crucial to success of the eradication initiative. Frequent contacts between laboratory staff and program epidemiologists are necessary to establish testing priorities and interpret findings. At the core of Network communications system is the timely flow of data from tests on more than 60,000 stool samples annually from countries worldwide. Computers, fax-machines, and dedicated telephone lines are available in all Network Laboratories to ensure unimpeded data flow. National Laboratories report isolation/identification test results to national programs and inform Regional Laboratories of impending shipments of poliovirus isolates. Regional Laboratories report ITD test results to the submitting National Laboratory, the national program, the WHO Regional Laboratory Coordinator, and the Regional program office. They also inform the Global Laboratory Coordinator and designated Specialized Laboratory of impending shipments of confirmed wild polioviruses and vaccine-derived polioviruses (VDPV).

The Specialized Laboratory reports the sequencing results and phylogenic relationships of isolates to the Global and Regional Laboratory Coordinators and respective programs. By taking advantage of the constant rate of poliovirus genetic mutation (molecular clock), sequencing makes it possible to track transmission pathways through analysis of genetic relatedness. The construction of phylogenic trees by Specialized Laboratories provides crucial programmatic information on individual chains of transmission for monitoring eradication progress, assessing surveillance sensitivity, and distinguishing among local endemic reservoirs. These valuable data are made possible only through close collaboration and open communications between the programs and all Laboratories of the Global Network.

Laboratory accreditation

Laboratory results are accepted only from WHO accredited poliovirus laboratories. Accreditation provides documentation that the laboratory has the capability and capacity to detect, identify, and promptly report wild polioviruses and VDPV that may be present in clinical and environmental specimens. The accreditation process also provides a learning opportunity, a mechanism for identifying resources and training needs, a measure of progress, and a link to the WHO Global Laboratory Network. Accreditation is renewed annually by WHO based on laboratory performance during the immediately preceding 12 months.

National Laboratories are accredited by the Regional Laboratory Coordinator. In addition to the performance requirements previously described, accreditation criteria include testing at least 150 stool specimens annually, achieving a 90% accuracy in detecting and identifying poliovirus, implementing the required internal quality controls, scoring at least 80% on the annual isolation/identification proficiency test, and achieving at least an 80% score on the annual on-site review. Major categories for the onsite checklist include laboratory staff and facilities, operating procedures and practices, biosafety measures, conditions of equipment and supplies, records keeping, and cooperation with program staff.

Regional Reference Laboratories are accredited by the Global Laboratory Coordinator. In addition to the performance requirements previously described, accreditation criteria include achieving a 90% or greater score on annual ITD and isolation/identification proficiency tests, and on the on-site review. Regional laboratories that also serve as National Laboratories must be fully accredited at that level as well.

Specialized Reference Laboratories serving as Regional or National Laboratories are accredited by the Global Laboratory Coordinator. No accreditation requirements have been established for research or genomic sequencing functions of these laboratories because of the complexity and evolving nature of these tasks.

In 2002, 133 (92%) Network Laboratories were fully accredited, 7 (5%) were provisionally accredited, and 5 (3%) were non-accredited. For the latter, arrangements are made for an accredited laboratory to perform duplicate tests on all specimens until accreditation status is resolved.

Support of the laboratory network

The assurance of high technical quality and consistent performance in a large number of widely dispersed laboratories requires constant support from the Global and Regional Network levels. Formal training courses are given annually in all endemic or recently endemic Regions. Such courses provide virologists and technicians with initial or refresher training in standard techniques, keep them informed of the aims and strategies of the eradication program, and reinforce their role in creating a strong and cohesive network of laboratories. Individual specialized training is provided by assignments for weeks to months to Regional or Specialized Laboratories and by providing trainers in country.

To ensure comparability of test results and confidence in Network performance, standard cell cultures of known quality and sensitivity and standard poliovirus typing antisera are provided centrally through each Regional Laboratory Coordinator. An annual global meeting of Regional and Specialized Laboratory directors develops recommendations on laboratory procedures and interpretations, Network performance standards, and ensuing revisions of the laboratory manual.

Each Laboratory must undergo one or more annual proficiency tests designed to reveal flaws in technique that may compromise poliovirus detection. Test panels consist of coded samples of varying quantities and mixtures of polioviruses and non-polio enteroviruses. A proficiency test score of less than 80% or an annual accreditation rating of less than 80% indicates that a laboratory is experiencing difficulty in maintaining the required standards of performance. If serious problems are suspected, a virologist visits the laboratory to review methods, needs and constraints, and makes recommendations as appropriate.

Regional Laboratory Coordinators must ensure that laboratories have functioning basic equipment, including microscopes, centrifuges, incubators, refrigerators, mechanical and nitrogen freezers, water purification units, and testing devices. Essential supplies must be available for processing samples, maintaining cell cultures, and identifying polioviruses. In many developing countries, inadequate national funding, high tariffs on imports, or excessive time required for approval of purchases shifts responsibility to the Network for meeting basic laboratory needs.

Conclusions

Ideally, the poliovirus laboratory in many countries and Regions will serve as the model for surveillance of additional diseases of public health importance. Many Network Laboratories already perform surveillance activities for other important virus diseases. Lessons from the Polio Laboratory Network demonstrate the value of a central coordinating body, an effective communication infrastructure, and full partnerships with peer epidemiology and medical sectors. The availability of essential equipment and supplies in all Network Laboratories assures capacity to meet investigative responsibilities. The provision of standard reagents and reference materials assures test comparability. High standards for annual accreditation assure confidence in test performance and reliability. The Poliovirus Network provides an existing laboratory infrastructure and human resources with the proven ability to achieve technology transfer and quality laboratory performance even in resource poor countries. Countries in polio non-endemic Regions are encouraged to explore the potential of the WHO Poliovirus Laboratory Network to serve as a model or an instrument for surveillance of emerging viral encephalitidies.

References

1. Andrus JK, de Quadros C, Olive JM, Hull HF (1992) Screening of cases of acute flaccid paralysis for poliomyelitis eradication: ways to improve specificity. Bull World Health Organ 70: 591–596
2. Hull BP, Dowdle WR (1997) Poliovirus surveillance: building the global polio laboratory network. J Infect Dis 175(Suppl 1): 113–116
3. Pallansch M, Roos R (2001) Enteroviruses: polioviruses, coxsackieviruses, echoviruses, and newer enteroviruses. In: Knipe D, Howley P, Griffin D, Lamb GA, Martin M, Roizman B, Straus S (eds) Field's virology, 4th edn. Lippincott Williams and Wilkins Philadelphia, pp 723–775

4. Pan American Health Organization (1985) Director announces campaign to eradicate poliomyelitis from the Americas by 1990. Bull PAHO 19: 21–35
5. World Health Assembly (1988) Global eradication of poliomyelitis by the year 2000: resolution of the 41st World Health Assembly. Resolution WHA 41.28. World Health Organization, Geneva, Switzerland
6. World Health Organization (1997) Polio: the beginning of the end. World Health Organization, Geneva, Switzerland
7. World Health Organization (2002) Progress toward the global eradication of poliomyelitis. Wkly Epidem Rec 77: 98–107
8. World Health Organization (2002) Manual for the virologic investigation of poliomyelitis. WHO/EPI/GEN/02.1. World Health Organization, Geneva, Switzerland

Author's address: Walter R. Dowdle, Task Force for Child Survival and Development, 750 Commerce Drive, Decatur, GA 30030, U.S.A.; e-mail: wdowdle@taskforce.org

Arch Virol (2004) [Suppl] 18: 21–33

Archives of Virology

Emergence and virulence of encephalitogenic arboviruses

D. E. Griffin, A. P. Byrnes, and **S. H. Cook**

W. Harry Feinstone Department of Molecular Microbiology and Immunology, Johns Hopkins Bloomberg School of Public Health, Baltimore, Maryland, U.S.A.

Summary. Each arbovirus that causes encephalitis is geographically restricted by the availability of appropriate vectors and reservoir hosts. These viruses evolve regionally by recombination, reassortment and point mutation and can "emerge" as causes of human encephalitis through extension to new geographic regions or by selection of more virulent or more efficiently transmitted virus variants. The properties of arboviruses that result in encephalitis involve efficient replication in peripheral tissues after initiation of infection, production of a viremia, entry into the central nervous system and efficient replication in neurons with spread to additional populations of neurons. Many of these steps are determined by properties of the envelope glycoproteins responsible for cellular attachment, but changes in noncoding regions of the genome, as well as in other structural and nonstructural proteins, also contribute to neurovirulence.

Introduction

Arthropod-borne viruses are responsible for a significant fraction of the cases of human encephalitis and encephalomyelitis in many regions of the world. All of these encephalitogenic arboviruses have neurons as primary target cells, but some may also be able to replicate in meningeal cells, ependymal cells or glial cells in the central nervous system (CNS). Arboviruses that cause encephalitis are found in the *Flaviviridae*, *Togaviridae* and *Bunyaviridae* families and specifically in the *Flavivirus*, *Alphavirus*, *Bunyavirus* and *Phlebovirus* genera. Encephalitogenic arboviruses vary in their transmitting vectors (mosquitoes, ticks), in their reservoir hosts (birds, mammals) and in their virulence for humans. These viruses evolve regionally in conjunction with hosts important for an endemic cycle of transmission and can "emerge" as causes of human encephalitis through extension to new geographic regions or by genetic mutation and selection in a region where the virus was present, but was not associated with encephalitis [22]. Genomic changes that result in increased human or animal disease can be the result of recombination or point mutations that improve replicative capacity in a target host or in a vector that transmits the virus more efficiently to a susceptible host.

The mechanisms by which genetic changes in arboviruses lead to new strains that cause encephalomyelitis are under intensive study. Successful infection of the nervous system requires the virus to first establish a peripheral site of replication, then spread to the CNS, infect and replicate in neurons and to spread within the CNS. Damage to neurons in the brain and spinal cord and the inflammatory response to infection lead to the clinical manifestations of encephalomyelitis. This review will provide an overview of the emergence of arboviruses and determinants of arbovirus neurovirulence.

Emergence

Extension to new regions

The geographic range of an arbovirus is generally determined by the mobility of the reservoir host and the availability of appropriate competent vectors. Thus, the ranges of viruses with migratory avian reservoir hosts are often more extensive than the ranges of arboviruses with small mammals as reservoir hosts [47, 82]. For instance, eastern equine encephalitis virus (EEEV), with migratory passerine birds as important reservoir hosts has an extensive range throughout North and South America [83], while Venezuelan equine encephalitis virus (VEEV), with mammals as reservoir hosts [21], evolves in focal regions of South and Central America [54, 58]. These natural determinants of geographic range continue to be influenced by human travel, with introductions of arboviruses to new continents by transport of vectors and/or hosts on boats and airplanes.

There are a number of current examples of encephalitogenic arboviruses that are expanding their ranges. The best-studied are the flaviviruses Japanese encephalitis virus (JEV) and West Nile virus (WNV). Over the last several decades, JEV has gradually spread from its probable origin in Indonesia, and initial recognition as a cause of encephalitis in Japan, to the Indian subcontinent, Pacific Islands and, most recently, the northern coast of Australia, where new outbreaks have been recognized [56, 61, 66]. WNV was introduced from its former extensive range in Africa, the Middle East and southern Europe to North America in 1999 [38]. WNV has now spread widely throughout the United States and Canada and continued extension of the range of this virus is likely since the reservoir hosts are migratory birds [59]. The source of the introduced virus (e.g. infected mosquito, human or bird) is unknown, but WNV is now well established on a new continent. In 2002 WNV caused thousands of cases of encephalitis in humans, birds, horses and other animals in North America [9].

Recombination and reassortment

Arboviruses circulating in the same region and infecting the same vectors or hosts have the opportunity to undergo recombination or reassortment. The alphaviruses and flaviviruses have a nonsegmented genome of message-sense RNA that can undergo recombination [67], while the bunyaviruses have a segmented genome that can also undergo reassortment [2]. In the laboratory, chimeric viruses are usually at a replicative disadvantage compared to the parent viruses [37], but recombination is occasionally successful in nature. Western equine encephalitis virus (WEEV) arose as a result of recombination between EEEV (an encephalitis virus) and a Sindbis-like virus (a virus that causes fever, rash and arthritis) and resulted in a new virus that causes encephalitis [23]. There are also examples of natural recombinations between genotypes of the flaviviruses JEV and St. Louis encephalitis virus [77]. As more sequence information becomes available for other encephalitogenic arboviruses, a better understanding of the contributions of recombination to arbovirus evolution will emerge.

Point mutation

The most common mutational strategy employed by all RNA viruses is the production of point mutations as a result of polymerase errors [15, 29]. The RNA-dependent RNA polymerases of the arboviruses do not have a proof-reading function and it is estimated that there is one error for each 5,000–10,000 nucleotides copied. This amounts to approximately 1 error per genome replicated. Most of these changes will be disadvantageous for replication and the genomes will not survive, others will be neutral and a few will provide a replicative advantage in the vector or reservoir host or may lead to increased virulence for humans or other susceptible hosts. Because of the constraints imposed by the necessity for arboviruses to maintain a cycle involving efficient replication in vertebrate and invertebrate hosts, evolution of arboviruses is slower than that of other plus-strand RNA viruses [6, 58]. This constraint can also be demonstrated in vitro. Fewer mutations accumulate when viruses are alternately passaged between vertebrate and invertebrate cells than when passage is in only one cell type [81]. Replication in ticks appears to be even more selective than replication in mosquitoes, resulting in the evolution of mosquito-borne flaviviruses at twice the rate of evolution of tick-borne flaviviruses [85].

As mutations are locally selected and maintained, the geographic evolution of arbovirus strains, variants and genotypes is influenced by the vertebrate hosts [13] and the insect vectors [85]. Viruses such as EEEV, Murray Valley encephalitis virus (MVEV) and WNV, using migratory birds as reservoir hosts, evolve as a few highly conserved genotypes, while viruses, such as Venezuelan equine encephalitis virus (VEEV) and Ross River virus (RRV) with small mammals as reservoir hosts evolve within multiple foci of distinct genotypes [1, 47, 82]. The short transmission season of many arboviruses will further constrain the rate of genetic evolution in a region [13].

Within this evolutionary background, viruses with increased virulence can arise. This is best documented for epizootics of VEE. Extensive study of the epizootic strains of VEEV that cause disease in humans and equines has demonstrated that these viruses are derived from enzootic strains that cause little disease. The epizootic variants have often acquired the ability to be more efficiently transmitted by mosquitoes and to replicate more efficiently in equines and humans [5, 20, 58, 60]. Differences in virulence and transmission patterns are also recognized for other arboviruses and these variations likely reflect the appearance of point mutations that have provided selective advantages for these viruses [4]. The specific mutations that facilitate epizootic or epidemic transmission and the mechanisms by which improvement occurs are not yet understood, but will be important for identifying the mechanisms of emergence and will improve our understanding of the determinants of virulence.

Neurovirulence

Closely related strains of arboviruses can differ significantly in their ability to cause human disease. For instance, the genotype of EEEV endemic in North America causes cases of fatal human and equine encephalitis along the eastern and Gulf coasts of the United States every year, while no cases of encephalitis due to EEEV caused by the 3 genotypes endemic in South America have been recognized [8, 80]. This is also true for WEEV with human and equine encephalitis recognized in North America, but only equine encephalitis in South America [43, 62]. However, for reasons that are not understood, human cases of WEE in the United States have been decreasing and the last reported case occurred in 1994. Likewise, WNV, throughout most of its extensive range, causes only mild febrile illness, but since 1996 outbreaks of human, avian and equine encephalitis have occurred in Europe and the Middle East [72]. The apparently more virulent strain from the Middle East was imported to North America in 1999 [38].

In order for an arbovirus to cause encephalitis it must be able to accomplish several pathogenetic steps. First, the virus must replicate efficiently in peripheral tissues after injection by the arthropod vector, then cause a viremia, enter the CNS and replicate efficiently in neurons (Fig. 1). Neuroinvasiveness, neurotropism and neurovirulence are distinct, separable properties of a virus and are influenced by the age, sex and genetic background of the host. Although for some encephalitogenic viruses, altered tropism for cells within the CNS accounts for changes in neurovirulence, this does not appear to be the case for arboviruses. Both virulent and avirulent strains of encephalitogenic arboviruses target neurons, but the efficiency of replication in neurons often differs dramatically without affecting replication in other types of cells [30]. Specific mechanisms by which changes in both coding and noncoding regions of the genome lead to more neurovirulent

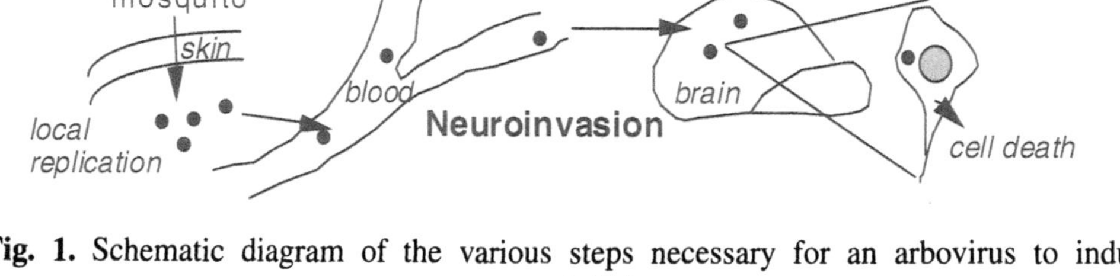

Fig. 1. Schematic diagram of the various steps necessary for an arbovirus to induce encephalitis

viruses are currently being identified for these different steps in pathogenesis, using animal models of arbovirus encephalomyelitis. An important issue to address in the future is why differences in replication are specific for neurons.

Viremia

A sustained high-titered viremia will favor CNS invasion, although it is not always required [31, 45, 51]. The level of virus in the blood reflects a balance between the amount of virus delivered to the circulation from peripheral sites of replication and the rapidity of clearance from the blood. For instance, amino acid changes that alter the efficiency of flavivirus fusion decrease replication in lymphoid tissue, lower the level of viremia and decrease the likelihood of subsequent CNS invasion [49]. Clearance is enhanced by virus binding to glycosaminoglycans, particularly heparan sulfate, which is abundant in the liver and in the extracellular matrix [3, 7, 39]. Glycosaminoglycans are highly sulfated, negatively charged, unbranched chains of repeating disaccharides that may or may not be linked to a protein core (proteoglycans). A common adaptation of arboviruses grown in tissue culture is enhanced binding to heparan sulfate [3, 36, 39, 48, 64]. It is thought the interaction of positively charged amino acids on the surface of the virus with the negatively charged glycosaminoglycans on the surface of the cell enhances infection and replication in vitro. These mutations are rapidly selected for during in vitro passage and often lead to decreased virulence in vivo [3, 7, 48].

We have studied encephalomyelitis caused by Sindbis virus infection of mice as a model system for understanding alphavirus neurovirulence. Viruses naturally selected for an ability to cause sustained viremia were sequenced and amino acid changes identified [7]. All coding changes were in the E2 glycoprotein and most amino acid substitutions resulted in the loss of positively charged amino acids in the surface binding domains of E2 (Table 1). These viruses bound heparin less well, were cleared from the blood more slowly, and produced higher titered

Table 1. Relationship of heparan sulfate (HS)-binding, viremia and virulence in recombinant strains of Toto1101 infecting 2 d-old mice

Virus[a]	HS-binding[b]	Viremia[c]	% Mortality
Toto1101	323	5.0	30
R157H	**289**	**7.8**	**100**
M62D	276	6.6	68
K159E	**261**	**7.7**	**25**
K230M	202	6.7	70
K76N	142	6.4	35
K76T	**97**	**6.4**	**25**
K76E	**66**	**6.4**	**70**

[a] – amino acid substitution in E2 of Toto1101
[b] – mM NaCl required for elution from heparin-Sepharose column
[c] – pfu/ml blood at 48 h after infection

viremias than did the parent strain Toto1101. Most of these viruses were more virulent than Toto 1101, but not all (e.g. K159E and K76T) indicating that other properties of the virus, in addition to high titered viremia, affect neurovirulence. These additional properties are likely to affect the ability of the virus to enter the CNS and the efficiency of replication in neurons, once the virus has reached the CNS. Similar observations have been made for VEEV [3] and the flaviviruses JEV, MVEV and tick-borne encephalitis virus (TBEV) [39] suggesting that similar principles may be operative for all encephalitogenic alphaviruses and flaviviruses.

Neuroinvasion

Viruses can enter the CNS by several routes and arboviruses have been suggested to use all of them in one or another animal model system. Viruses in the blood can infect cerebrocapillary endothelial cells and spread directly to parenchymal cells in the brain or spinal cord [33, 65] or can infect choroid plexus epithelial cells and spread through the cerebrospinal fluid to ependymal cells and then to the parenchyma [42]. In addition, viruses can enter through neural routes either from the periphery (e.g. rabies virus) or the olfactory epithelium (e.g. St. Louis encephalitis virus, MVE and VEEV) [12, 50, 51, 79]. Olfactory sensory neurons have dendritic knobs that protrude into the nasal cavity, cell bodies in the nasal epithelium, and an axon that extends to the brain. These neurons can be infected by virus present in the blood or by virus in the respiratory tract. Host responses to infection may also influence the likelihood of neuroinvasion [41].

Often it is difficult to determine the route of entry when amounts of virus in the blood are high. To facilitate studies of virus spread to the CNS, we have developed an in vivo imaging system that allows tracking of virus spread from one location to another [14]. Sindbis virus was engineered to express the reporter gene luciferase. Sites of virus replication were identified by inoculating mice infected with a luciferase-expressing virus, with luciferin as a substrate for the luciferase. Infected cells will produce luciferase that will convert the luciferin to produce light. The light that is emitted from infected cells is captured by a highly sensitive CCD camera (Xenogen Corp., Alameda, CA) at various times after infection. Because production of light requires cellular ATP, virus in the blood does not interfere with localization of the sites of virus replication. When Sindbis virus-infected mice were followed after inoculation in a hind foot, we found that virus appeared to enter the CNS both by retrograde axonal transport (appearing first in the lumbar spinal cord in some animals) and by infection of the nasal epithelium, with subsequent transport to the brain in others (Fig. 2). It is likely that specific features of virus attachment or replication will determine the route of entry. A single neurotropic virus may use more than one route of entry and this may account for some variability in outcome of infection.

Genetic dissection of the determinants of neuroinvasion for arboviruses has shown that cell attachment regions of the surface glycoproteins are important determinants of this component of virulence. Regions of the virus genome that affect neuroinvasion have been mapped to the envelope (E) glycoprotein of JEV,

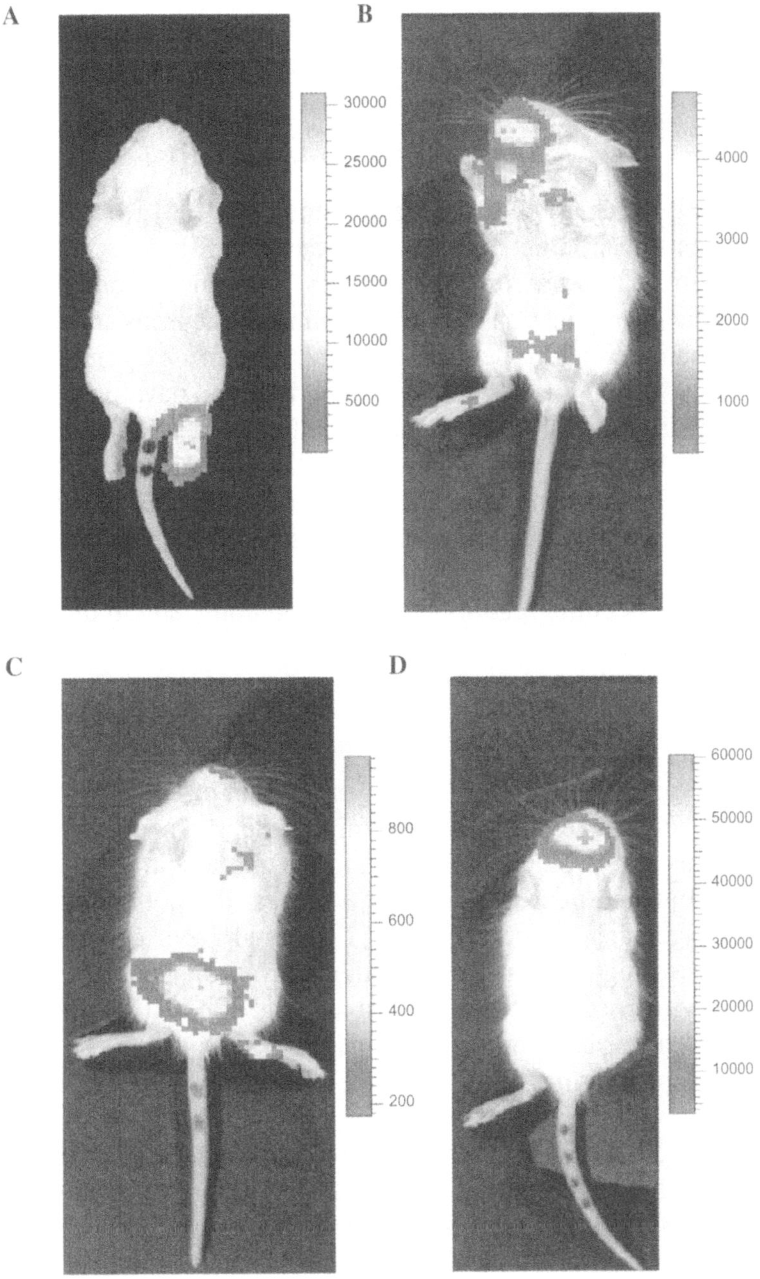

Fig. 2. Illustration of the light emitted by cells infected with a luciferase-expressing virus. Sindbis virus expressing firefly luciferase from a second subgenomic promoter was inoculated into the right hind foot of white B6 mice (**A**). Mice were then imaged daily to track spread of the virus from the foot to the CNS. In some mice virus appeared first in the lumbar spinal cord (**C**) and in other mice virus was detected first in the nose (**B**) before appearing in the brain (**D**), illustrating more than one route of entry into the CNS. Reproduced with permission from [14]

MVEV, WNV and TBEV [10, 11, 49], to the 5′ NTR and E2 glycoprotein of Sindbis virus [17] and to the G1 glycoprotein of LaCrosse virus [32].

Replication in neurons

Most encephalitogenic arboviruses are neuronotropic [25, 34, 52] and neurovirulence correlates with efficient replication in neurons and rapid spread from initially infected neurons to new neuronal populations. Properties of both the host and the virus affect efficient replication in neurons. Immature neurons are more susceptible to infection than are mature neurons [55] and arboviruses generally replicate to higher titer in young animals, which are more likely to develop fatal disease than are older animals [33, 35, 84]. In human infections, older adults, as well as young children, are more likely to develop encephalitis than young and middle-aged adults [43, 44]. In addition, there are genetic determinants of susceptibility and strains of mice vary in susceptibility to fatal flavivirus and alphavirus encephalitis [69, 70].

Changes in noncoding regions of the virus genome and in the coding regions of both structural and nonstructural proteins of the virus can influence replication in neurons, often without a perceptible effect on replication in cells that are not neurons [27, 46, 49, 73, 75]. As with neuroinvasiveness, amino acid changes in the glycoprotein responsible for binding to cells are commonly associated with changes in neurovirulence [53, 75], suggesting that efficient binding to neurons and entry into neurons are often the determining steps in neurovirulence. Changes in other proteins may promote replication in the unique intracellular environment of fully differentiated neurons. For bunyaviruses, the polymerase protein is an important determinant of neurovirulence [18]. For alphaviruses, mutations in nsP1 that affect processing of the nsPs affect the neurovirulence of Sindbis virus [68] and the neurovirulence of Semliki Forest virus is affected by mutations in nsP3 and in the nuclear localization signal of nsP2 [19, 76].

Neurovirulent arboviruses cause severe cytopathic effects in infected neurons [24, 26, 52, 71] and the ability of the virus to induce cell death correlates with neurovirulence [57, 78]. One important change that determines virulence of Sindbis virus for adult mice is a glutamine to histidine change at position 55 in the E2 glycoprotein [75]. This change improves efficiency of infection of neurons at an early step in replication associated with binding to neurons and efficient fusion of the viral membrane with the neuronal plasma membrane [16, 40, 74]. However, the precise mechanism of improved neuron infection has not yet been determined. Even severely damaged neurons do not always die and may recover function as the virus is cleared [25]. Identifying the factors that contribute to neuronal cell death will be important for a full understanding of arbovirus neurovirulence.

Acknowledgments

Work from the authors' laboratory was supported by research grant R01 NS18596 and training grant T32 AI07417 from the National Institutes of Health.

References

1. Beasley DW, Davis CT, Guzman H, Vanlandingham DL, Travassos da Rosa AP, Parsons RE, Higgs S, Tesh RB, Barrett AD (2003) Limited evolution of West Nile virus has occurred during its southwesterly spread in the United States. Virology 309: 190–195
2. Beaty BJ, Sundin DR, Chandler LJ, Bishop DHL (1985) Evolution of bunyaviruses by genome reassortment in dually infected mosquitoes *(Aedes triseriatus)*. Science 230: 548–550
3. Bernard KA, Klimstra WB, Johnson RE (2000) Mutations in the E2 glycoprotein of Venezuelan equine encephalitis virus confer heparan sulfate interaction, low morbidity, and rapid clearance from blood of mice. Virology 276: 93–103
4. Bianchi TI, Aviles G, Monath TP, Sabattini MS (1993) Western equine encephalomyelitis: virulence markers and their epidemiologic significance. Am J Trop Med Hyg 49: 322–328
5. Brault AC, Powers AM, Weaver SC (2002) Vector infection determinants of Venezuelan equine encephalitis virus reside within the E2 envelope glycoprotein. J Virol 76: 6387–6392
6. Burness AHT, Pardoe I, Faragher SG, Vrati S, Dalgarno L (1988) Genetic stability of Ross River virus during epidemic spread in nonimmune humans. Virology 167: 639–643
7. Byrnes AP, Griffin DE (2000) Large-plaque mutants of Sindbis virus show reduced binding to heparan sulfate, heightened viremia and slower clearance from the circulation. J Virol 74: 644–651
8. Calisher CH (1994) Medically important arboviruses of the United States and Canada. Clin Microbiol Rev 7: 89–116
9. CDC (2002) Provisional surveillance summary of the West Nile virus epidemic–United States, January–November 2002. MMWR Morb Mortal Wkly Rep 51: 1129–1133
10. Cecilia D, Gould EA (1991) Nucleotide changes responsible for loss of neuroinvasiveness in Japanese encephalitis virus neutralization-resistant mutants. Virology 181: 70–77
11. Chambers TJ, Halevy M, Nestorowicz A, Rice CM, Lustig S (1998) West Nile virus envelope proteins: nucleotide sequence analysis of strains differing in mouse neuroinvasiveness. J Gen Virol 79: 2375–2380
12. Charles PC, Walters E, Margolis F, Johnston RE (1995) Mechanism of neuroinvasion of Venezuelan equine encephalitis virus in the mouse. Virology 208: 662–671
13. Clinis MJ, Kang W, Weaver SC (1996) Genetic conservation of Highlands J viruses. Virology 218: 343–351
14. Cook S, Griffin DE (2003) Luciferase imaging of a neurotropic viral infection in intact animals. J Virol 77: 5333–5338
15. Drake JW, Holland JJ (1999) Mutation rates among RNA viruses. Proc Natl Acad Sci USA 96: 13910–13913
16. Dropulic LK, Hardwick JM, Griffin DE (1997) A single amino acid change in the E2 glycoprotein of Sindbis virus confers neurovirulence by altering an early step of virus replication. J Virol 71: 6100–6105
17. Dubuisson J, Lustig S, Ruggli N, Akov Y, Rice CM (1997) Genetic determinants of Sindbis virus neuroinvasiveness. J Virol 71: 2636–2646
18. Endres MJ, Griot C, Gonzalez-Scarano F, Nathanson N (1991) Neuroattenuation of an avirulent bunyavirus variant maps to the L RNA segment. J Virol 65: 5465–5470
19. Fazakerley JK, Boyd A, Mikkola ML, Kaariainen L (2002) A single amino acid change in the nuclear localization sequence of the nsP2 protein affects the neurovirulence of Semliki Forest virus. J Virol 76: 392–396
20. Gonzalez-Salazar D, Estrada-Franco JG, Carrara AS, Aronson JF, Weaver SC (2003) Equine amplification and virulence of subtype IE Venezuelan equine encephalitis

viruses isolated during the 1993 and 1996 Mexican epizootics. Emerg Infect Dis 9: 161–168
21. Grayson MA, Galindo P (1968) Epidemiologic studies of Venezuelan equine encephalitis virus in Almirante, Panama. Am J Epidemiol 88: 80–96
22. Gubler DJ (2002) The global emergence/resurgence of arboviral diseases as public health problems. Arch Med Res 33: 330–342
23. Hahn CS, Lustig S, Strauss EG, Strauss JH (1988) Western equine encephalitis virus is a recombinant virus. Proc Natl Acad Sci USA 85: 5997–6001
24. Hase T (1993) Virus-neuron interactions in the mouse brain infected with Japanese encephalitis virus. Virchows Arch B Cell Pathol Incl Mol Pathol 64: 161–170
25. Hase T, Summers PL, Dubois DR (1990) Ultrastructural changes of mouse brain neurons infected with Japanese encephalitis virus. Int J Exp Pathol 71: 493–505
26. Havert M, Schofield B, Griffin D, Irani D (2000) Activation of divergent neuronal cell death pathways in different target cell populations during neuroadapted Sindbis virus infection of mice. J Virol 74: 5352–5356
27. Heise MT, Simpson DA, Johnston RE (2000) A single amino acid change in nsP1 attenuates neurovirulence of the Sindbis-group alphavirus A.A.AR86. J Virol 74: 4207–4213
28. Hirayama N, Senda M, Nakashima N, Takagi M, Sugiyama M, Yoshikawa Y, Yamanouchi K (1991) Protective effects of monoclonal antibodies against lethal canine distemper virus infection in mice. J Gen Virol 72: 2827–2830
29. Holland J, Spindler K, Horodyski F, Grabau E, Nichol S, VandePol S (1982) Rapid evolution of RNA genomes. Science 215: 1577–1585
30. Jackson AC, Moench TR, Trapp BD, Griffin DE (1988) Basis of neurovirulence in Sindbis virus encephalomyelitis of mice. Lab Invest 58: 503–509
31. Jahrling PB, Heisey GB, Hesse RA (1977) Evaluation of vascular clearance as a marker for virulence of alphaviruses: disassociation of rapid clearance with low virulence of Venezuelan encephalitis virus strains in guinea pigs. Infect Immunol 17: 356–360
32. Janssen RS, Nathanson N, Endres MJ, Gonzalez-Scarano F (1986) Virulence of LaCrosse virus is under polygenic control. J Virol 59: 1–7
33. Johnson KP, Johnson RT (1968) California encephalitis. II. Studies of experimental infection in the mouse. J Neuropathol Exp Neurol 27: 390–400
34. Johnson RT, Burke DS, Elwell M, Leake CJ, Nisalak A, Hoke CH, Lorsomrudee W (1985) Japanese encephalitis: Immunocytochemical studies of viral antigen and inflammatory cells in fatal cases. Ann Neurol 18: 567–573
35. Johnson RT, McFarland HF, Levy SE (1972) Age-dependent resistance to viral encephalitis: Studies of infections due to Sindbis virus in mice. J Infect Dis 125: 257–262
36. Klimstra WB, Ryman KD, Johnston RE (1998) Adaptation of Sindbis virus to BHK cells selects for use of heparan sulfate as an attachment receptor. J Virol 72: 7357–7366
37. Kuhn RJ, Griffin DE, Owen KE, Niesters HGM, Strauss JH (1996) Chimeric Sindbis-Ross River viruses to study interactions between alphavirus nonstructural and structural regions. J Virol 70: 7900–7909
38. Lanciotti RS, Roehrig JT, Deubel V, Smith J, Parker M, Steele K, Crise B, Volpe KE, Crabtree MB, Scherret JH, Hall RA, Mackenzie JS, Cropp CB, Panigrahy B, Ostlund E, Schmitt B, Malkinson M, Banet C, Weissman J, Komar N, Savage HM, Stone W, McNamara T, Gubler DJ (1999) Origin of the West Nile virus responsible for an outbreak of encephalitis in the northeastern United States. Science 286: 2333–2337
39. Lee E, Lobigs M (2002) Mechanism of virulence attenuation of glycosaminoglycan-binding variants of Japanese encephalitis virus and Murray valley encephalitis virus. J Virol 76: 4901–4911

40. Lee P, Knight R, Smit JM, Wilschut J, Griffin DE (2002) A single mutation in the E2 glycoprotein important for neurovirulence influences binding of Sindbis virus to neuroblastoma cells. J Virol 76: 6302–6310
41. Licon Luna RM, Lee E, Mullbacher A, Blanden RV, Langman R, Lobigs M (2002) Lack of both Fas ligand and perforin protects from flavivirus-mediated encephalitis in mice. J Virol 76: 3202–3211
42. Liu C, Voth DW, Rodina P, Shauf LR, Gonzalez G (1970) A comparative study of the pathogenesis of western equine and eastern equine encephalomyelitis viral infections in mice by intracerebral and subcutaneous inoculations. J Infect Dis 122: 53–63
43. Longshore WA, Stevens IM, Hollister AC, Gittelsohn A, Lennette EH (1956) Epidemiologic observations on acute infectious encephalitis in California with special reference to the 1952 outbreak. Am J Hyg 63: 69–86
44. Luby JP, Miller G, Gardner P, Pigford CA, Henderson BE, Eddins D (1967) The epidemiology of St. Louis encephalitis in Houston, Texas, 1964. Am J Epidemiol 86: 584–597
45. Lustig S, Halevy M, Ben-Nathan D, Akov Y (1992) A novel variant of Sindbis virus is both neurovirulent and neuroinvasive in adult mice. Arch Virol 122: 237–248
46. Lustig S, Jackson AC, Hahn CS, Griffin DE, Strauss EG, Strauss JH (1988) The molecular basis of Sindbis virus neurovirulence in mice. J Virol 62: 2329–2336
47. Mackenzie JS, Poidinger M, Lindsay MD, Hall RA, Sammels LM (1995) Molecular epidemiology and evolution of mosquito-borne flaviviruses and alphaviruses enzootic in Australia. Virus Genes 11: 225–237
48. Mandl CW, Kroschewski H, Allison SL, Kofler R, Holzmann H, Meixner T, Heinz FX (2001) Adaptation of tick-borne encephalitis virus to BHK-21 cells results in the formation of multiple heparan sulfate binding sites in the envelope protein and attenuation in vivo. J Virol 75: 5627–5637
49. McMinn PC (1997) The molecular basis of virulence of the encephalitogenic flaviviruses. J Gen Virol 78: 2711–2722
50. McMinn PC, Dalgarno L, Weir RC (1996) A comparison of the spread of Murray Valley encephalitis viruses of high or low neuroinvasiveness in the tissues of Swiss mice after peripheral inoculation. Virology 220: 414–423
51. Monath TP, Cropp CP, Harrison AK (1983) Mode of entry of a neurotropic arbovirus into the central nervous system. Reinvestigation of an old controversy. Lab Invest 48: 399–410
52. Murphy FA, Harrison AK, Gary GW Jr, Whitfield SG, Forrester FT (1969) St. Louis encephalitis virus infection of mice electron microscopic studies of central nervous system. Lab Invest 19: 652–662
53. Ni H, Barrett AD (1998) Attenuation of Japanese encephalitis virus by selection of its mouse brain membrane receptor preparation escape variants. Virology 241: 30–36
54. Oberste MS, Schmura SM, Weaver SC, Smith JF (1999) Geographic distribution of Venezuelan equine encephalitis virus subtype IE genotypes in Central America and Mexico. Am J Trop Med Hyg 60: 630–634
55. Ogata A, Nagashima K, Hall WW, Ichikawa M, Kimura-Kuroda J, Yasui K (1991) Japanese encephalitis virus neurotropism is dependent on the degree of neuronal maturity. J Virol 65: 880–886
56. Paul WS, Moore PS, Karabatsos N, Flood SP, Yamada S, Jackson T, Tsai TF (1993) Outbreak of Japanese encephalitis on the island of Saipan, 1990. J Infect Dis 167: 1053–1058

57. Pekosz A, Phillips J, Pleasure D, Merry D, Gonzalez-Scarano F (1996) Induction of apoptosis by La Crosse virus infection and role of neuronal differentiation and human bcl-2 expression in its prevention. J Virol 70: 5329–5335
58. Powers AM, Oberste MS, Brault AC, Rico-Hesse R, Schmura SM, Smith JF, Kang W, Sweeney WP, Weaver SC (1997) Repeated emergence of epidemic/epizootic Venezuelan equine encephalitis from a single genotype of enzootic subtype ID virus. J Virol 71: 6697–6705
59. Rappole JH, Derrickson SR, Hubalek Z (2000) Migratory birds and spread of West Nile virus in the Western Hemisphere. Emerg Infect Dis 6: 319–328
60. Rico-Hesse R, Weaver SC, de Siger J, Medina G, Salas RA (1995) Emergence of a new epidemic/epizootic Venezuelan equine encephalitis virus in South America. Proc Natl Acad Sci USA 92: 5278–5281
61. Ritchie SA, Rochester W (2001) Wind-blown mosquitoes and introduction of Japanese encephalitis into Australia. Emerg Infect Dis 7: 900–903
62. Sabattini MS, Monath TP, Mitchell CJ, Daffner JF, Bowen GS, Pauli R, Contigiani MS (1985) Arbovirus investigations in Argentina, 1977–1980. I. Historical aspects and description of study sites. Am J Trop Med Hyg 34: 937–944
63. Schlesinger Y, Tebas P, Gaudreault-Keener M, Buller RS, Storch GA (1995) Herpes simplex virus type 2 meningitis in the absence of genital lesions: improved recognition with use of the polymerase chain reaction. Clin Infect Dis 20: 842–848
64. Smit JM, Waarts B-L, Kimata K, Klimstra WB, Bittman R, Wilschut J (2002) Adaptation of alphaviruses to heparan sulfate: Interaction of sindbis and semliki forest viruses with liposomes containing lipid-conjugated heparin. J Virol 76: 10128–10137
65. Soilu-Hanninen M, Eralinna JP, Hukkanen V, Roytta M, Salmi AA, Salonen R (1994) Semliki Forest virus infects mouse brain endothelial cells and causes blood-brain barrier damage. J Virol 68: 6291–6298
66. Solomon T, Ni H, Beasley DW, Ekkelenkamp M, Cardosa MJ, Barrett AD (2003) Origin and evolution of Japanese encephalitis virus in southeast Asia. J Virol 77: 3091–3098
67. Strauss JH, Strauss EG (1997) Recombination in alphaviruses. Semin Virol 8: 85–94
68. Susin SA, Daugas E, Ravagnan L, Samejima K, Zamzami N, Loeffler M, Costantini P, Ferri KF, Irinopoulou T, Prevost M-C, Brothers G, Mak TW, Penninger J, Earnshaw WC, Kroemer G (2000) Two distinct pathways leading to nuclear apoptosis. J Exp Med 192: 571–579
69. Thach DC, Kleeberger SR, Tucker PC, Griffin DE (2001) Genetic control of neuroadapted Sindbis virus replication in female mice maps to chromosome 2 and associates with paralysis and mortality. J Virol 75: 8674–8680
70. Thach D, Kimura T, Griffin D (2000) Differences between C57BL/6 and BALB/cBy mice in mortality and virus replication after intranasal infection with neuroadapted Sindbis virus. J Virol 74: 6156–6161
71. Thompson WH, Kalfayan B, Anslow RO (1965) Isolation of California encephalitis group virus from a fatal human illness. Am J Epidemiol 81: 245
72. Tsai TF, Popovici F, Cernescu C, Campbell GL, Nedelcu NI (1998) West Nile encephalitis epidemic in southeastern Romania. Lancet 352: 767–771
73. Tucker PC, Griffin DE (1991) The mechanism of altered Sindbis virus neurovirulence associated with a single amino acid change in the E2 glycoprotein. J Virol 65: 1551–1557
74. Tucker PC, Lee SH, Bui N, Martinie D, Griffin DE (1997) Amino acid changes in the Sindbis virus E2 glycoprotein that increase neurovirulence improve entry into neuroblastoma cells. J Virol 71: 6106–6112
75. Tucker PC, Strauss EG, Kuhn RJ, Strauss JH, Griffin DE (1993) Viral determinants of age-dependent virulence of Sindbis virus in mice. J Virol 67: 4605–4610

76. Tuittila M, Hinkkanen AE (2003) Amino acid mutations in the replicase protein nsP3 of Semliki Forest virus cumulatively affect neurovirulence. J Gen Virol 84: 1525–1533
77. Twiddy SS, Holmes EC (2003) The extent of homologous recombination in members of the genus *Flavivirus*. J Gen Virol 84: 429–440
78. Ubol S, Tucker PC, Griffin DE, Hardwick JM (1994) Neurovirulent strains of alphavirus induce apoptosis in Bcl-2-expressing cells; role of a single amino acid change in the E2 glycoprotein. Proc Natl Acad Sci USA 91: 5202–5206
79. Vogel P, Abplanalp D, Kell W, Ibrahim S, Downs MB, Pratt WD, Davis KJ (1996) Venezuelan equine encephalitis in BALB/c mice: Kinetic analysis of central nervous system infection following aerosol or subcutaneous inoculation. Arch Pathol Lab Med 120: 164–172
80. Walder R, Jahrling PB, Eddy GA (1980) Differentiation markers of eastern equine encephalitis (EEE) viruses and virulence. Zbl Bact Microbiol Hyg Suppl 9: 237–250
81. Weaver SC, Brault AC, Kang W, Holland JJ (1999) Genetic and fitness changes accompanying adaptation of an arbovirus to vertebrate and invertebrate cells. J Virol 73: 4316–4326
82. Weaver SC, Hagenbaugh A, Bellew LA, Gousset L, Mallampalli V, Holland JJ, Scott TW (1994) Evolution of alphaviruses in the eastern equine encephalomyelitis complex. J Virol 68: 158–169
83. Weaver SC, Scott TW, Rico-Hesse R (1991) Molecular evolution of eastern equine encephalomyelitis virus in North America. Virology 182: 774–784
84. Weiner LP, Cole GA, Nathanson N (1970) Experimental encephalitis following peripheral inoculation of West Nile virus in mice of different ages. J Hyg 68: 435–446
85. Zanotto PM, Gao GF, Gritsun T, Marin MS, Jiang WR, Venugopal K, Reid HW, Gould EA (1995) An arbovirus cline across the northern hemisphere. Virology 210: 152–159

Author's address: Diane E. Griffin, MD PhD, Department of Molecular Microbiology and Immunology, Johns Hopkins Bloomberg School of Public Health, 615 N. Wolfe St., Rm E5132, Baltimore, MD 21205, U.S.A.; e-mail: dgriffin@jhsph.edu

Arch Virol (2004) [Suppl] 18: 35–41

Archives of Virology

Molecular determinants of virulence of West Nile virus in North America

D. W. C. Beasley[1], **C. T. Davis**[1], **M. Whiteman**[1], **B. Granwehr**[1], **R. M. Kinney**[2], and **A. D. T. Barrett**[1]

[1]Center for Biodefense and Emerging Infectious Diseases, Sealy Center for Vaccine Development and Department of Pathology, University of Texas Medical Branch, Galveston, Texas, U.S.A.
[2]Division of Vector-Borne Infectious Diseases, Centers for Disease Control and Prevention, Fort Collins, Colorado, U.S.A.

Summary. West Nile virus (WNV) is a mosquito-borne flavivirus that until very recently had not been found in the Americas. In 1999, there was an outbreak of West Nile encephalitis in New York and surrounding areas, involving 62 human cases, including 7 fatalities. The virus has subsequently become established in the United States of America (U.S.) with 4156 human cases, including 284 deaths, in 2002. The WNV strains found in the U.S. are members of "lineage I", a genetic grouping that includes viruses from Europe, Asia and Africa. Molecular epidemiologic studies indicate that two genetic variants of WNV emerged in 2002. The major genetic variant is found in most parts of the U.S., while the minor genetic variant has been identified only on the southeast coast of Texas. Investigation of WNV in mouse and hamster models demonstrated that strains from the U.S. are highly neurovirulent and neuroinvasive in these laboratory rodents. Other strains, such as Ethiopia 76a from lineage I, are not neuroinvasive and represent important viruses which can be used to elucidate the molecular basis of virulence and attenuation of WNV. To identify putative molecular determinants of virulence and attenuation, we have undertaken comparative nucleotide sequencing of Ethiopia 76a and strains from the U.S. The results show that the two viruses differ by 5 amino acids in the envelope (E) protein, including loss of the glycosylation site. Comparison of our panel of 27 WNV strains suggests that E protein glycosylation is a major determinant of the mouse neuroinvasive phenotype.

Introduction

West Nile virus (WNV) is a member of the Japanese encephalitis antigenic complex of the genus *Flavivirus*, family *Flaviviridae*. Until recently, human infections with WNV were associated with a mild undifferentiated fever [8]. However, recent outbreaks in Europe, Israel and North America involving humans and animals have been associated with significant rates of neurological disease [5, 6, 10, 11]. The most important event was the introduction of WNV into the western hemisphere in the summer of 1999 when the virus was first isolated in New York City before spreading along the eastern seaboard of the U.S. The initial outbreak in 1999 involved 62 human cases, including 7 fatalities (a case:fatality rate of 12%) and 25 clinical cases in equines, including 9 deaths (a case:fatality rate of 36%). The virus has continued to spread in subsequent years and in 2002 was found in 44 states and the District of Columbia in the U.S. and 5 provinces in Canada. At least 4156 human cases, including 284 fatalities, and more than 14,000 equine cases were reported in 2002, representing the largest recorded epidemic of arboviral meningoencephalitis in the western hemisphere. West Nile virus is expected to continue to be a public health and veterinary problem in the western hemisphere in 2003.

There is extensive knowledge of the molecular biology and antigenic relationships of WNV. Early studies with polyclonal antisera and more recent studies with monoclonal antibodies have identified three antigenic complexes that are geographically distinct: viruses predominantly from Africa; viruses from India, and viruses predominantly from Europe and the Middle East [2, 4]. The genome of a Nigerian strain was sequenced by Castle et al. [3] and subsequently led to phylogenetic studies that defined genotypes that broadly agreed with previous antigenic studies. Two major lineages were defined that contained several subtypes and clades [7, 12]. Lineage II contained only strains from Africa, while Lineage I was large and contained two major subtypes (representing strains from India and Kunjin virus), plus additional clades that included strains from North America, Europe and the Middle East. In comparison to the detailed studies on epidemiology, serology, and genetics of WNV, there is very limited understanding of the molecular determinants of virulence of WNV.

Phylogenetic analysis of WNV strains

The World Arbovirus Reference Center at the University of Texas Medical Branch at Galveston retains 27 low passage strains of WNV and Kunjin virus that were used in these studies: these represent strains isolated between 1950 and 2002 from different hosts and geographical locations (see Table 1). Genomic RNA was subjected to RT-PCR using mosquito-borne flavivirus primers (vd8 and emf1) that amplify a region in NS5 and the 3′ untranslated region [9] and a phylogenetic tree was constructed (Fig. 1). Our phylogenetic analysis gave comparable results to those of Lanciotti et al. [7] and identified two major lineages. Lineage II contained 10 strains; 9 from east, central and west Africa and one from Cyprus, while lineage I contained 17 strains. Within lineage I, two of the three Indian strains were members of the Indian subtype, the two Kunjin strains represented a second subtype. The remaining 13 strains could be grouped into two major clades; clade A contained strains from the Middle East, India and East Africa, while clade B contained strains from the United States and east, west and central Africa.

Table 1. Properties of West Nile and Kunjin virus strains used in this study

Designation	Strain	Origin	Year	Lineage[a]	CHO[b]	3- to 4-week-old mouse i.p. LD_{50} (pfu)
CAR67	ArB-310/67	Central African	1967	I	−S	12.6
NIG65	IbAn7019	Nigeria	1965	I	+	3.2
SEN79	ArD-27875	Senegal	1979	I	+	0.2
KEN98	KEN3829	Kenya	1998	I	+	<1.0
USA99a	31A	United States	1999	I	+	0.5
USA99b	385-99	United States	1999	I	+	3.2
IND68	68856	India	1968	I	+	3.2
EGY50	Egypt101	Egypt	1950	I	−S	50*
EGY52	Ar248	Egypt	1952	I	−N	795*
ISR52	TL443	Israel	1952	I	−S	8000*
ISR53	Goldblum	Israel	1953	I	−S	≥10,000
ETH76a	EthAn4766	Ethiopia	1976	I	−N	2000*
ETH76b	EthAn4767	Ethiopia	1976	I	−N	≥10,000
AUS60	MRM16 (Kunjin)	Australia	1960	KUN (I)	−S	≥10,000
AUS91	K6453 (Kunjin)	Australia	1991	KUN (I)	+	≥10,000
IND57	IG-15578	India	1957	IND (I)	−S	8000*
IND80	804994	India	1980	IND (I)	−S	50*
SEN90	ArD-76104	Senegal	1990	II	−Δ	50*
CAR82	ArB3573/82	Central African	1982	II	+	0.8
SA58a	SAH-442	South Africa	1958	II	+	3.2
CON58	Eyoku	Congo	1958	II	−S	0.8
SA89	SPU116-89	South Africa	1989	II	+	5.0
UGA37	B956 (prototype)	Uganda	1937	II	−Δ	7.9
MAD88	ArMg-979	Madagascar	1988	II	+	500*
SA58b	SAAn2842	South Africa	1958	II	+	125*
CYP68	Q3574-5	Cyprus	1968	II	−N	≥10,000
MAD78	DakAnMg798	Madagascar	1978	II	−S	≥10,000

*LD_{50} values could not be reliably calculated for these strains due to inconsistent mortality across the range of challenge doses

[a]Based on sequencing of a 3′NCR fragment; see Fig. 1

[b]Indicates presence or absence of potential N-linked glycosylation site in the E protein sequence (E154–156 NYS): +, present; −, Absent; N, asparagine mutant (N → S); S, serine mutant (S → P/A); Δ, deletion mutant

E protein glycosylation

Most flaviviruses have an envelope (E) protein that is glycosylated. In the case of WNV, the majority of strains are glycosylated but some are not, due to modifications in the NYS motif at residues E154-156. Five (UGA37, CON58, SEN90, CYP68 and MAD78) of the 10 lineage II strains were not glycosylated, while 10 (the two Indian subtype strains [IND57 and IND80], AUS60, CAR67, ETH76a,

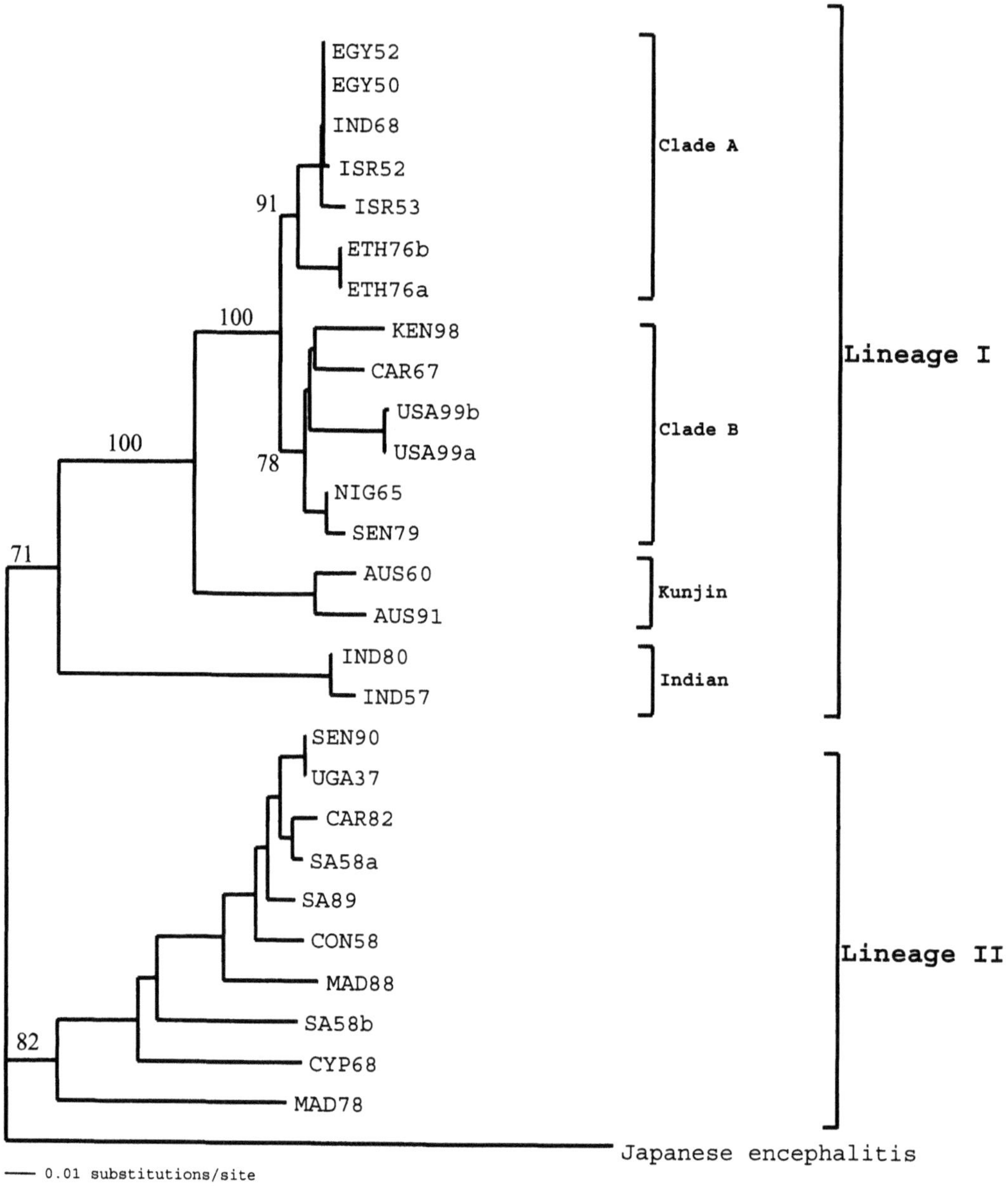

Fig. 1. Phylogenetic tree of WNV strains described in Table 1. Numbers next to branches refer to bootstrap values

ETH76b, ISR52, ISR53, EGY52, and EGY50) of 17 lineage I strains were not glycosylated (Table 1).

Mouse neuroinvasiveness

We used 3- to 4-week-old weaned NIH Swiss mice as a model to investigate virulence of WNV. All strains of WNV that we have tested were lethal for mice inoculated intracerebrally, indicating that all strains are able to replicate in the brain. In comparison, there were significant differences in the neuroinvasiveness of strains following intraperitoneal inoculation (Table 1). The four ancestral strains in lineage II (MAD78, CYP68, SA58b and MAD88) each required at least 125 pfu to induce a lethal infection while the other strains required 50 pfu or less for a lethal infection. Similarly, AUS60, AUS91, ETH76b and ISR53 in lineage I were attenuated and would not kill at any dose. All the strains in clade B of lineage I were virulent, with fewer than 12 pfu causing a lethal infection, while the majority in clade A were attenuated requiring at least 2000 pfu to induce a lethal infection. Thus, the evolution of both lineage I and lineage II strains shows increased neuroinvasiveness in the mouse model. Similar results have been found for hamster neuroinvasiveness [1]. There was no direct correlation between mouse neuroinvasiveness and E protein glycosylation, although the majority of the attenuated strains were nonglycosylated.

Genetic variation of WNV in the United States

During 2002, WNV spread throughout much of the U.S. and was introduced to many states for the first time, including Texas. Sampling of isolates made in the area surrounding Houston revealed two genetic variants. The major variant was found in much of Texas and differed from the prototype North American strain NY99 (385-99) by 0.18% at the nucleotide level. The minor variant has only been found on the southeast coastal area of Texas and differs by 0.35% from NY99. Significantly, the major and minor genetic variants differ by 0.5% from each other. Our subsequent studies have shown that the major variant was found through much of the U.S. during 2002, whereas NY99 typified isolates prior to 2002 [7]. Interestingly, the major variant has an amino acid substitution at E159 that was found in some strains from the Middle East and Europe prior to the introduction of the virus into North America in 1999 [7].

Compared to 1999, the minor variant was found to have a total of 30 nucleotide differences scattered throughout the genome encoding a total of five amino acid substitutions (one each in E, NS1, NS2A, NS4B and NS5). Interestingly, all of the amino acid substitutions were unique and not found in other isolates from North America that had been sequenced. The mouse neuroinvasive phenotypes of the major and minor genetic variants were compared to NY99 but no significant differences were found, indicating that there has been no major virulence phenotype difference in North America since 1999.

Attenuated phenotype of strain ETH76a

As described above, strain ETH76a is attenuated for mouse neuroinvasiveness (Table 1), yet is closely related phylogenetically to USA99 strains (Fig. 1). Comparison of replication of ETH76a and USA99b in mice following intraperitoneal inoculation shows that strain USA99b causes a viremia followed by invasion of the brain, while ETH76 causes a brief, low level viremia that is only detectable

Table 2. Virus titers in the sera and brains of mice, determined at daily intervals post-inoculation with neuroinvasive (USA99b) or non-invasive (ETH76a) WNV strains[a]

Day post-inoculation	Animal	USA99b inoculated		ETH76a inoculated	
		Serum titer (pfu/mL)	Brain titer (pfu/brain)	Serum titer (pfu/mL)	Brain titer (pfu/brain)
1	1	300	–*	350	–
	2	600	–	–	–
	3	200	–	–	–
2	1	2,000	–	50	–
	2	15,000	–	–	–
	3	35,000	–	–	–
3	1	3,000	–	–	–
	2	70,000	–	–	–
	3	25,000	–	–	–
4	1	2,000	–	–	–
	2	300	25	–	–
	3	150	2,000	–	–
5	1	–	–	–	–
	2	–	50	–	–
	3	–	750	–	–

*Indicates no virus detected; limits of detection were 50 pfu/mL of serum and 25 pfu/brain

[a]Groups of mice were inoculated with 1000 pfu by the ip route and three mice sampled per day and infectivity determined by plaque assay in Vero cells

in some animals on days 1 and 2 post-infection, and no detectable invasion of the brain (Table 2). Thus, attenuation is associated with low-level replication following infection and no invasion of the brain. Comparing the nucleotide sequences of the structural protein genes of the two virus strains reveals 128 nucleotide differences, encoding 6 amino acid substitutions; one in prM and five in the E protein. Current studies are utilizing infectious clone technology to make chimeric ETH76a/USA99 viruses to precisely map the molecular determinants of attenuation and virulence of WNV.

Conclusion

Although our knowledge of WNV has greatly increased in recent years, our understanding on the molecular basis of virulence of the virus is still rudimentary. Nonetheless, reverse genetics should help elucidate the molecular determinants of virulence and attenuation of WNV.

Acknowledgements

This work was funded in part by the State of Texas Advanced Research Program and the U.S. Centers for Disease Control and Prevention, Atlanta, Georgia.

References

1. Beasley DWC, Li L, Suderman MT, Barrett ADT (2002) Mouse neuroinvasive phenotype of West Nile virus strains varies depending upon virus genotype. Virology 296: 17–23
2. Blackburn NK, Thompson DL, Jupp PG (1987) Antigenic relationship of West Nile strains by titre ratios calculated from cross-neutralization test results. Epidem Infect 99: 551–557
3. Castle E, Leidner U, Nowak T, Wengler G, Wengler G (1986) Primary structure of the West Nile flavivirus genome region coding for all nonstructural proteins. Virology 149: 10–26
4. Hammam HM, Clarke DH, Price WH (1965) Antigenic variation of West Nile virus in relation to geography. Am J Epidemiol 82: 40–55
5. Hubálek Z, Halouzka J (1999) West Nile fever: a reemerging mosquito-borne viral disease in Europe. Emerg Infect Dis 5: 643–650
6. Lanciotti RS, Roehrig JT, Deubel V, Smith J, Parker M, Steele K, Crise B, Volpe KE, Crabtree MB, Scherret JH, Hall RA, MacKenzie JS, Cropp CB, Panigrahy B, Ostlund E, Schmitt B, Malkinson M, Banet C, Weissman J, Komar N, Savage HM, Stone W, McNamara T, Gubler DJ (1999) Origin of the West Nile virus responsible for an outbreak of encephalitis in the northeastern United States. Science 286: 2333–2337
7. Lanciotti RS, Ebel GD, Deubel V, Kerst AJ, Murri S, Meyer R, Bowen M, McKinney N, Morrill WB, Crabtree MB, Kramer LD, Roehrig JT (2002) Complete genome sequences and phylogenetic analysis of West Nile virus strains isolated from the United States, Europe, and the Middle East. Virology 298: 96–105
8. Monath TP, Heinz FX (1996) Flaviviruses. In: Fields BN, Knipe DM, Howley PM, et al. (eds) Fields virology, Lippincott-Raven Publishers, Philadelphia, pp 961–1025
9. Pierre V, Drouet MT, Deubel V (1994) Identification of mosquito-borne flavivirus sequences using universal primers and reverse transcription/polymerase chain reaction. Res Virol 145: 93–104
10. Solomon T, Vaughn DW (2002) Pathogenesis and clinical features of Japanese encephalitis and West Nile virus infections. Curr Topics Micro Immun 267: 171–194
11. Tsai TF, Popovici F, Cernescu C, Campbell GL, Nedelcu NI (1998) West Nile encephalitis epidemic in southeastern Romania. Lancet 352: 767–771
12. Scherret JM, Poidinger M, MacKenzie JS, Broom AK, Deubel V, Lipkin WI, Briese T, Gould EA, Hall RA (2001) Studies on the relationships between West Nile and Kunjin viruses. Emerg Infect Dis 7: 697–705

Author's address: Alan D. T. Barrett, Department of Pathology, University of Texas Medical Branch, 301 University Blvd, Galveston, TX 77555-0609, U.S.A.; e-mail: abarrett@utmb.edu

Arch Virol (2004) [Suppl] 18: 43–64

Archives of Virology

Genetic determinants of Venezuelan equine encephalitis emergence

S. C. Weaver[1], **M. Anishchenko**[1], **R. Bowen**[2], **A. C. Brault**[1], **J. G. Estrada-Franco**[1], **Z. Fernandez**[1], **I. Greene**[1], **D. Ortiz**[1], **S. Paessler**[1], and **A. M. Powers**[1]

[1]Center for Biodefense and Emerging Infectious Diseases and Department of Pathology, University of Texas Medical Branch, Galveston, Texas, U.S.A.
[2]Animal Reproduction and Biotechnology Laboratory, Colorado State University, Ft. Collins, Colorado, U.S.A.

Summary. Following a period of inactivity from 1973–1991, Venezuelan equine encephalitis (VEE) reemerged during the past decade in South America and Mexico. Experimental studies of VEE virus (VEEV) infection of horses with virus strains isolated during these outbreaks have revealed considerable variation in the ability of equine-virulent, epizootic strains to exploit horses as efficient amplification hosts. Subtype IC strains from recent outbreaks in Venezuela and Colombia amplify efficiently in equines, with a correlation between maximum viremia titers and the extent of the outbreak from which the virus strain was isolated. Studies of enzootic VEEV strains that are believed to represent progenitors of the epizootic subtypes support the hypothesis that adaptation to efficient replication in equines is a major determinant of emergence and the ability of VEEV to spread geographically. Correlations between the ability of enzootic and epizootic VEEV strains to infect abundant, equiphilic mosquitoes, and the location and extent of these outbreaks, also suggest that specific adaptation to *Ochlerotatus taeniorhynchus* mosquitoes is a determinant of some but not all emergence events. Genetic studies imply that mutations in the E2 envelope glycoprotein gene are major determinants of adaptation to both equines and mosquito vectors.

History of VEE

Venezuelan equine encephalitis (VEE) is a disease of humans and equines that has affected many parts of Latin America for nearly a century. Because equines remain important in Latin America for agriculture and transportation, VEE outbreaks have had both indirect (social and economic) as well as direct effects on human health [29, 38, 77, 81]. First recognized as a disease of horses, mules and donkeys in

Colombia and Venezuela during the 1930's, VEE outbreaks continued sporadically for most of the 20th century, primarily in northern South America but occasionally in Central America, Mexico and the U.S. Some of these outbreaks involved up to hundreds of thousands of human and equine cases, and spread over wide geographic areas.

From 1938–1956, VEE virus (VEEV) strains were only isolated during equine epizootics and human epidemics in Venezuela, Colombia, Trinidad and Peru. However, beginning in the late 1950's, antigenically related virus strains were detected in forest and swamp habitats in Central America [30, 58], South America [11, 62], Mexico [57] and Florida, U.S.A. [13] in the absence of equine disease. Further studies of these viruses indicated that they utilized mosquito vectors, as did their epizootic VEEV relatives, but they primarily infected small mammalian reservoir hosts, circulated continuously, and did not cause equine disease [29]. However, humans were shown to become infected with these enzootic VEEV and VEE complex strains when they entered enzootic foci of transmission, with occasional fatal cases [30, 90]. Later antigenic [89] and phylogenetic studies [48, 83] demonstrated that these enzootic viruses, along with the epizootic variants, comprise a complex of related alphaviruses.

The VEE complex of alphaviruses

The VEE complex consists of 13 subtypes and varieties, and includes 7 different virus species, in the genus *Alphavirus* of the family *Togaviridae* (Table 1) [85]. The VEEV strains isolated during major outbreaks are referred to as epizootic or epidemic (henceforth referred to as epizootic in this review) and belong to subtypes IAB and IC. The remaining subtypes (ID-IF, II-VI) are considered enzootic strains. A practical definition of epizootic is the association with equine outbreaks and equine virulence, while enzootic strains are isolated in situations not associated with equine epizootics and are generally avirulent for horses, donkeys and mules. However, VEEV strains recovered from recent equine outbreaks in Mexico, which produce encephalitis but little or no viremia [25], defy these traditional definitions and underscore the complex relationship between serotype, virulence, and the ability of VEEV strains to cause widespread human disease by exploiting equines as amplification hosts.

Structure and replication of VEEV

Alphaviruses have single stranded, positive sense RNA genomes of ca. 11.5 kB. At its 5′ end, the VEEV genome encodes four nonstructural proteins called nsP1-4, which participate in genome replication and viral protein processing in the host cell cytoplasm (Fig. 1). The 3′ one-third of the genome is colinear with a subgenomic message that is translated to produce the 3 major structural proteins: the capsid, and the E1 and E2 envelope glycoproteins. VEEV can bind to the laminin binding protein [39] for entry into cells via receptor-mediated endocytosis and E2 protein

Table 1. VEE antigenic complex viruses

Subtype	Species	Variety	Transmission pattern	Equine virulence	Location	Vector
I	VEE virus	AB	epizootic	yes	C., S., N. America	mammalophilic mosquitoes
	VEE virus	C	epizootic	yes	S. America	mammalophilic mosquitoes
	VEE virus	D	enzootic	no	Central, S. America	*Culex (Mel.) aikenii s.l (ocossa, panocossa)* [23, 24]; *vomerifer, pedroi, adamesi* [21]
	VEE virus	E	enzootic	variable	C. America, Mexico	*Culex (Mel.) taeniopus* [15]
	Mosso das Pedras virus*	F	enzootic	unknown	Brazil	unknown
II	Everglades virus		enzootic	no	Southern Florida	*Culex (Mel.) cedecei* [12, 13, 88]
III	Mucambo virus	A	enzootic	no	S. America	*Culex (Mel.) portesi* [2, 16]
	Tonate virus	B (also Bijou Bridge virus)	enzootic	unknown	S., N. America	Unknown, *Oeciacus vicarius*** (cliff swallow bug) [42]
	Mucambo virus	C (strain 71D1252)	enzootic	unknown	Western Peru	Unknown
	Mucambo virus	D (strain V407660)	enzootic	unknown	Western Peru	Unknown
IV	Pixuna virus		enzootic	unknown	Brazil	Unknown
V	Cabassou virus		enzootic	unknown	French Guiana	Unknown
VI	Rio Negro* virus		enzootic	unknown	Northern Argentina	*Cx. (Mel.) delpontei***

*Revision approved by the International Committee on the Taxonomy of Viruses
**Preliminary vector incrimination based only on virus isolation

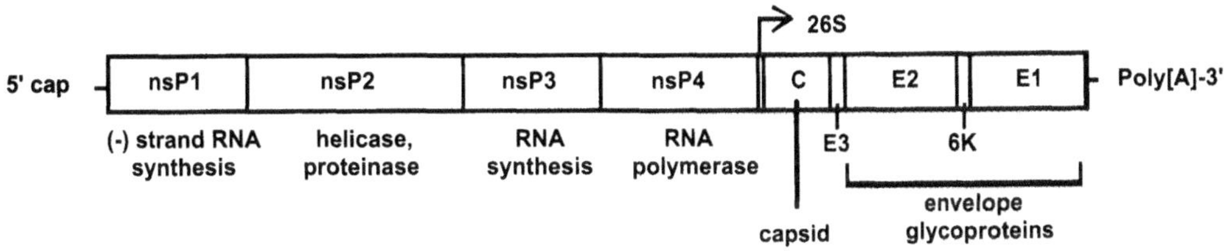

Fig. 1. Organization of the VEEV genome showing encoded proteins and their major functions

interactions. Other studies have shown that passage in some vertebrate cell cultures selects for binding to glycosaminoglycans via mutations on the surface of the E2 protein [5, 80]. However, these interactions with highly conserved cell surface molecules do not explain the restricted vector host range exhibited by some VEEV strains. After fusion of virions with the membrane of endosomes at low pH via a hydrophobic amino acid sequence in the E1 protein, the virion disassembles and the genome is translated in the cytoplasm to generate the nonstructural polyprotein. Viral genome replication takes place on the cytoplasmic surface of endosomes to produce minus strand copies that serve as the templates for genomic plus strands and an excess of subgenomic 26S messages that encode the structural proteins. Following interaction of 240 copies of the capsid protein with genomic RNA to form a nucleocapsid and processing of the envelope glycoproteins in the secretory pathway, maturation of virus particles occurs on the surface of cells, where nucleocapsids interact with the envelope glycoproteins to initiate budding of the 70 nm icosahedral

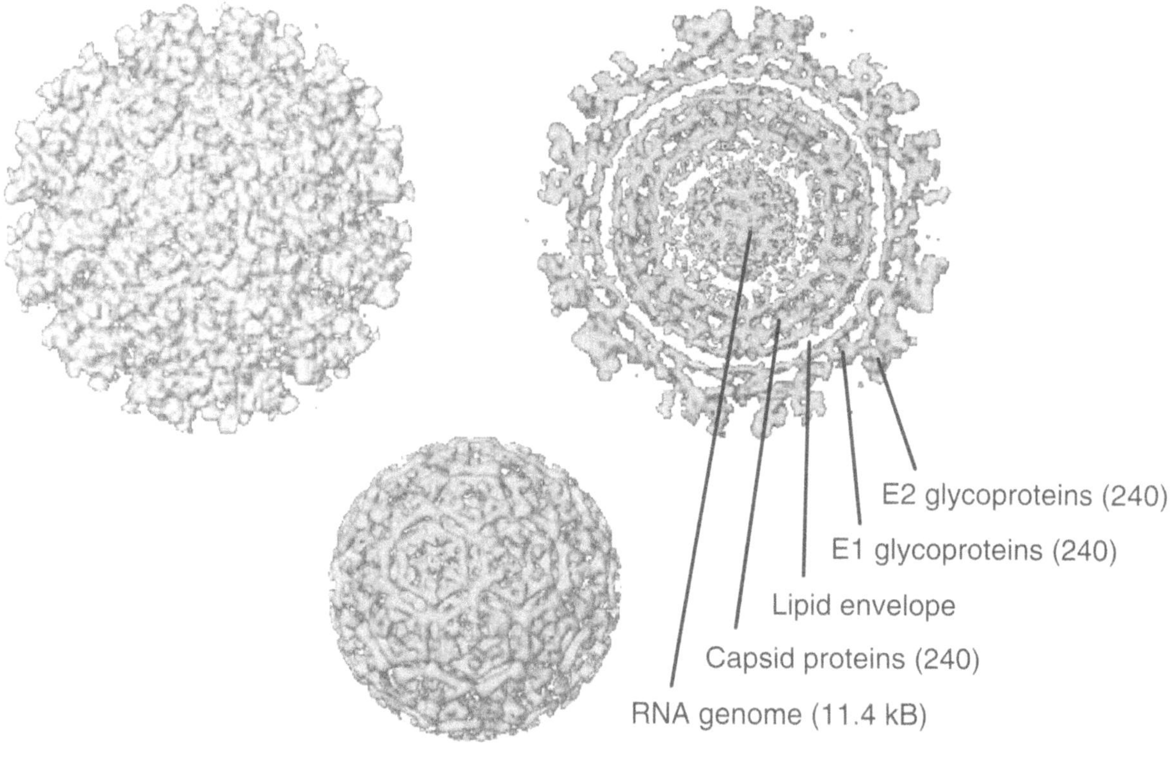

Fig. 2. Cryo-electron microscopic reconstruction of the VEE virion at 14 Å resolution. Upper left, external view of the virion showing the envelope glycoproteins on the surface; Upper right, cross section of the virion showing the RNA genome at the center surrounded by 240 copies of the capsid protein, the bilipid envelope derived from the host cell plasma membrane, the E1 envelope glycoprotein lying parallel to the envelope, and the E2 envelope glycoprotein forming spikes on the surface; Lower, nucleocapsid showing the capsid proteins in T = 4 icosahedral symmetry. Courtesy of Angel Paredes and Wah Chiu, Baylor College of Medicine, and Stanley Watowich and Scott Weaver, University of Texas Medical Branch

virion with T = 4 symmetry [45]. Structural studies of other alphaviruses [46] indicate that the E2 protein forms spikes on the surface of the virion, while the E1 protein lies adjacent to the host cell-derived lipid envelope (Fig. 2).

Recent VEE outbreaks

Between 1973 and 1992, no confirmed epizootic VEE activity was documented, prompting speculation that these strains had become extinct [77]. However, three recent epizootics have underscored the continuing threat of VEEV to human and equine health. The first began in December of 1992, in Trujillo State of western Venezuela, soon after the opening of a nearby dam and reservoir [50]. Several dozen human and equine cases were documented in December 1992 and January 1993. Then, in July of 1993, the outbreak reappeared on the western shore of Lake Maracaibo in Zulia State. Viruses isolated from humans and horses during this small outbreak represented the evolution of a new subtype IC strain genetically distinct from those isolated during all previous epizootics. The high degree of nucleotide (less than 0.5%) and amino acid (less than 0.3%) sequence divergence in comparison to enzootic, subtype ID strains from the same region of western Venezuela, as well as phylogenetic relationships, suggested that this outbreak resulted from the recent evolution of the IC strain from enzootic, subtype ID progenitors in western Venezuela [48, 50, 78].

In July of 1993, a small equine epizootic was reported in Pacific coastal areas of Chiapas State in Southern Mexico; human cases were not documented, although recent human serosurveys indicate that VEEV is endemic in the region (J. Estrada-Franco, S. C. Weaver, unpublished). Three years later, a similar outbreak involving several dozen equine cases occurred just to the north in coastal Oaxaca State [43]. Virus isolates from both outbreaks belong to subtype IE, not previously associated with equine epizootics. Genetic studies again supported the hypothesis that these epizootic viruses had recently gained the equine virulence phenotype from enzootic progenitors, but in this case without a change in serotype [7]. Surprisingly, experimental equine infections with these strains indicated that, although they can produce encephalitis, they do so without high serum viremia titers [25]. Thus, the Mexican subtype IE epizootic strains differ fundamentally from epizootic strains (IAB and IC) isolated during more extensive outbreaks, which exploit equines as amplification hosts. The lack of equine amplification during the 1993 and 1996 Mexican outbreaks probably prevented VEEV movement into Texas, as occurred 32 years ago from the same Mexican regions [38].

The most recent VEE outbreak began in April of 1995 in Falcon State of northern Venezuela and quickly spread into adjoining states. After movement across the mouth of Lake Maracaibo and into the Guajira Peninsula during the summer, a major equine epizootic and human epidemic occurred in both Venezuela and Colombia, with an estimated 75,000–100,000 human cases in total. Viruses isolated during this outbreak were identical serologically (subtype IC) and nearly identical genetically to those isolated from a 1962–64 epizootic with a nearly identical distribution. The genetic identity of a reference strain isolated in 1963 with the phylogenetically predicted ancestral sequence of the 1995 outbreak suggested a possible laboratory source for the 1995 epizootic [8].

Potential sources of epizootic VEEV strains

Because of the sporadic nature of outbreaks, a seminal question in VEE research has been the source of the IAB and IC strains isolated from all major epizootics. Although several hypotheses have been presented over the years [29], direct

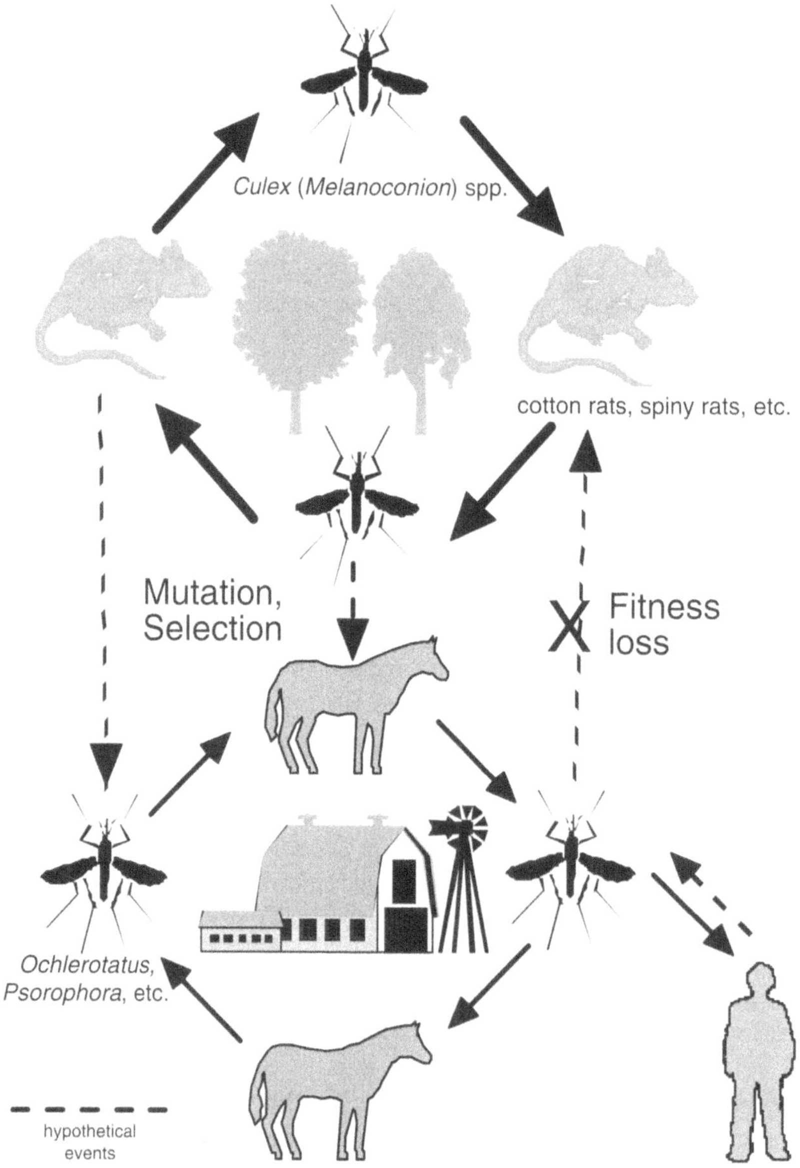

Fig. 3. Cartoon depicting enzootic (above) and epizootic/epidemic (below) transmission cycles of VEEV. Arrows show hypothetical adaptation to new hosts that mediate VEE emergence from enzootic progenitors

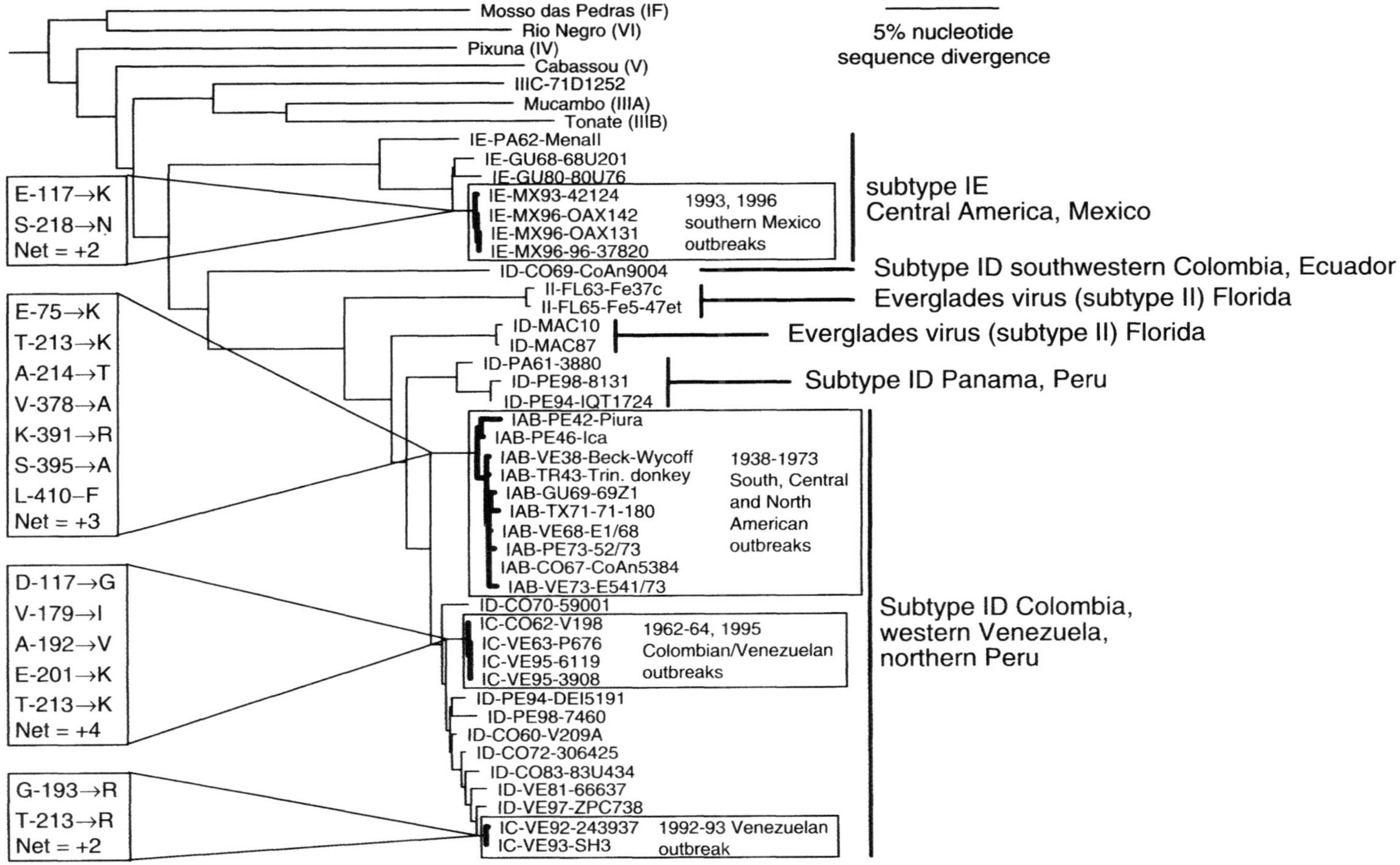

Fig. 4. Phylogenetic tree derived from nucleotide sequences encoding the PE2 envelope glycoprotein precursors using the neighbor joining method and the HK distance formula as implemented in the PAUP 4.0 software package [67]. Strains are indicated by subtype, followed by country (FL = Florida, USA) and strain designation. Major lineages of VEEV are indicated on the right, and strains isolated from epizootics are boxed and branches printed bold. Boxes on the left indicate E2 envelope glycoprotein changes accompanying the branches representing evolution of epizootic strains from enzootic progenitors, with net charge changes summarized for each epizootic clade. Bootstrap values for all major lineages and epizootic clades were $\geq$98%

evidence has been generated in support of only two: 1) genetic studies of subtype IAB strains isolated during epizootics [32, 86] as well as the isolation of live virus from some human vaccinees after administration of "inactivated" vaccine preparations made from IAB strains [66], indicate that some of the early outbreaks may have been initiated by the inoculation of equines with incompletely inactivated vaccines. However, all recent VEE outbreaks have involved subtypes IC and IE, which, to our knowledge, have never been used to make vaccines, indicating a non-vaccine source for all outbreaks since 1973; 2) genetic similarity as determined first using RNA fingerprinting [49] and later using nucleotide sequencing of enzootic and epizootic VEEV strains [31] suggested that the IAB and IC subtypes might have evolved from subtype ID. Later, phylogenetic analyses of the entire VEE complex indicated that the subtype IAB and IC strains had evolved at least three times from one of 6 principal enzootic lineages, subtype ID strains that occur in Colombia, western Venezuela and northern Peru [48, 53, 84]. The IAB stains probably evolved once, in the early 20th century, while IC strains evolved independently in 1962 and emerged again in 1992 (Fig. 4) [8]. The strongest biogeographical and epidemiological link supporting the evolution of epizootic strains from enzootic progenitors is between subtype IC strains isolated during the 1992–93 Venezuelan epizootic, and closely related ($<0.5\%$ nucleotide sequence divergence and only 7 amino acids associated with emergence of the IC strains) subtype ID strains isolated in 1997 in the same region of western Venezuela [50, 78]. Similar evidence implicates enzootic subtype IE strains that circulate on the Pacific coast of Guatemala or Mexico as the progenitors of equine-virulent, subtype IE strains isolated during the 1993 and 1996 Mexican epizootics [7, 43, 44].

The above mentioned data imply that enzootic VEEV strains undergo a small number of critical mutations that allow them to increase their virulence and ability to induce viremia in equines, resulting in highly efficient epizootic amplification and widespread human disease (Fig. 3). Because the vectors implicated during outbreaks also differ fundamentally from those that transmit enzootic strains [*Culex* (*Melanoconion*) spp.], adaptation to epizootic vectors could also be a critical mechanism of VEE emergence. These two kinds of adaptations are the subject of ongoing research in our laboratories, as described below.

Transmission cycles of VEEV: epizootic strains

Transmission cycles during VEE epizootics have received considerable attention since the 1930's and many aspects are relatively well understood. Subtype IAB and IC strains generally produce high titered viremia in naturally and experimentally infected equines [2–4 days of up to 10^8 infectious units per ml serum [18, 28, 29, 33, 34, 40, 41, 52, 54, 76]], and serve as excellent amplification hosts due to their attractiveness to mosquitoes and their limited defensive behavior. These viremic equines therefore infect large numbers of mosquitoes in agricultural settings, resulting in transmission to other equines, humans, and a variety of domestic animals (Fig. 3).

Humans develop similar levels of viremia (up to 10^8 infectious units/ml) to those of equines [6, 10, 54, 63, 87], and transmission studies with IAB viruses [64, 65] indicate that these titers exceed infection thresholds of some epizootic vector species. Experimental transmission from naturally infected humans to mice by *Ae. aegypti* was demonstrated during a Venezuelan epidemic [63]. However, humans are probably not major contributors to epidemic amplification, since major epidemics in the absence of equine amplification have never been reported. During epizootics, most people probably become infected tangentially when mosquitoes, infected by biting viremic equines, feed on people following extrinsic incubation [29, 77, 81]. Direct human-to-human transmission of VEE viruses has also been suggested by the sudden appearance, rapid increase and brief occurrence of human disease within affected communities during epidemics [63, 87]. This rapid epidemic transmission is surprising considering the requirement for extrinsic incubation in the vector of at least several days prior to transmission, in addition to an incubation period of one to a few days in the vertebrate host. However, household surveys conducted in epidemic regions of Colombia during 1995 indicated low secondary attack rates, and secondary cases among attending Venezuelan medical personnel were not identified [51]. Thus, there is no evidence of direct human-human transmission during epidemics.

Because infected equines develop high titered viremias, many different mosquitoes, some only marginally susceptible, can probably become infected by biting equines. Field studies have indicated that more than one vector species can be involved in transmission [54, 65]. Susceptibility to oral infection is a prerequisite for transmission, and some species appear to be nearly completely refractory [70]. However, ecological and behavioral traits such as longevity, host preference, survival and population size are probably more important than susceptibility differences in vectorial capacity for VEEV, and some species that appear to be only moderately susceptible to infection have been incriminated as important vectors during outbreaks.

Based on traditional criteria [3], the following species are proven epizootic vectors: *Psorophora confinnis* [54] and *Ps. columbiae* [65] (some previous publications did not distinguish these species), *Ochlerotatus sollicitans* [65, 73] and *Oc. taeniorhynchus* [35, 65, 73]. The latter species is probably the most important epizootic vector in South America, especially in coastal areas including the Guajira Peninsula where the largest outbreaks on record have occurred. *Culex (Deinocerites)* spp. may also be VEEV vectors in coastal areas [26].

Non-mosquito arthropods have also been implicated as VEEV vectors during outbreaks. Mechanical transmission (infection via contaminated vector mouth parts, without viral replication) by black flies has probably contributed to epidemic transmission during a Colombian epizootic [54]. Mites are also capable of experimental mechanical transmission of VEEV [19]. Ticks including *Amblyomma cajennense* (Acari: Ixodidae) [36] and *Hyalomma truncatum* [37] are susceptible to infection by enzootic and epizootic VEEV strains, but transstadial transmission, required for biological transmission, is generally inefficient.

Transmission cycles of enzootic VEEV strains

Enzootic VEE complex virus (subtypes ID-IF, II-VI) transmission cycles differ fundamentally from that of epizootic strains (Fig. 3). Enzootic viruses circulate nearly continuously in discrete, wet tropical forest or shaded swamp habitats. Transmission can be remarkably focal, varying dramatically within distances of only meters [17, 22]. Sylvatic rodents in the genera *Sigmodon*, *Oryzomys*, *Zygodontomys*, *Heteromys*, *Peromyscus* and *Proechimys* are believed to be the principle vertebrate amplifying hosts of enzootic VEE viruses because they are frequently infected in nature, have high rates of immunity, and develop moderate to high titered viremias after experimental infection [29, 77]. Other mammals such as opossums (*Didelphis marsupialis*) are also frequently infected [4, 23, 74, 75], and bats and shore birds may occasionally be involved in dispersal of enzootic viruses. Natural reservoir hosts are believed to show little adverse effect from enzootic VEE virus infection. However, unlike equines, people become ill after natural or laboratory-acquired (aerosol exposure) infection with enzootic VEE serotypes ID, IE, II and IIIA; natural enzootic infections occur sporadically when people enter enzootic transmission foci, but outbreaks involving tens to hundreds of people have also been described [55] as well as fatal cases.

Enzootic mosquito vectors of 4 different VEE serotypes have been identified in several locations. All are members of the subgenus *Culex* (*Melanoconion*) (Table 1), and all belong to the Spissipes Section of this subgenus, which contains only 23 species. Experimental studies indicate that these vectors tend to be highly susceptible to experimental infection, even with small oral doses of virus [23, 24, 56, 71, 72, 88], and a high degree of specificity has been observed between the vectors and VEE serotypes [59, 60].

The ecology of most *Culex* (*Melanoconion*) species is poorly understood, partly due to taxonomic difficulties. The immature stages of many species have never been described, and ecological knowledge is generally restricted to that of adult females. Most species occur primarily in tropical humid forests, and several occupy swamps [1] in association with aquatic plants like *Pistia* [23]. Many adult females, especially sylvatic vectors in the Spissipes Section, feed primarily on small sylvatic mammals [1, 68]. However, some species, including proven VEEV vectors, exhibit more opportunistic feeding behavior, and readily bite humans [14]. Blood feeding periodicity is generally crepuscular or nocturnal.

Equine amplification as a determinant of VEE emergence

Because epizootic mosquito vectors are often not extremely susceptible to infection with low oral doses of epizootic VEEV strain, the magnitude of equine viremia may be an important determinant of the efficiency of transmission and the temporal and geographic extent of VEE outbreaks. To test this hypothesis, data on equine viremia after experimental infection of equines with strains isolated during several different outbreaks can be compared from the literature and from

Table 2. Equine viremia after experimental infection with Venezuelan equine encephalitis viruses*

Virus strain(s)	Outbreak(s) (subtype)	Estimated magnitude of outbreak(s) (number of equine or human cases)	Peak mean equine viremia titer(s) ($\log_{10}$ $SMICLD_{50}$/ml serum or equivalent)**	Reference
V-38	1938 Venezuela (IAB)	Unknown, but extensive	6.1	[40]
P676	1963 Venezuela (IC)	32,000 human	7.0	[28, 40]
MF-8	Honduras 1969 (IAB)	20,000 equine	5.6	[18]
GJ9-1BJ	Guatemala 1969 (IAB)	20,000 equine	6.3	[28]
MF-8, Sal Pelayo, P-676***	Honduras 1969 (IAB); Colombia, 1963; Venezuela (IC) 1967; (IAB)	20,000 equine; 200,000 equine; 32,000 human, respectively, for each strain	6.1	[76]
243937, SH3	1992–93 (IC)	26 equine, 4 human (confirmed cases only)	6.0, 5.1	[79]
3908	1995 (IC)	100,000 human	6.7	R. A. Bowen, A. C. Brault, S. C. Weaver, unpublished
CPA201, I-290-93, OAX131, OAX142	1993, 1996 Mexico (IE)	125, 32 (equine only)	1.4, <0.6, 2.8, <0.6	[25]

*Results are summarized only for experiments with inocula titers of approximately $\log_{10}$ 2–5 $SMICLD_{50}$
**1,000 $SMICLD_{50}$/ml = ca. 1 Vero cell plaque forming unit
***Results were reported in a single graph summarizing experiments with all three strains

unpublished results (Table 2). A strong correlation between the magnitude of equine viremia and the size of the epizootics can be seen. The subtype IAB strains, responsible for the earliest documented VEE outbreaks, generate peak equine viremia titers of ca. 6 $\log_{10}$ $SMICLD_{50}$/ml (Table 2). Later IAB strains, probably related to vaccine-initiated transmission, also generate high mean peak titers of 5.6–6.3 $\log_{10}$ $SMICLD_{50}$/ml. Subtype IC strains from the major 1962–64 and 1995 epizootics also generate very high peak viremia titers of 6–7 $\log_{10}$ $SMICLD_{50}$/ml serum. Viruses isolated from the small 1992–93 Venezuelan epizootic generate slightly lower viremia titers (peak titers of 5.1–6.0 $\log_{10}$ $SMICLD_{50}$/ml) that may have resulted in slightly less efficient transmission. Strains isolated from the 1993 and 1996 Mexican epizootics produce very little viremia (<0.6–2.8 $\log_{10}$ $SMICLD_{50}$/ml) consistent with the limited spread and small size of these outbreaks despite their occurrence in the same region through which passed the major 1969–71 subtype IAB outbreak. These data support the hypothesis that equine viremia is a major determinant of VEEV amplification and the size of epizootics.

Table 3. Evidence for epizootic VEEV adaptation to mosquitoes from comparisons of the susceptibility of four species to closely related epizootic and enzootic strains*

Species	Subtype, outbreak and genetic determinants of adaptation							
	IAB	Determinant	1962/95 IC	Determinant	1992 IC	1993 IE	Determinant	Reference
Oc. taeniorhynchus	yes	E2 protein	yes	E2 protein	no	yes	E2-S218N	[9], A. C. Brault, S. C. Weaver, unpublished
Ps. confinnis	NT**	NA	no	NA	no	no	NA	D. Ortiz, S. C. Weaver, unpublished
Ae. aegypti	NT	NA	no	NA	no	NT	NA	D. Ortiz, S. C. Weaver, unpublished
Ae. albopictus	NT	NA	NT	NA	no	NT	NA	[20]

*Subtype IAB and IC strains were compared with closely related subtype ID strains; subtype IE Mexican strains were compared with closely related Guatemalan subtype IE strains not associated with equine outbreaks

**NT – not tested; NA – not applicable

Genetic determinants of equine amplification

The strong association of human disease epidemics with equine outbreaks, and the ability of epizootic subtype IAB and IC, but not enzootic VEEV strains to generate high titer viremia in equines, provide strong evidence supporting the idea that adaptation to equines is critical to VEE emergence. The ancestral relationship of the Colombia/western Venezuelan ID lineage to the IAB and IC clades indicated in phylogenetic studies provides hypotheses regarding the genetic determinants of equine amplification. Using maximum parsimony methods to predict nucleotide and amino acid changes that accompanied evolution of the epizootic strains, relatively small numbers of mutations can be tested using a reverse genetic approach and infectious cDNA clones derived from epizootic and enzootic strains representing their ancestors [78]. The first such studies used infectious clones derived from enzootic subtype IE and epizootic subtype IAB strains, which show different virulence patterns in guinea pigs. Chimeric viruses including the nonstructural proteins of an epizootic IAB strain and structural proteins of an enzootic IE strain (IAB/IE), or reciprocal constructs (IE/IAB), showed intermediate levels of virus replication and virulence for guinea pigs compared to the parental strains [47]. The IE/IAB chimera produced slightly higher viremia and an average survival time 2 days less than the VE/IAB-IE virus, suggesting that the structural proteins are more important determinants of virulence than the nonstructural proteins. Similar results implicating the E2 envelope glycoprotein as an important determinant have been obtained with subtype ID and IC strains, and chimeric recombinants with swapped E2 genes (I. Greene, S. Paessler and S. C. Weaver, unpublished). Initial studies with individual E2 positive charge mutations placed into a subtype ID backbone support the hypothesis that E2 mutations are critical to the acquisition of equine virulence and the ability of the virus to induce high serum viremia (R. Bowen, S. Paessler, M. Anishchenko, E. Brault, E. Wang, S. C. Weaver, unpublished). However, nonstructural protein mutations also may be involved.

Vector susceptibility as a determinant of VEE emergence

Because the mosquito host range must also change with the emergence of epizootic strains from enzootic progenitors, evidence has been sought in support of the hypothesis that selection for efficient infection of epizootic vectors mediates VEE emergence, using experimental infections mosquito infections. Although their studies were not designed to test this hypothesis, Kramer et al. [35] produced the first evidence that an established epizootic vector, *Oc. taeniorhynchus*, is more susceptible to epizootic (subtype IAB) than enzootic (subtype IE) VEEV strains. More recent work with this mosquito species has focused on subtype ID strains, which are believed to be the progenitors of epizootic VEEV. Like the IAB strains, a subtype IC strain (3908) isolated during the 1995 outbreak also infects *Oc. taeniorhynchus* more efficiently than does a closely related ID strain [9]. These findings support the hypothesis that adaptation to epizootic vector species can mediate VEE emergence.

More recent studies with *Oc. taeniorhynchus* have tested the ability of subtype IE strains from the 1993 and 1996 Mexican and 1992–93 Venezuelan epizootics to infect this species. The Mexican outbreaks occurred in coastal areas, and mosquito surveillance indicates that *Oc. taeniorhynchus* is the predominant mosquito species feeding on equines in the region. Experimental infections of *Oc. taeniorhynchus* with Mexican IE strains isolated from equines, and closely related IE isolates from nearby coastal Guatemala, not associated with equine disease, show a greater susceptibility to the Mexican strains (A. C. Brault, A. M. Powers, S. C. Weaver, unpublished). These results support the hypothesis that the Mexican outbreaks were enhanced by the ability of *Oc. taeniorhynchus* to efficiently transmit the virulent viruses to equines.

The 1992–93 Venezuelan subtype IC outbreak occurred in an inland region of western Venezuela and did not affect the Guajira Peninsula, the epicenter of many IAB outbreaks, as well as the 1962–64 and 1995 IC epizootics. The regions affected in 1992–93 were generally not inhabited by *Oc. taeniorhynchus*. Interestingly, 2 strains from this recent outbreak appear to be no more infectious for this species than for closely related ID strains, indicating that adaptation to *Oc. taeniorhynchus* did not mediate the 1992 emergence (D. Ortiz, S. C. Weaver, unpublished). These findings may help to explain why the 1992–93 epizootic did not spread into coastal areas such as the Guajira Peninsula where more extensive VEE outbreaks have occurred.

To examine further the role of VEEV adaptation to mosquito vectors in epizootic emergence, another proven epizootic vector, *Ps. confinnis*, has been tested with both enzootic ID and epizootic IAB and IC strains. Unlike *Oc. taeniorhynchus*, this species exhibits no differential susceptibility to the enzootic vs. epizootic strains (D. Ortiz, S. C. Weaver, unpublished). Two other species never implicated as important vectors, *Ae. albopictus* and *Ae. aegypti*, have also been evaluated. In addition to serving as controls due to their lack of involvement as epizootic vectors, they allowed us to examine the specificity of the vector adaptation associated with the IAB and IC (1962–64 and 1995) outbreaks. When susceptibility was evaluated by feeding *Ae. albopictus* blood meals containing a 1992 subtype IC strain vs. a closely related ID strain from Venezuela, both were found to be highly infectious for this species [20]. Similar experiments with *Ae. aegypti* showed slightly lower susceptibility but no difference between epizootic vs. enzootic VEEV strains (D. Ortiz, S. C. Weaver, unpublished). In total, these results suggest that adaptation to *Oc. taeniorhynchus* of the IAB epizootic strains, as well as the 1962–64 and 1995 IC strains, contributed to the ability of these viruses to cause extensive coastal epizootics in Latin America. Strains from the small 1992–93 subtype IC outbreak do not appear to infect mosquitoes any better than closely related subtype ID strains. However, the adaptation, mediated by mutations in the E2 envelope glycoprotein gene, appears to be species- or genus-specific and does not involve another proven vector, *Ps. confinnis*.

Another interesting feature of VEEV-mosquito interactions is that enzootic vectors are, in some cases, more susceptible to infection by sympatric enzootic than epizootic or allopatric enzootic variants. *Cx. taeniopus*, the enzootic subtype

IE vector in Guatemala, is highly susceptible to infection with IE strains but relatively refractory to subtype IAB, IC, ID, II (Everglades), III (Mucambo) and IV (Pixuna) viruses [59, 60]. The refractoriness of *Cx. taeniopus* to epizootic strains may explain the inability of the subtype IAB virus to persist in Guatemala following the 1969 outbreak. Additional experimental infections of subtype ID enzootic vectors (Table 1) with epizootic and enzootic strains are needed to test the hypothesis that adaptation to epizootic vectors reduces the fitness of epizootic strains for the ancestral enzootic vectors (Fig. 3).

Genetic determinants of adaptation to epizootic vectors

To map mosquito infection determinants of these subtype IAB, IC, ID and IE VEEV strains, reciprocal chimeric viruses generated from epizootic and enzootic strains were tested for mosquito infectivity. Chimeras containing the epizootic structural gene regions and, more specifically, the PE2 envelope glycoprotein E2 precursor gene, efficiently infected *Oc. taeniorhynchus*. Introduction of the PE2 gene from an enzootic subtype ID virus into an epizootic (IAB or IC) genetic backbone resulted in lower infection rates than those of the epizootic parent. These results demonstrate that the E2 envelope glycoprotein, the site of epitopes that define the enzootic and epizootic subtypes, also encodes mosquito infection determinants [9]. Similar reverse genetic studies with the Guatemalan and Mexican subtype IE strains implicate a single mutation in the E2 protein as the determinant of enhanced *Oc. taeniorhynchus* infection exhibited by strains isolated during the epizootics (A. C. Brault, A. M. Powers, S. C. Weaver, unpublished).

VEE virus as a biological weapon

In addition to being a natural human pathogen via its natural emergence mechanisms described above, VEEV is a potential agent of biological warfare or terrorism (BWT) for several reasons [27]:

(a) VEEV was highly developed as a BWT agent in the U.S. and the former USSR.
(b) VEEV is readily isolated from natural sources throughout much of the New World [81].
(c) VEEV replicates to very high titer in a variety of cell cultures (10^8–10^9 infectious units per ml) and is more amenable to culture and handling compared with many other viral BWT agents [69].
(d) VEEV is highly stable when lyophilized.
(e) VEEV is highly infectious by aerosol; for example, over 150 laboratory infections have been documented, and most were presumed to have occurred via aerosols [61].
(f) VEEV produces a highly debilitating and sometimes fatal disease, with permanent neurological sequelae in many human cases. It also produces temporary immunosuppression [69].

(g) If introduced into a location with equines and mosquitoes, VEEV can produce a widespread epidemic [82].
(h) No licensed human vaccine exists for VEEV; current experimental vaccines produced by the U.S. Army (TC-83 live-attenuated, C84 inactivated vaccines) have poor efficacy and/or a high rate of adverse reactions (TC-83), and are unavailable to many non-military investigators.
(i) No effective antivirals have been developed against VEEV [69].
(j) The genomes of VEEV can be placed into infectious cDNA clones to allow virtually any genetic manipulation, including the introduction of foreign genes such as toxins or human cytokines. It therefore has the potential to be developed into an even more virulent pathogen.

For the above reasons, VEEV is a select agent and a category B pathogen on both the CDC Critical Biological Agents for Public Health Preparedness the NIAID Virus Threat List.

Conclusions

Sequencing and phylogenetic studies led to the hypothesis that epizootic strains of VEEV arise via mutations in enzootic, subtype ID and IE variants that mediate changes in host range. Adaptation for efficient replication and high titer viremia in equines is supported by experimental infections with enzootic and most epizootic variants, and phylogenetic and reverse genetic studies have implicated mutations in the E2 envelope glycoprotein, some of which increase the charge on the virion surface, as determinants of this host range change. Some but not all epizootic variants also show evidence of adaptation for efficient infection of *Oc. taeniorhynchus*, one of the most important epizootic mosquito vectors. As with adaptation to equine infection, the determinants of *Oc. taeniorhynchus* infection are mutations in the E2 glycoprotein. More detailed, mechanistic studies are needed to elucidate the cellular and molecular interactions between VEEV and its hosts and vectors that are responsible for host range mediated VEE emergence.

Acknowledgments

The authors' research is supported by National Institutes of Health grants AI39800 and AI48807, and by the National Aeronautics and Space Administration (NASA). ACB was supported by the James L. McLaughlin Infection and Immunity Fellowship Fund and the NIH Emerging Tropical Diseases T32 Training grant AI-107526. AMP, SP and DO were supported by the James W. McLaughlin Fellowship Fund.

References

1. Aitken TH, Worth CB, Tikasingh ES (1968) Arbovirus studies in Bush Bush Forest, Trinidad, W. I., September 1959–December 1964. 3. Entomologic studies. Am J Trop Med Hyg 17: 253–268
2. Aitken THG (1972) Habits of some mosquito hosts of VEE (Mucambo) virus from northeastern South America, including Trinidad. In: Proc, Workshop-Symposium on

Venezuelan encephalitis virus. Pan American Health Organization, Washington DC, Scientific Publ 243, pp 254–256
3. Barnett HC (1960) The incrimination of arthropods as vectors of disease. In: Proc. 11th Intl Congr Entomol 2: 341–345
4. Barrera R, Ferro C, Navarro JC, Freier J, Liria J, Salas R, Ahumada M, Vasquez C, Gonzalez M, Kang W, Boshell J, Weaver SC (2002) Contrasting sylvatic foci of Venezuelan equine encephalitis virus in northern South America. Am J Trop Med Hyg 67: 324–334
5. Bernard KA, Klimstra WB, Johnston RE (2000) Mutations in the E2 glycoprotein of Venezuelan equine encephalitis virus confer heparan sulfate interaction, low morbidity, and rapid clearance from blood of mice. Virology 276: 93–103
6. Bowen GS, Calisher CH (1976) Virological and serological studies of Venezuelan equine encephalomyelitis in humans. J Clin Microbiol 4: 22–27
7. Brault AC, Powers AM, Holmes EC, Woelk CH, Weaver SC (2002) Positively charged amino acid substitutions in the E2 envelope glycoprotein are associated with the emergence of Venezuelan equine encephalitis virus. J Virol 76: 1718–1730
8. Brault AC, Powers AM, Medina G, Wang E, Kang W, Salas RA, De Siger J, Weaver SC (2001) Potential sources of the 1995 Venezuelan equine encephalitis subtype IC epidemic. J Virol 75: 5823–5832
9. Brault AC, Powers AM, Weaver SC (2002) Vector infection determinants of Venezuelan equine encephalitis virus reside within the E2 envelope glycoprotein. J Virol 76: 6387–6392
10. Briceno Rossi AL (1967) Rural epidemic encephalitis in Venezuela caused by a group A arbovirus (VEE). Progr Med Virol 9: 176–203
11. Causey OR, Causey CE, Maroja OM, Macedo DG (1961) The isolation of arthropod-borne viruses, including members of hitherto undescribed serological groups, in the Amazon region of Brazil. Am J Trop Med Hyg 10: 227–249
12. Chamberlain RW (1972) Discussion comments Proc. Workshop-Symposium on Venezuelan Encephalitis Virus. Pan American Health Organization, Washington DC, Scientific Publ 243, pp 257–258
13. Chamberlain RW, Sudia WD, Work TH, Coleman PH, Newhouse VF, Johnston JG Jr (1969) Arbovirus studies in south Florida, with emphasis on Venezuelan equine encephalomyelitis virus. Am J Epidemiol 89: 197–210
14. Cupp EW, Scherer WF, Lok JB, Brenner RJ, Dziem GM, Ordonez JV (1986) Entomological studies at an enzootic Venezuelan equine encephalitis virus focus in Guatemala, 1977–1980. Am J Trop Med Hyg 35: 851–859
15. Cupp EW, Scherer WF, Ordonez JV (1979) Transmission of Venezuelan encephalitis virus by naturally infected *Culex* (*Melanoconion*) *opisthopus*. Am J Trop Med Hyg 28: 1060–1063
16. Davies JB, Martinez R (1970) Observations on the population dynamics, behavior and maintenance of a laboratory colony of *Culex* (*Melanoconion*) *portesi* Senevet and Abonnenc, 1941 (Diptera: *Culicidae*). J Med Entomol 7: 179–188
17. Dickerman RW, Cupp EW, Groot H, Alarcon AM, Cura E, Dickerman AW, Ibagos AL, Ricco-Hesse R, Taylor CA, Weaver SC (1986) Venezuelan equine encephalitis virus activity in northern Colombia during April and May 1983. Bull Pan Am Health Organ 20: 276–283
18. Dietz WH Jr, Alvarez O Jr, Martin DH, Walton TE, Ackerman LJ, Johnson KM (1978) Enzootic and epizootic Venezuelan equine encephalomyelitis virus in horses infected by peripheral and intrathecal routes. J Infect Dis 137: 227–237

19. Durden LA, Linthicum KJ, Turell MJ (1992) Mechanical transmission of Venezuelan equine encephalomyelitis virus by hematophagous mites (Acari). J Med Entomol 29: 118–121
20. Fernández Z, Moncayo AC, Carrara AS, Forattini OP, Weaver SC (2003) Vector competence of *Aedes (Stegomyia) albopictus* (Diptera: Culicidae) from São Paulo State, Brazil for IC, ID and IF subtypes of Venezuelan equine encephalitis virus. J Med Entomol 40: 522–527
21. Ferro C, Boshell J, Moncayo AC, Gonzalez M, Ahumada ML, Kang W, Weaver SC (2003) Natural enzootic vectors of Venezuelan equine encephalitis virus, Magdalena Valley, Colombia. Emerg Infect Dis 9: 49–54
22. Franck PT, Johnson KM (1970) An outbreak of Venezuelan encephalitis in man in the Panama Canal Zone. Am J Trop Med Hyg 19: 860–865
23. Galindo P (1972) Endemic vectors of Venezuelan encephalitis. In: Proc. Workshop-Symposium on Venezuelan Encephalitis Virus, Scientific Publ. 243. Pan American Health Organization, Washington DC, pp 249–253
24. Galindo P, Grayson MA (1971) *Culex (Melanoconion) aikenii*: Natural vector in Panama of endemic Venezuelan encephalitis. Science 172: 594–595
25. Gonzalez-Salazar D, Estrada-Franco JG, Carrara AS, Aronson JF, Weaver SC (2003) Equine amplification and virulence of subtype IE Venezuelan equine encephalitis viruses isolated during the 1993 and 1996 Mexican epizootics. Emerg Infect Dis 9: 161–168
26. Grayson MA, Galindo P (1972) Experimental transmission of Venezuelan equine encephalitis virus by *Deinocerites pseudes* Dyar and Knab, 1909. J Med Entomol 9: 196–200
27. Hawley RJ, Eitzen EM Jr (2001) Biological weapons – a primer for microbiologists. Annu Rev Microbiol 55: 235–253
28. Henderson BE, Chappell WA, Johnston JG Jr, Sudia WD (1971) Experimental infection of horses with three strains of Venezuelan equine encephalomyelitis virus. I. Clinical and virological studies. Am J Epidemiol 93: 194–205
29. Johnson KM, Martin DH (1974) Venezuelan equine encephalitis. Adv Vet Sci Comp Med 18: 79–116
30. Johnson KM, Shelokov A, Peralta PH, Dammin GJ, Young NA (1968) Recovery of Venezuelan equine encephalomyelitis virus in Panama. A fatal case in man. Am J Trop Med Hyg 17: 432–440
31. Kinney RM, Tsuchiya KR, Sneider JM, Trent DW (1992) Genetic evidence that epizootic Venezuelan equine encephalitis (VEE) viruses may have evolved from enzootic VEE subtype I-D virus. Virology 191: 569–580
32. Kinney RM, Tsuchiya KR, Sneider JM, Trent DW (1992) Molecular evidence for the origin of the widespread Venezuelan equine encephalitis epizootic of 1969 to 1972. J Gen Virol 73: 3301–3305
33. Kissling RE, Chamberlain RW (1967) Venezuelan equine encephalitis. Adv Vet Sci 11: 65–84
34. Kissling RE, Chamberlain RW, Nelson DB, Stamm DD (1956) Venezuelan equine encephalomyelitis in horses. Am J Hyg 63: 274–282
35. Kramer LD, Scherer WF (1976) Vector competence of mosquitoes as a marker to distinguish Central American and Mexican epizootic from enzootic strains of Venezuelan encephalitis virus. Am J Trop Med Hyg 25: 336–346
36. Linthicum KJ, Gordon SW, Monath TP (1992) Comparative infections of epizootic and enzootic strains of Venezuelan equine encephalomyelitis virus in *Amblyomma cajennense* (Acari: *Ixodidae*). J Med Entomol 29: 827–831

37. Linthicum KJ, Logan TM (1994) Laboratory transmission of Venezuelan equine encephalomyelitis virus by the tick *Hyalomma truncatum*. Trans R Soc Trop Med Hyg 88: 126
38. Lord RD (1974) History and geographic distribution of Venezuelan equine encephalitis. Bull PAHO 8: 100–110
39. Ludwig GV, Kondig JP, Smith JF (1996) A putative receptor for Venezuelan equine encephalitis virus from mosquito cells. J Virol 70: 5592–5599
40. Mackenzie RM, de Siger J, Parra D (1976) Venezuelan equine encephalitis virus: comparison of infectivity and virulence of strains V-38 and P676 in donkeys. Am J Trop Med Hyg 25: 494–499
41. Martin DH, Dietz WH, Alvaerez O Jr, Johnson KM (1982) Epidemiological significance of Venezuelan equine encephalomyelitis virus in vitro markers. Am J Trop Med Hyg 31: 561–568
42. Monath TP, Lazuick JS, Cropp CB, Rush WA, Calisher CH, Kinney RM, Trent DW, Kemp GE, Bowen GS, Francy DB (1980) Recovery of Tonate virus ("Bijou Bridge" strain), a member of the Venezuelan equine encephalomyelitis virus complex, from cliff swallow nest bugs (*Oeciacus vicarius*) and nestling birds in North America. Am J Trop Med Hyg 29: 969–983
43. Oberste MS, Fraire M, Navarro R, Zepeda C, Zarate ML, Ludwig GV, Kondig JF, Weaver SC, Smith JF, Rico-Hesse R (1998) Association of Venezuelan equine encephalitis virus subtype IE with two equine epizootics in Mexico. Am J Trop Med Hyg 59: 100–107
44. Oberste MS, Schmura SM, Weaver SC, Smith JF (1999) Geographic distribution of Venezuelan equine encephalitis virus subtype IE genotypes in Central America and Mexico. Am J Trop Med Hyg 60: 630–634
45. Paredes A, Alwell-Warda K, Weaver SC, Chiu W, Watowich SJ (2001) Venezuelan equine encephalomyelitis virus structure and its divergence from Old World alphaviruses. J Virol 75: 9532–9537
46. Pletnev SV, Zhang W, Mukhopadhyay S, Fisher BR, Hernandez R, Brown DT, Baker TS, Rossmann MG, Kuhn RJ (2001) Locations of carbohydrate sites on alphavirus glycoproteins show that E1 forms an icosahedral scaffold. Cell 105: 127–136
47. Powers AM, Brault AC, Kinney RM, Weaver SC (2000) The use of chimeric Venezuelan equine encephalitis viruses as an approach for the molecular identification of natural virulence determinants. J Virol 74: 4258–4263
48. Powers AM, Oberste MS, Brault AC, Rico-Hesse R, Schmura SM, Smith JF, Kang W, Sweeney WP, Weaver SC (1997) Repeated emergence of epidemic/epizootic Venezuelan equine encephalitis from a single genotype of enzootic subtype ID virus. J Virol 71: 6697–6705
49. Rico-Hesse R, Roehrig JT, Trent DW, Dickerman RW (1988) Genetic variation of Venezuelan equine encephalitis virus strains of the ID variety in Colombia. Am J Trop Med Hyg 38: 195–204
50. Rico-Hesse R, Weaver SC, de Siger J, Medina G, Salas RA (1995) Emergence of a new epidemic/epizootic Venezuelan equine encephalitis virus in South America. Proc Natl Acad Sci USA 92: 5278–5281
51. Rivas F, Diaz LA, Cardenas VM, Daza E, Bruzon L, Alcala A, De la Hoz O, Caceres FM, Aristizabal G, Martinez JW, Revelo D, De la Hoz F, Boshell J, Camacho T, Calderon L, Olano VA, Villarreal LI, Roselli D, Alvarez G, Ludwig G, Tsai T (1997) Epidemic Venezuelan equine encephalitis in La Guajira, Colombia, 1995. J Infect Dis 175: 828–832

52. Roberts ED, Sanmartin C, Payan J, Mackenzie RB (1970) Neuropathologic changes in 15 horses with naturally occurring Venezuelan equine encephalomyelitis. Am J Vet Res 31: 1223–1229
53. Salas RA, Garcia CZ, Liria J, Barrera R, Navarro JC, Medina G, Vasquez C, Fernandez Z, Weaver SC (2001) Ecological studies of enzootic Venezuelan equine encephalitis in north-central Venezuela, 1997–1998. Am J Trop Med Hyg 64: 84–92
54. Sanmartin C, Mackenzie RB, Trapido H, Barreto P, Mullenax CH, Guteirrez E, Lessnes C (1967) Encefalitis equina Venezolana en Colombia, 1967. Bol Sanit Panam 74: 108–137
55. Sanmartin-Barberi C, Osorno-Mesa E (1954) Human epidemic in Colombia caused by the Venezuelan equine encephalomyelitis virus. Am J Trop Med Hyg 3: 283–291
56. Scherer WF, Cupp EW, Lok JB, Brenner RJ, Ordonez JV (1981) Intestinal threshold of an enzootic strain of Venezuelan equine encephalomyelitis virus in *Culex* (*Melanoconion*) *taeniopus* mosquitoes and its implication to vector competency and vertebrate amplifying hosts. Am J Trop Med Hyg 30: 862–869
57. Scherer WF, Dickerman RW, Chia CW, Ventura A, Moorhouse A, Geiger R, Diaz-Najera A (1963) Venezuelan equine encephalitis virus in Veracruz, Mexico, and the use of hamsters as sentinels. Science 145: 274–275
58. Scherer WF, Dickerman RW, Ordonez JV (1970) Discovery and geographic distribution of Venezuelan encephalitis virus in Guatemala, Honduras, and British Honduras during 1965–68, and its possible movement to Central America and Mexico. Am J Trop Med Hyg 19: 703–711
59. Scherer WF, Weaver SC, Taylor CA, Cupp EW (1986) Vector incompetency: its implication in the disappearance of epizootic Venezuelan equine encephalomyelitis virus from Middle America. J Med Entomol 23: 23–29
60. Scherer WF, Weaver SC, Taylor CA, Cupp EW, Dickerman RW, Rubino HH (1987) Vector competence of *Culex* (*Melanoconion*) *taeniopus* for allopatric and epizootic Venezuelan equine encephalomyelitis viruses. Am J Trop Med Hyg 36: 194–197
61. US Dept. Health and Human Services (1999) Biosafety in microbiological and biomedical laboratories, 4th edn. US Government Printing Office, Washington DC
62. Shope RE, Causey OR, Homobono Paes de Andrade A, Theiler M (1964) The Venezuelan equine encephalomyelitis complex of group A arthropod-borne viruses, including Mucambo and Pixuna from the Amazon region of Brazil. Am J Trop Med Hyg 13: 723–727
63. Suarez OM, Bergold GH (1968) Investigations of an outbreak of Venezuelan equine encephalitis in towns of eastern Venezuela. Am J Trop Med Hyg 17: 875–880
64. Sudia WD, Newhouse VF (1975) Epidemic Venezuelan equine encephalitis in North America: a summary of virus-vector-host relationships. Am J Epidemiol 101: 1–13
65. Sudia WD, Newhouse VF, Beadle ID, Miller DL, Johnston JG Jr, Young R, Calisher CH, Maness K (1975) Epidemic Venezuelan equine encephalitis in North America in 1971: vector studies. Am J Epidemiol 101: 17–35
66. Sutton LS, Brooke CC (1954) Venezuelan equine encephalomyelitis due to vaccination in man. J Am Med Assoc 155: 1473–1476
67. Swofford DL (1998) PAUP*. Phylogenetic Analysis Using Parsimony (* and other methods). Version 4. Sinauer Associates, Sunderland, Massachusetts
68. Tempelis CH, Galindo P (1975) Host-feeding patterns of *Culex* (*Melanoconion*) and *Culex* (*Aedinus*) mosquitoes collected in Panama. J Med Entomol 12: 205–209
69. Tsai TF, Weaver SC, Monath TP (2002) Alphaviruses. In: Richman DD, Whitley RJ, Hayden FG (eds) Clinical virology. ASM Press, Washington DC, pp 1177–1210

70. Turell MJ (1999) Vector competence of three Venezuelan mosquitoes (Diptera: Culicidae) for an epizootic IC strain of Venezuelan equine encephalitis virus. J Med Entomol 36: 407–409
71. Turell MJ, Barth J, Coleman RE (1999) Potential for Central American mosquitoes to transmit epizootic and enzootic strains of Venezuelan equine encephalitis virus. J Am Mosq Control Assoc 15: 295–298
72. Turell MJ, Jones JW, Sardelis MR, Dohm DJ, Coleman RE, Watts DM, Fernandez R, Calampa C, Klein TA (2000) Vector competence of Peruvian mosquitoes (Diptera: Culicidae) for epizootic and enzootic strains of Venezuelan equine encephalomyelitis virus. J Med Entomol 37: 835–839
73. Turell MJ, Ludwig GV, Beaman JR (1992) Transmission of Venezuelan equine encephalomyelitis virus by *Aedes sollicitans* and *Aedes taeniorhynchus* (Diptera: Culicidae). J Med Entomol 29: 62–65
74. Walder R, Suarez OM, Calisher CH (1984) Arbovirus studies in southwestern Venezuela during 1973–1981. II. Isolations and further studies of Venezuelan and eastern equine encephalitis, Una, Itaqui, and Moju viruses. Am J Trop Med Hyg 33: 483–491
75. Walder R, Suarez OM, Calisher CH (1984) Arbovirus studies in the Guajira region of Venezuela: activities of eastern equine encephalitis and Venezuelan equine encephalitis viruses during an interepizootic period. Am J Trop Med Hyg 33: 699–707
76. Walton TE, Alvarez O, Buckwalter RM, Johnson KM (1973) Experimental infection of horses with enzootic and epizootic strains of Venezuelan equine encephalomyelitis virus. J Infect Dis 128: 271–282
77. Walton TE, Grayson MA (1988) Venezuelan equine encephalomyelitis. In: Monath TP (ed) The arboviruses: epidemiology and ecology, vol IV. CRC Press, Boca Raton, pp 203–231
78. Wang E, Barrera R, Boshell J, Ferro C, Freier JE, Navarro JC, Salas R, Vasquez C, Weaver SC (1999) Genetic and phenotypic changes accompanying the emergence of epizootic subtype IC Venezuelan equine encephalitis viruses from an enzootic subtype ID progenitor. J Virol 73: 4266–4271
79. Wang E, Bowen RA, Medina G, Powers AM, Kang W, Chandler LM, Shope RE, Weaver SC (2001) Virulence and viremia characteristics of 1992 epizootic subtype IC Venezuelan equine encephalitis viruses and closely related enzootic subtype ID strains. Am J Trop Med Hyg 65: 64–69
80. Wang E, Brault AC, Powers AM, Kang W, Weaver SC (2003) Glycosaminoglycan binding properties of natural venezuelan equine encephalitis virus isolates. J Virol 77: 1204–1210
81. Weaver SC (1998) Recurrent emergence of Venezuelan equine encephalomyelitis. In: Scheld WM, Hughes J (eds) Emerging infections I. ASM Press, Washington DC, pp 27–42
82. Weaver SC (2001) Venezuelan equine encephalitis. In: Service MW (ed) The encyclopedia of arthropod-transmitted infections. CAB International, Wallingford, UK, pp 539–548
83. Weaver SC, Bellew LA, Rico-Hesse R (1992) Phylogenetic analysis of alphaviruses in the Venezuelan equine encephalitis complex and identification of the source of epizootic viruses. Virology 191: 282–290
84. Weaver SC, Bellew LA, Rico-Hesse R (1992) Phylogenetic analysis of alphaviruses in the Venezuelan equine encephalitis complex and identification of the source of epizootic viruses. Virology 191: 282–290
85. Weaver SC, Dalgarno L, Frey TK, Huang HV, Kinney RM, Rice CM, Roehrig JT, Shope RE, Strauss EG (2000) Family *Togaviridae*. In: van Regenmortel MHV, Fauquet CM, Bishop DHL, Carstens EB, Estes MK, Lemon SM, Maniloff J, Mayo MA, McGeogh

DJ, Pringle CR, Wickner RB (eds) Virus taxonomy. Classification and nomenclature of viruses. Seventh Report of the International Committee on Taxonomy of Viruses. Academic Press, San Diego, pp 879–889
86. Weaver SC, Pfeffer M, Marriott K, Kang W, Kinney RM (1999) Genetic evidence for the origins of Venezuelan equine encephalitis virus subtype IAB outbreaks. Am J Trop Med Hyg 60: 441–448
87. Weaver SC, Salas R, Rico-Hesse R, Ludwig GV, Oberste MS, Boshell J, Tesh RB (1996) Re-emergence of epidemic Venezuelan equine encephalomyelitis in South America. VEE Study Group. Lancet 348: 436–440
88. Weaver SC, Scherer WF, Taylor CA, Castello DA, Cupp EW (1986) Laboratory vector competence of *Culex (Melanoconion) cedecei* for sympatric and allopatric Venezuelan equine encephalomyelitis viruses. Am J Trop Med Hyg 35: 619–623
89. Young NA, Johnson KM (1969) Antigenic variants of Venezuelan equine encephalitis virus: their geographic distribution and epidemiologic significance. Am J Epidemiol 89: 286–307
90. Zarate ML, Scherer WF, Dickerman RW (1970) Venezuelan equine encephalitis virus as a human infection determinant. Description of a fatal case occurring in Jaltipan, Ver., in 1965. Rev Invest Salud Publ 30: 296–302

Author's address: Scott C. Weaver, Department of Pathology, University of Texas Medical Branch, Galveston, Texas 77555-0609, U.S.A.; e-mail: sweaver@utmb.edu

Arch Virol (2004) [Suppl] 18: 65–84

Archives of Virology

Evolution and dispersal of encephalitic flaviviruses

E. A. Gould, S. R. Moss, and **S. L. Turner**

Centre for Ecology and Hydrology, Oxford, U.K.

Summary. There are two major groups of encephalitic flaviviruses, those that infect and are transmitted by ticks, particularly *Ixodes* spp. and those that infect and are transmitted by mosquitoes, particularly *Culex* spp. The tick-borne encephalitic flaviviruses exhibit evolutionary characteristics that are largely determined by the protracted life cycle of the tick, its habitat and the prevailing climatic conditions. These viruses appear to have evolved gradually from non-encephalitic viruses that radiated eastwards and north eastwards out of Africa into Asia and the southern islands, then northwards to far east Asia and finally westwards across Eurasia to western Europe, during the past two to four thousand years. Only one of these recognized species has found its way to North America viz. Powassan virus. In contrast, the evolution of the recognized mosquito-borne encephalitic flaviviruses reflects the wide range of mosquito species that they infect. They emerged out of Africa relatively recently and at roughly the same time, i.e., probably during the past few centuries. Although many of these mosquito-borne viruses are geographically widely dispersed, with the exception of West Nile virus, they are found either in the Old World or the New World, never in both, and we are now beginning to understand the reasons. Phylogenetic trees will be used here to describe the evolution, epidemiology and dispersal characteristics of these viruses, taking into account the importance of virus persistence and recombination.

Introduction

The genus *Flavivirus* in the family *Flaviviridae*, comprises approximately 70 single-stranded, positive sense RNA viruses that encode the viral structural proteins in a single open reading frame [16]. The translated polyprotein is processed co- and post-translationally to produce three structural proteins, the capsid (C) protein, the membrane (M) protein which is cleaved from a membrane pre-cursor (PrM), and the envelope (E) protein, the latter two of which are glycosylated, and seven non-structural proteins (NS1, NS2a, NS2b, NS3, NS4a, NS4b and NS5) that effect the replicative, proteolytic and maturation functions [31, 41, 42]. The open reading frame of all flaviviruses is flanked by 5′ (about 130 nucleotides) and 3′ (400–700 nucleotides) untranslated regions (UTR). RNA in these regions forms secondary stem-loop structures that probably serve as *cis*-acting elements for genome amplification, translation or packaging [13, 35, 38].

Approximately two thirds of the flaviviruses infect and are known to be transmitted to wildlife vertebrate species by ticks or mosquitoes and amongst these arthropod-transmitted viruses approximately 60% produce encephalitis in the hosts that they infect. The encephalitic flaviviruses can be divided into two groups, viz., the Mammalian tick-borne virus group, and the Mosquito-borne viruses that are mainly, but not exclusively, associated with *Culex* spp., including those of the Japanese encephalitis virus group, the Ntaya virus group, and the Kokobera virus group [16]. In terms of encephalitis, these groupings are not absolute, for example three viruses in the tick-borne encephalitis (TBE) complex, ie Omsk haemorrhagic fever virus (OHFV), Kyasanur Forest disease virus (KFDV) and the closely related subtype Alkhurma virus (ALKV), are known primarily to cause haemorrhagic disease although they may cause encephalitis [14, 21]. The *Aedes* spp.-associated dengue viruses, that usually only cause dengue fever (DF) or dengue haemorrhagic fever/shock syndrome (DHF/DSS) also occasionally cause fatal encephalitis [47] but will not be included in this review.

There are now available several comprehensive phylogenetic analyses of the flaviviruses, mostly based on the non-structural NS5 and NS3 genes, the structural envelope (E) gene or complete genome sequences [1, 11, 23, 29, 63]. Regardless of the genes analysed, in general, the viruses occupy the same relative phylogenetic

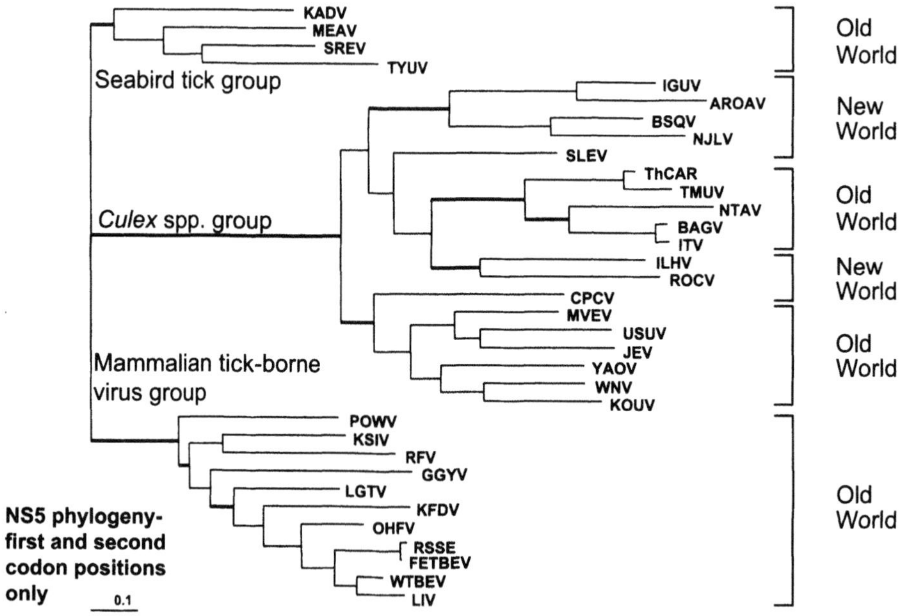

Fig. 1. ML phylogenetic tree based on partial NS5 gene sequence (first and second codon positions only) [11, 23] of representative viruses in the Mammalian tick-borne virus group. The tree was developed in PAUP*. Variable parameters were estimated using MODEL TEST. Parameters were estimated from the data, the optimal model was TrN + I + Γ. The tree is rooted on the Seabird tick-borne virus cluster

positions in these trees although some exceptions will be highlighted in this review. In all these trees the Mammalian tick-borne virus group and the *Culex* spp.-associated viruses form clades that are distinct from each other and from the other viruses in the monophyletic genus *Flavivirus*. This is illustrated in Fig. 1 which presents a maximum likelihood phylogenetic tree that focuses on these two encephalitic groups of virus. An additional group of viruses, the Seabird tick-borne virus group, none of which is known to cause encephalitis in humans or livestock, is included in the tree as a basal cluster. The tree also identifies the geographic position of the viruses on the basis of whether or not they were first identified in the Old World or the New World. The evolution and dispersal of the viruses most frequently associated with human disease will form the focus of attention in this review.

Encephalitic tick-borne viruses

This group currently consists of nine recognized species and eight closely related viruses or subtypes [16] although in the light of recent phylogenies [11] the current inclusion of Kadam virus (KADV) may not be justified. Examination of the available phylogenies that describe these viruses reveals some disagreement in the relative degrees of divergence. However, in many trees, the most divergent virus in the Mammalian tick-borne virus group is Powassan virus (POWV), closely followed by Royal Farm virus (RFV), Karshi virus (KSIV) and Gadgets Gully virus (GGYV). It has long been recognized that the Mammalian tick-borne viruses are antigenically and genetically closely related and distributed, in the main, across the northern hemisphere. These factors led to the original proposal that the TBE complex viruses had evolved in a continuous or gradual manner across Asia and Europe [9]. Subsequent studies revealed that the tick-borne and mosquito-borne viruses had diverged relatively early in their recent evolutionary history [29, 63, 64]. The evolution of the tick-borne viruses showed a direct correlation between genetic and geographic distance between the viruses [63]. In phylogenetic trees the branching pattern was continuous and asymmetric or ladder-like, inferring gradual evolution through time. It was estimated that the tick-borne viruses had evolved at a rate one half or one third of that of the mosquito-borne viruses [63] and this has been explained largely by the fact that ticks have a relatively long life cycle, lasting from two to five years, during which the tick takes only three blood meals. Since the level of virus infectivity in the tick remains relatively low, virus turnover is constrained. Moreover, transmission of tick-borne flaviviruses occurs between ticks when they co-feed on vertebrate hosts. Thus, whilst the vertebrate host is important for virus transmission between ticks, it probably plays a small role in total virus turnover and generating genetic variation.

Most of the mammalian tick-borne flaviviruses can infect and be transmitted by a wide range of tick species. For example, in far east Asia TBE viruses have been isolated from 18 different tick species associated with forest animals [46, 65] whereas Gadgets Gully virus (GGYV) was isolated from *Ixodes uriae*, which is associated with seabirds. Powassan virus has been isolated from several different species of ticks, including *Ixodid* spp., and also mosquitoes, including *Culex* spp. (D. K. Lvov, Ivanovsky Institute, Moscow, personal communication). It is interesting that POWV overlaps geographically with far-eastern TBE viruses in the Primorsk region of Russia, whereas in North and South America, where POWV is known to be present and to cause encephalitis in humans, there are no TBE viruses. The most logical interpretation of this is that POWV originated in Asia and was subsequently introduced into the Americas. Presumably this occurred recently, otherwise more recent tick-borne lineages would have emerged in the New World. The tree (Fig. 1) shows that POWV is relatively closely related to GGYV, which was isolated from *Ixodes* uriae ticks collected on Macquarie Island, South of Tasmania. It is presumed that GGYV was carried to this island by seabirds or their associated ticks. Because the Storm Petrel (*Oceanites oceanicus*) and the Arctic Tern (*Sterna paradisaea*) migrate between the Arctic and Antarctic Circles, and they are found on Macquarie Island, it was suggested that GGYV could provide a genetic link between the subarctic and subantarctic TBE viruses [49].

Kyasanur Forest disease virus was discovered in 1957 in the Shimoga District, Karnataka State, India [59, 60] when monkeys were found dying with a disease that resembled haemorrhagic fever in humans. The virus has been isolated from several species of ticks including *Haemaphysalis*, *Dermacentor* and *Ixodes*. Until recently, KFDV appeared to be geographically isolated from other members of the Mammalian tick-borne virus group but the discovery of ALKV in Saudi Arabia in 1995 [62] now provides a rational explanation for the presence of KFDV in the Indian forests.

Alkhurma virus is genetically very closely related to KFDV [4] and produces a haemorrhagic disease in humans that closely resembles the disease produced by KFDV. Estimates of the times of divergence between these two viruses are about 500–600 years ago. Therefore, although they have not yet been discovered, other related strains of haemorrhagic and encephalitic tick-borne viruses almost certainly exist in other parts of India and in those countries that connect Saudi Arabia with India. Further support for this idea comes from the isolation and phylogenetic analyses of KSIV from Uzbekistan and RFV from Afghanistan. These viruses infect and are transmitted by Argasid ticks. The phylogenetic tree shows that they are also close relatives of KFDV.

Langat virus (LGTV), which is found in southern Asia, also is closely related to these Middle Eastern and Asian haemorrhagic fever viruses but is also closely related to tick-borne encephalitic viruses found in Japan, far east Russia, central and western Europe, Scandinavia and the British Isles. *Langat virus* is found in *Ixodes* and *Haemaphysalis* ticks in Malaysia and Thailand and, like TBE viruses, is transmitted between ticks co-feeding on rodent hosts (M. Labuda, Institute of Virology, Bratislava, personal communication).

Omsk haemorrhagic fever virus occurs in Siberia and at first glance appears geographically out of place. Whilst this virus is related to TBEV, it is primarily a haemorrhagic disease virus and therefore in this sense it resembles KFDV and ALKV, although cases of encephalitis due to OHFV have been recorded. Based on detailed studies of OHFV in the lowland forest-steppe of western Siberia it appears that this virus has circulated harmlessly amongst indigenous small mammals (including various rodent species) for many years [14, 21]. The virus came to the attention of local inhabitants and also scientists following the introduction of muskrats and other species from North America in the mid 1930's. Many of these introduced species died in an intense epizootic of unknown origin at the time. Investigators in western Siberia noticed haemorrhagic disease in humans associated with the muskrats and named the disease "muskrat disease". Subsequently it was observed that no natural virus infection was detected in any wild or domesticated animals where OHFV was known to be present in muskrats and it was experimentally demonstrated that water voles, which are very common in this region, were not susceptible to OHFV [21]. The fact that OHFV is highly virulent for the North American muskrat but not for indigenous mammal species in Siberia adds further support to the proposed Old World origin for this virus.

The clinal biogeographic evolution and dispersal of the Mammalian tick-borne virus group has been described in detail [11, 63]. Whilst small genetic differences between individual strains have been identified [15] the TBE viruses in far east Russia and Japan are extremely closely related but phylogenetically distinct from those of Siberia and central Europe, where they are typified by the Vasilchenko strain [12] and in western Europe by the European subtype [16]. These viruses are all associated with ticks and rodents in forest environments. The clinal pattern of evolution for these tick-borne flaviviruses infers that birds are unlikely to contribute significantly to their geographic dispersal. Nevertheless, TBE viruses have been isolated from birds [7] and isolates that resemble strains of far east Asian TBE virus have recently been observed in central Europe (unpublished data). The possibility that birds could introduce far-east Asian strains of TBE virus into western Europe should therefore be taken seriously. This is of importance to government health departments because it is recognized that far-east Asian strains of TBE virus are more virulent for humans than west European strains [14].

Perhaps two of the most interesting and important modern discoveries regarding TBE viruses and related viruses were the observations that these viruses are transmitted between ticks when they co-feed in close proximity on individual rodent hosts and that this can occur even if the hosts have antibodies to the virus [20, 24]. It is not necessary for the vertebrate host either to be susceptible to virus infection or for it to develop a viraemia. Further research has shown that the focal distribution of TBE viruses in western Europe may be determined by the geographically variable degree of synchrony in the seasonal questing activity of larval and nymphal *Ixodes ricinus* [36]. The perpetuation of these viruses in the forest appears to be ensured if the following criteria are satisfied. Firstly, non-infected larvae must co-feed with infected nymphs. Secondly, there must be a high ratio of larvae to nymphs, which is largely dependent on high humidity. Thirdly, large numbers of unfed larvae need to be available to emerge synchronously with infected nymphs in the spring. This condition is highly dependent on the climatic conditions of the previous late summer to autumn period.

Only a relatively limited area of geographic zones in Eurasia will fulfill these requirements and this explains the highly defined areas in which tick-borne encephalitis occurs. By augmenting ground collected climatic information and vector habitat distribution with remotely sensed data, detailed information may be obtained. Using GIS satellite data risk maps have been produced which predict the pan-European distribution of TBE viruses. The prediction maps fit well with the recorded disease distribution [36].

Not all of the encephalitic tick-borne flaviviruses are found on the Eurasian forest floor. Louping ill virus is the name given to a tick-borne flavivirus that was first observed to produce encephalomyelitis in sheep, goats and grouse on the sheep-rearing hillsides of Scotland, north Wales, and in south-west and northern England [39]. Subtypes of LIV that produce similar disease in sheep and goats have been identified in Turkey (Turkish sheep encephalomyelitis virus TSEV) [8], Greece (Greek Goat encephalomyelitis virus GGEV) [16] and Spain (Spanish sheep encephalomyelitis virus SSEV) [28].

A plausible explanation for the appearance of TSEV/GGEV on the Turkish and Greek hillsides can be deduced by combining the phylogenetic data with the ideas that arose from a review by Vesenjak-Hirjan [56] of studies in Croatia and Bosnia. Migrating seminomadic flocks of sheep were found to be carrying ticks infected with TBE virus. Thus, the seasonal introduction of non-immune sheep or goats onto grazing land that skirts the forests of Eurasia exposes them to forest ticks infected with TBE viruses. The presence of these animals on the hillsides would provide the ideal environment for amplification of the tick population and a viraemic host for transmission of the disease. The phylogenetic data [11, 30] infer that TSEV/GGEV emerged earliest of the sheep encephalomyelitic viruses and were probably transported to France and Spain with ticks of sheep and goats. It then appears that SSEV or a descendant lineage was transported to Ireland where the Irish louping ill virus evolved and was eventually transported to the British mainland, possibly through north Wales [30]. Louping ill virus then emerged and was dispersed along the drover's routes to northern England and Scotland. This dispersal pattern demonstrates the very significant influence that agricultural practices and transportation can have on processes in the natural environment. It is worth noting that the recorded appearance of LIV as a veterinary disease on the British hillsides coincides closely with the estimates for the emergence of Scottish LIV, based on comparative analysis of sequence data and mutation rate estimations [30].

Encephalitic mosquito-borne viruses

Viruses in the *Japanese encephalitis virus* (JEV) species (8 viruses, and 2 subtypes), the *Ntaya virus* (NTAV) species (5 viruses, and one subtype) and the *Kokobera virus* (KOKV) species (1 virus, and one subtype), are primarily maintained in the natural environment in bird-*Culex* spp. cycles. Their original grouping was based largely on antigenic cross-reactivity tests in neutralization, haemagglutination-inhibition and complement fixation tests [3, 34] and in general phylogenetic analyses support this scheme of classification. Many but not all of these viruses may cause encephalitis in humans and other animals but, with the exception of Japanese encephalitis virus (JEV), St. Louis encephalitis virus (SLEV), West

Nile virus (WNV), Kunjin virus (KUNV) and Murray Valley encephalitis virus (MVEV), the sequence data are limited. Phylogenetic analyses of these viruses based on partial NS5 gene sequence [11, 23] revealed four groups based on their geographic distribution, viz. two groups exclusively found in the Old World and two groups exclusively found in the New World and these can be seen in the maximum likelihood (ML) phylogenetic analysis presented in Fig. 1. Following its inadvertent introduction into North America in 1999, WNV became an exception to this rule and is now widely distributed in both the Old and the New World [18]. Phylogenies that include all recognized flaviviruses show that the *Culex* spp.-associated viruses are less divergent than the *Aedes* spp.-associated viruses, all of which are found in the Old World. Therefore, all *Culex* spp.-associated viruses have their evolutionary origins in the Old World and this should be taken into account when studying the evolution and dispersal of the New World *Culex* spp.-associated viruses.

St. Louis encephalitis virus is widely distributed throughout North and South America. It circulates between *Culex* spp. [40] and wild birds, primarily passeriformes and columbiformes [53]. This virus occurs in both rural and urban areas, in a primary enzootic cycle, depending on the accessibility of *Culex* species. In the Caribbean, and central America and South America, SLEV has also been isolated from mosquitoes of other genera, including *Mansonia*, *Haemagogus* and *Sabethes*. Detailed phylogenetic analyses based on the complete E gene sequence of 62 isolates of SLEV from different regions of the Americas showed that the viruses formed a monophyletic group and a tendency to cluster according to geographic origin, i.e., one group in Panama, one in South America and two distinct groups in the USA. However, viruses of the latter grouping were not strictly defined, indicating that although SLEV is predominantly maintained locally, it may also be transported large distances, probably by birds [22].

The tree in Fig. 1, which is based on partial NS5 gene sequence, shows that SLEV diverges with both Old and New World viruses in the *Culex* spp.-associated group. It is interesting to note that in all published trees based on NS5 gene data, SLEV is positioned in a clade that contains New World Ilheus virus (ILHV) and Rocio virus (ROCV), and some Old World viruses, including Tembusu virus (TMUV). In contrast, on the basis of E gene data, SLEV is placed nearest to a clade that contains members of the JEV complex, i.e., mostly Old World viruses [10]. Tree incongruency of this sort usually implies recombination. Therefore, SLEV could be a recombinant of either an Old World and a New World virus or two Old World viruses. In either case, since all the related Old World viruses are found in Africa, the phylogenies imply that SLEV or a closely related lineage circulated in Africa. It is therefore reasonable to suppose that SLEV or an unidentified ancestral lineage crossed the Atlantic Ocean, perhaps in a manner similar to that of Yellow fever virus (YFV), i.e., on the slave ships [11, 50]. It is unlikely that SLEV or other related *Culex* spp.-associated viruses crossed the Atlantic Ocean earlier, carried by wind-blown birds, because if this had happened then other ornithophilic mosquito-borne flaviviruses, such as WNV, JEV, Usutu virus, and KOUV might also have done the same thing – but there is no evidence that this has occurred.

Moreover, of the *Culex* spp.-associated flaviviruses, which are less divergent than the Aedes spp.-associated viruses, i.e. they emerged relatively recently; all are very closely related genetically, implying that the African and American viruses have not been separated for many years. Whether or not recombination took place before SLEV appeared in the New World may become clear when genes other than the E gene of a selection of SLEV strains are sequenced.

The ML tree (Fig. 2) based on E gene sequence of SLEV isolates, which is similar to the tree published previously [22] shows that the most divergent viruses were isolated in South America, specifically Argentina, Panama and Brazil, whereas all the least divergent lineages of SLEV are isolates from North America inferring that SLEV was first successfully introduced into and became established in South America and then evolved and dispersed northwards. This is further supported by the fact that all of the other New World *Culex* spp.-associated viruses

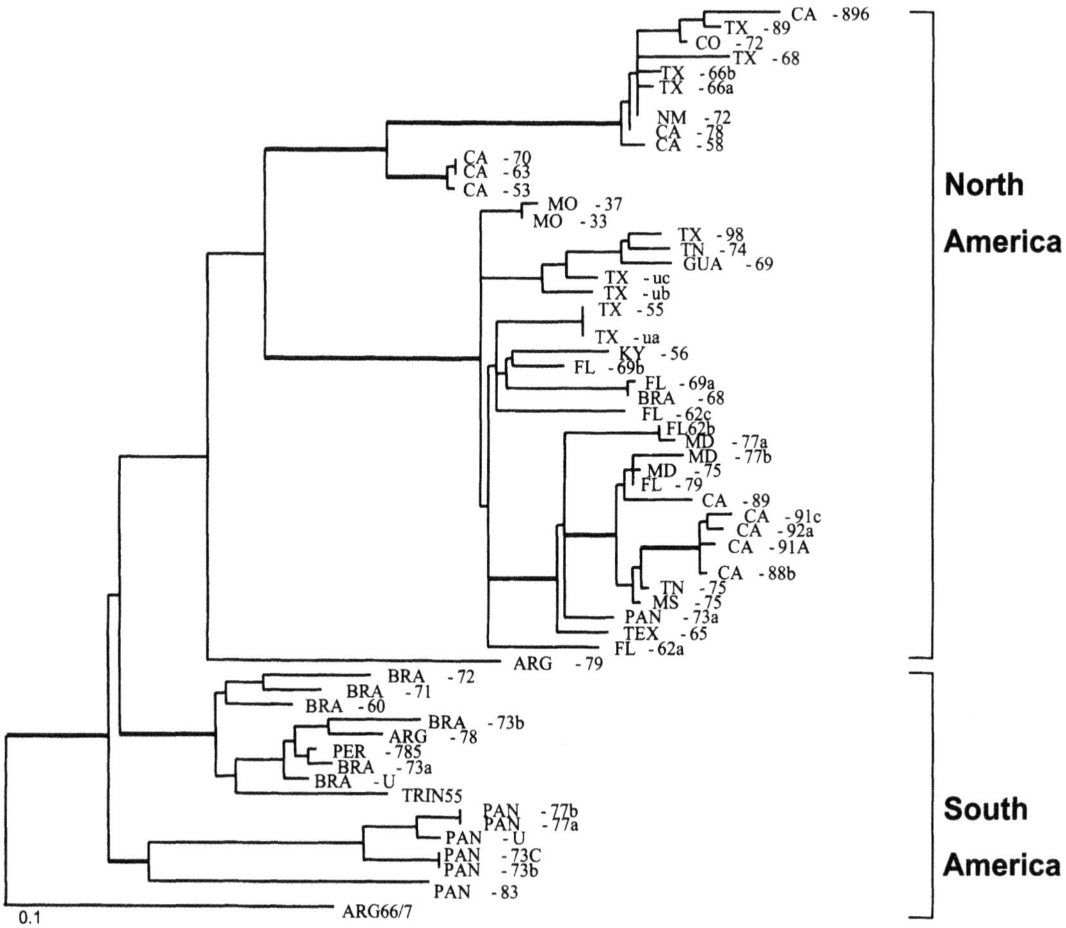

Fig. 2. ML phylogenetic tree based on complete E gene sequence of representative strains of SLEV [22]. The topology of the tree shown is based on a tree rooted with a virus closely related to the basal cluster. Emboldened branches correspond to bootstrap values >75%. The tree was developed as described for Fig. 1. All codon positions were included

(ROCV, ILHV, Naranjal virus [NJLV], Bussuquara virus [BSQV], Iguape virus [IGUV], Aroa virus [AROAV] and SLEV) are found in South America but few are found in North America. Also, each of the clades of *Culex* spp.-associated viruses contains members from both the Old World and the New World. The most likely interpretation of this is that during the past centuries several different viruses have been introduced into Latin America, from the Old World, rather than a single virus having been introduced with subsequent divergence to generate the different viruses found in the Americas.

The New World viruses, ROCV, ILHV, NJLV, BSQV, IGUV, AROAV, are not as well studied as is SLEV but, with the possible exception of ROCV and ILHV, they appear to be less widely distributed. This probably reflects firstly, the local animal species with which they are principally associated, for example, local rodents. Secondly, we suspect that they were introduced very recently and have therefore had very little opportunity to disperse widely across the Americas.

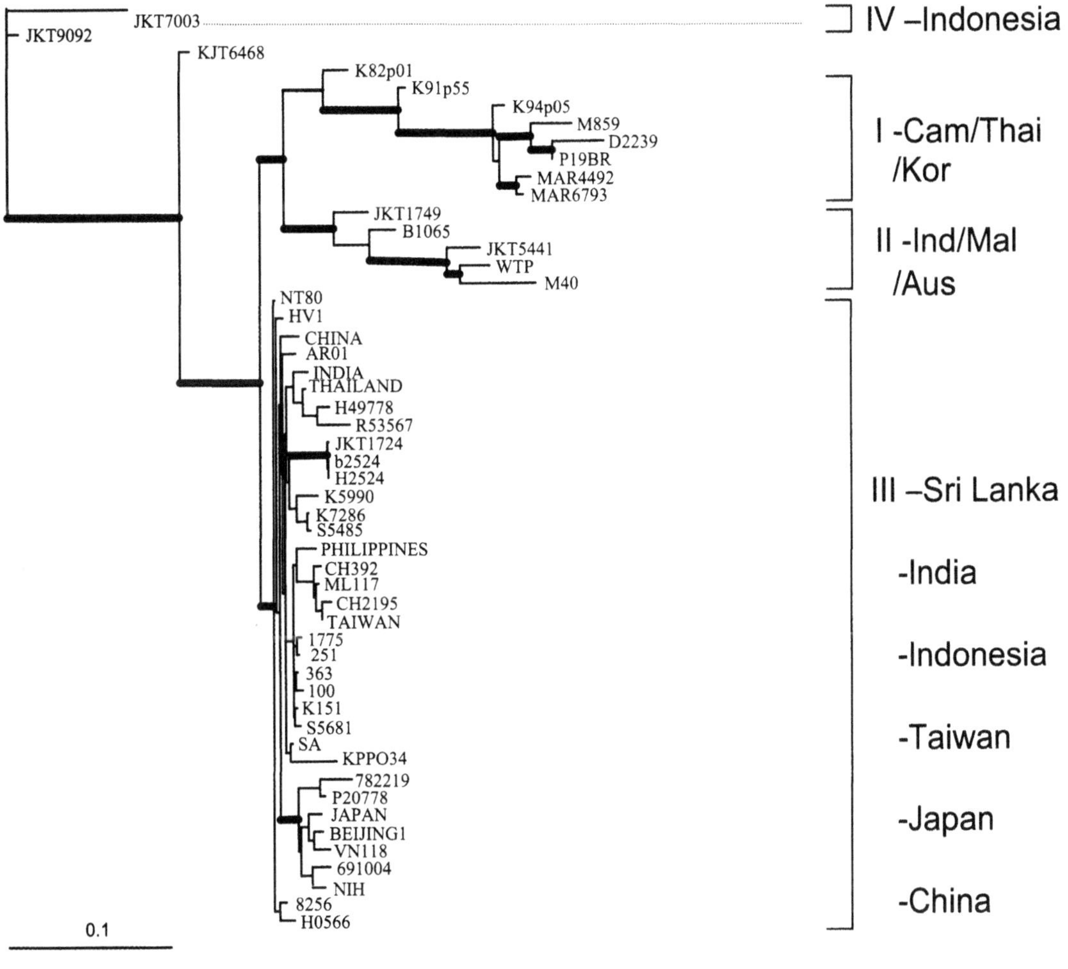

Fig. 3. ML phylogenetic tree based on complete E gene sequence of representative strains of JEV [48, 54]. The tree was developed as described for Fig. 2

A large number of strains of JEV have been sequenced and many phylogenetic trees constructed on the basis of these sequences [5, 6, 19, 32, 54, 57]. Figure 3 presents an ML phylogenetic tree based on the E gene of strains of JEV selected to represent each of the four genotypes that are described in most of the publications cited above. As can be seen, our analysis shows that the viruses can be assigned to five genotypes, without the inclusion of the Muar strain, which will be discussed separately below. Although the disease in humans and horses was first recognized in Japan, the most divergent viruses, i.e., those that we have assigned to genotype IV and V, but which were previously assigned to genotypes III or IV [48, 54], respectively, have been isolated only in Indonesia. The more recent lineages (the least divergent) of JEV appear to have radiated north, east and westward from Indonesia, presumably having been carried by migratory birds, and commercial transportation of animals such as pigs and poultry.

One can speculate on the origin of JEV, based on the phylogenetic data. This virus is closely related to other African and Australian *Culex* spp., associated viruses such as USUV and MVEV. Figure 1 shows that these viruses emerged at roughly similar times, which for JEV was within the last 300 years [55]. Previous estimates for the times of emergence of flaviviruses [64] infer that the *Culex* spp., associated viruses emerged more recently than the *Aedes* spp., associated viruses (not shown in Fig. 1), i.e. those that include Yellow fever virus (YFV) which is most likely of African origin [11, 50]. Thus, JEV probably emerged in south east Asia following the introduction of an ancestral virus of African origin. The newly emergent virus became established in Indonesia and was then dispersed radially and quite rapidly reaching Japan, where epidemics in horses and humans brought the virus to our attention. It is clear from many phylogenetic trees that these viruses are frequently moved long distances. In some cases this can occur very rapidly. For example, closely related viruses can be isolated thousands of miles apart and more distantly related viruses have been isolated from the same geographic areas. Whilst birds may provide some of this movement it is likely that commercial transportation of poultry and pigs, both of which can be infected by JEV, contributes significantly to the dispersal of JEV. Figure 3 also shows that the viruses assigned to genotype III which are all very closely related, span the widest geographic range, whereas the viruses in all other genotypes show more restricted distributions. Presumably this reflects the particular ecological niches into which each genotype has been introduced.

Anomalies in genotyping due to sequencing errors?

Recent phylogenetic analyses of JEV isolates describe at least four genotypes, each of which shows distinct geographic distributions [48, 54]. It has been suggested that the Muar strain of JEV, isolated in Singapore, represents a fifth and the most divergent JEV genotype [48, 54]. We have performed comparative alignments of the E gene sequence of Muar virus with E gene sequences of other strains of JEV and other flaviviruses. The alignment reveals substitutions in the Muar strain amongst nucleotides that are otherwise totally conserved throughout the entire genus. In other words, these substitutions are almost certainly sequencing errors. Analyses of other structural genes also appear to have been based on sequencing errors at totally conserved positions. Comparative alignments also revealed several other nucleotide positions that are conserved in all JEV strains and other *Culex* spp.-associated viruses except the Muar strain, although they are not totally conserved in every virus within the genus *Flavivirus*. In view of these potential sequencing errors, a re-analysis of the Muar strain of JEV seems justified and until this has taken place it is not considered appropriate to include this virus in phylogenetic analyses.

Anomalies in JEV phylogenies due to recombination?

Recombination amongst strains of JEV and SLEV was recently reported [53]. Three homologous recombinant strains of JEV and one homologous recombinant strain of SLEV were detected by comparative analyses of the envelope gene. This adds to the previous reports of recombination seen among the dengue viruses [17, 51, 55, 61]. We have carried out additional comparative analyses of the capsid and E gene phylogenies of JEV isolates and have discovered several further examples of tree incongruency that may impact on our understanding of the evolution, dispersal and even the pathogenicity of these viruses (manuscript in preparation). The data reveal that homologous recombination is occurring much more frequently in JEV than we had hitherto imagined. Figure 4 presents the bootscan analyses of two examples of recombinant viruses. These recombinants were initially detected by constructing trees based on sequential one-third segments of the E gene sequence of each virus. The bootscan illustrates the breakpoint within the E gene where recombination occurs. In the first example, there is one crossover point, whereas in the second example two crossover points can be seen. A recent report [53] emphasises the apparent ease with which genome switching by the polymerase may occur during RNA replication. These data appear to invalidate the assumption that genetic diversity in the flaviviruses is largely generated through mutational changes. Caution is required in constructing phylogenies of closely related viruses to avoid the possibility of anomalies arising due to recombination, i.e., the viruses may appear in different genotypes depending on the region of the genome used to develop the phylogenetic tree. The impression is often given in the published literature that whole genome sequencing provides a more accurate phylogenetic representation of viruses. This is a reliable conclusion if recombination between the viruses under analysis has not occurred. If on the other hand recombination has occurred and particularly if there are several recombinant crossovers or breakpoints throughout the genome, these events have to be identified before any phylogenetic data can be interpreted meaningfully. Another important implication of recombination is the possible impact on virus virulence. In view of the fact that live attenuated vaccines based on chimeric viruses are currently being developed, the potential risks that recombination might pose require serious consideration.

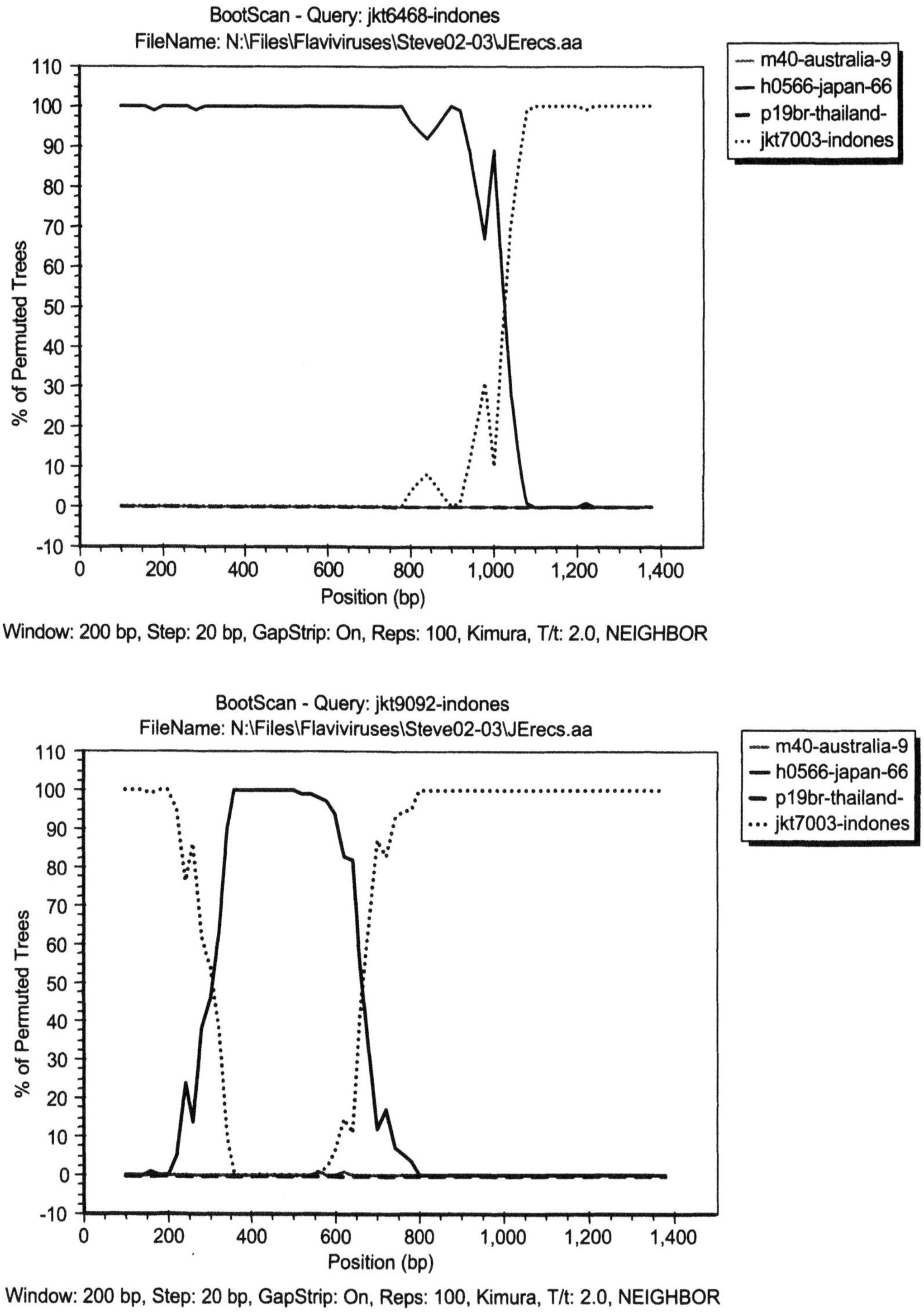

Fig. 4. Bootscanning analysis [44] of jkt6468 and jkt9092 sequence against four strains of JEV (m40-australia-9, h0566-japan-66, p19br-thailand and jkt7003), representing each of the main genotypes observed in the Maximum Likelihood phylogeny

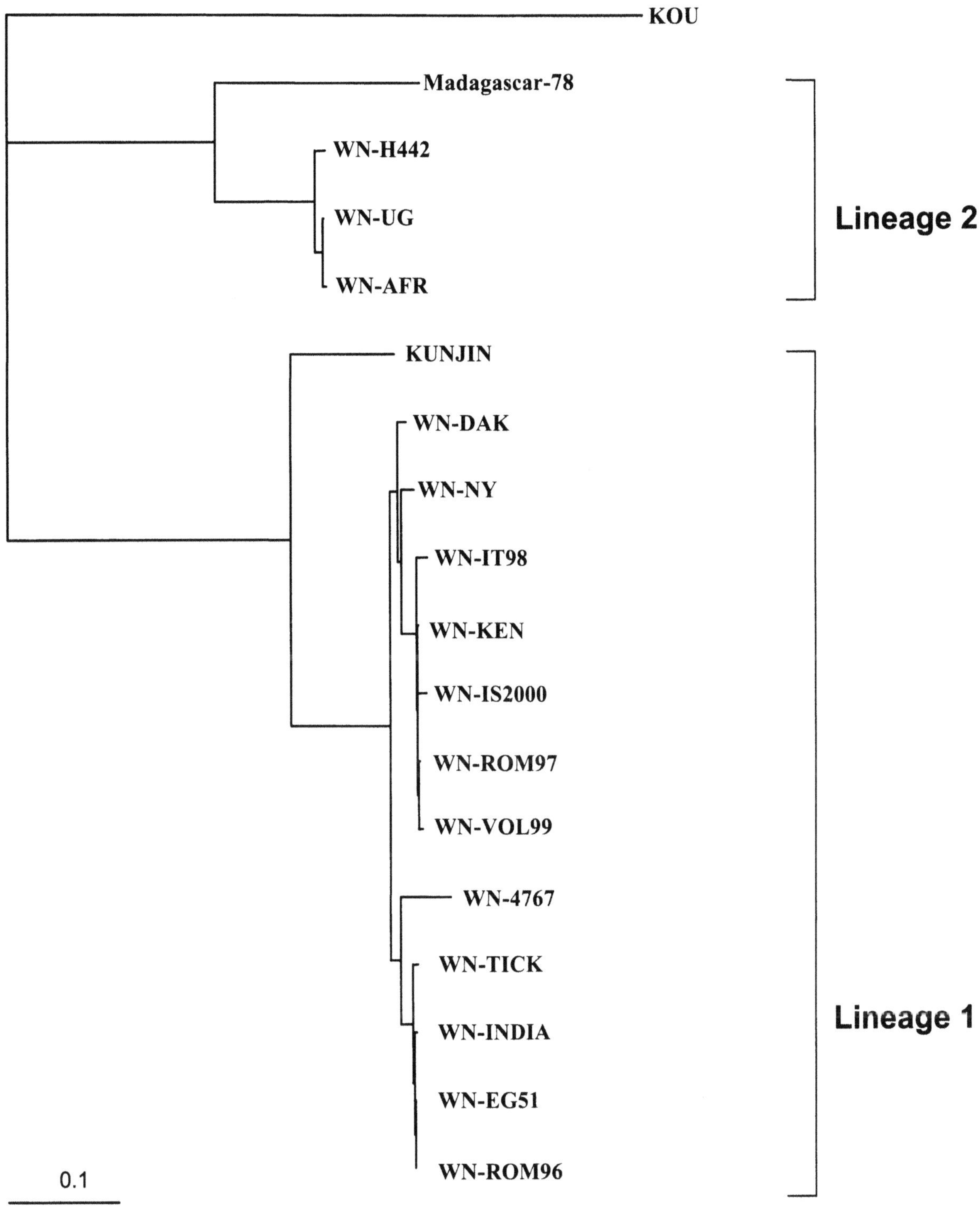

Fig. 5. ML phylogenetic tree based on partial E gene sequence of representative strains of WNV [25] (and unpublished data). The tree was developed as described for Fig. 2

Whilst we can therefore tentatively conclude that JEV emerged in Asia, probably from African roots, and then radiated widely, largely as the result of human commercial practices, we need to show caution in our interpretation of the relationships between flavivirus strains until more detailed comparative analyses of the sequence data have been performed.

Despite the facts that WNV and JEV are genetically closely related, share overlapping geographic distributions and circulate between *Culex* spp. and birds, they show significantly different evolutionary and dispersal characteristics. In terms of its global distribution WNV has clearly been the most successful of the encephalitic flaviviruses. This virus infects and is transmitted by a wide range of ornithophilic *Culex* spp., other mosquito species, and even ticks [31]. Its close association with migratory birds is well documented [27, 37]. The ML phylogenetic tree (Fig. 5) based on partial E gene sequence of representative global WNV isolates shows that the most divergent strains are of African origin, in lineage 2. These viruses, which circulate amongst birds in Africa, do not appear anywhere outside Africa and are rarely associated with epidemic outbreaks. The data therefore infer an African origin for WNV. If this assumption is correct the less divergent lineage 1 viruses probably dispersed out of Africa being vectored by *Culex* spp. and moved by migratory birds. West Nile virus appears to have dispersed to the Mediterranean countries, northern Europe, eastern Europe, southern Asia including India, Malaysia and finally Australia where it is recognized as Kunjin virus (KUNV). As more sequence data accumulate and the extent of the phylogenetic analyses increases, it will become clear whether or not this pattern of evolution and dispersal is correct.

Several phylogenetic analyses of WNV have shown that there are two distinct lineages [2, 26, 33, 45] and these are also evident in Fig. 5. The lineage 1 viruses have previously been divided into three clades [43] corresponding to strains that were isolated in Africa, Europe, the Middle East, Russia or North America (clade 1a), viruses that were isolated only in Australia (clade 1b), known as KUNV, and viruses isolated only in India (clade 1c). This sub-division does not correlate precisely with the topology shown in Fig. 5. As more data become available we presume these anomalies will be resolved. The lineage 2 viruses are found only in Africa and it has been suggested that these viruses may be less virulent for humans than the lineage 1 viruses that have been isolated during epidemics in which encephalitis and mortality were prominent features [43]. At a recent CDC meeting on WNV held in New Orleans, evidence was presented to show that the virulence of the North American strain of WNV for some North American birds was higher than that of a Kenyan isolate of WNV or a strain of KUNV. This work was reported as ongoing (A. Brault, CDC, Fort Collins, personal communication). The division of viruses into those that have dispersed out of Africa and those that apparently have not dispersed out of Africa raises an important question. Do the lineage 1 and lineage 2 viruses really show different dispersal characteristics and if so, why? It is generally believed that WNV has dispersed out of Africa as the result of being transported to Europe and elsewhere by migrating birds [27]. Does this mean that the lineage 2 viruses do not use this dispersal mechanism?

If we assume that this difference is not simply explained by sampling errors, i.e. the viruses are not recognized because they do not cause significant disease, one possible explanation is that in Africa the lineage 2 viruses occupy different ecological niches, ones that do not normally coincide with migratory bird flyways. Alternatively, the lineage 1 viruses could represent a more virulent mutant lineage that "broke out of Africa". Both possibilities are difficult to accept because lineage 2 viruses are widespread throughout Africa and appear to overlap geographically with lineage 1 viruses. Alternatively, the lineage 2 viruses may be transmitted less efficiently than lineage 1 viruses by ornithophilic mosquitoes. Whatever the reasons for the different distribution of the lineage 1 and 2 viruses, it would be important to know whether or not they do differ significantly in virulence for humans, equines and birds. If indeed the lineage 2 viruses are of lower virulence, they could circulate relatively harmlessly in Europe and other regions outside Africa and in so doing provide a background of immunity to more virulent strains of WNV. This would explain why WNV in the Old World rarely causes epidemics that are characterized by high mortality.

Until 1999, when WNV was first identified in New York, there were no extant Old World *Culex* spp.-associated viruses in the New World. It was not too surprising therefore that WNV was initially thought to be SLEV, the only closely related mosquito-borne flavivirus previously observed in north east America [43]. It is not known precisely how WNV was introduced into North America but it was almost certainly the result of human transportation. The possibility that the virus was introduced by a wind-blown infected bird, from the Old World is very low since WNV and other African *Culex* spp.-associated viruses such as USUV and Yaounde virus (YAOV) would presumably have been introduced many times to the New World and this does not appear to have been the case. Perhaps an infected bird or mosquito was inadvertently transported to New York, from the Middle East, on a commercial airliner. Once the virus had succeeded in entering North America it rapidly dispersed southwards along the east coast on the recognized bird migratory routes and gradually moved westwards, reaching the east side of the Rocky Mountains by the late summer of 2002. Although one case of WNV infection in a human was reported in California in 2002, it is believed to have been introduced, i.e., it did not arise from a local source. There is now considerable discussion concerning how long it will take before WNV becomes established on the west side of the Rocky Mountains. This may depend on how long it takes for WNV to become established in South America. At the time of writing, WNV antibodies had just been reported in birds in the Dominican Republic, Jamaica, El Salvador, Nicaragua, and Guatemala, and in horses in Mexico. The apparent tardiness of WNV to reach South America after dispersing so rapidly through North America, may merely indicate that there has been insufficient investigation, and this probably applies to many Latin American countries, but could be due to competitive interference by the presence of SLEV, ILHV, ROCV and other *Culex* spp.-associated flaviviruses which have been established in South America for some considerable time. Once WNV is established in South America, one might then expect birds migrating from South America to the west coast of North America to deliver the virus to the west side of the Rockies.

Occasional isolates of Murray Valley encephalitis virus (MVEV) and Alfuy virus (ALFV) have been identified in south-east Asia, but they do not appear to have spread widely out of Australia in a westerly direction. These viruses are phylogenetically close relatives of JEV [11, 23] and it is therefore possible that one or the other emerged at about the same time as JEV, possibly in Indonesia, dispersing towards Australia more rapidly than JEV. Why have these viruses not radiated out of Australia and spread westerly? Firstly, the virulence for humans of MVEV and ALFV appears to be relatively low compared with viruses such as JEV and, therefore, their presence in a background of JEV could easily be missed. Secondly, more divergent but related flaviviruses are already established in Asia, providing a form of natural immunity in nearby countries. Thirdly, they may not use ornithophilic mosquitoes with such efficiency as some of the other related flaviviruses and therefore have not had time to disperse out of Australia. Fourthly, the disease incidence of any flavivirus in Australia is generally low compared with that of many Asian countries, and Australian control measures across borders are well organized. These factors are therefore likely to restrict the dispersal of MVEV and ALFV. Presumably the same arguments apply to KOKV and Stratford virus (STRV) both of which are found in Australia and which seem to have diverged independently of the JEV complex viruses.

In Africa, Usutu virus (USUV) has a very similar econiche to that of WNV. Although USUV is not known to be encephalitic for humans, it came to the attention of virologists in August and September 2001 when a considerable die-off of Eurasian Blackbirds (*Turdus merula*) was observed in and around Vienna, Austria [58]. The aetiological agent was originally presumed to be WNV but was identified by RT-PCR sequencing as USUV. The virus reappeared twelve months later, in July 2002, causing more bird die-offs and is therefore assumed to have become established in this region of Austria (N Nowotny, Institute of Veterinary Medicine, Vienna, personal communication).

As there are no reports of other Old World *Culex* spp.-associated viruses such as YAOV and Koutango virus (KOUV), in Europe, they are currently assumed to circulate only in Africa. It seems surprising that USUV has not been observed in any country other than Austria, since significant bird die-offs would normally be expected to be noticed and recorded. However, other African viruses may not be associated with bird die-offs and could therefore understandably go unnoticed.

Conclusions

As we acquire a greater understanding of the evolution and dispersal of the flaviviruses it is becoming increasingly clear that these viruses are only now being recognized as causing emerging diseases and that there are many more viruses to discover. Thinking logically, there is no reason to believe that the alphaviruses and the bunyaviruses are significantly different from the flaviviruses. Perhaps the arbovirologists of the twenty first century will enjoy the re-birth of this subject knowing that they are better equipped to understand the epidemiology of these viruses than were the excellent pioneers of arbovirology in the last century.

References

1. Billoir F, de Chesse R, Tolou H, de Micco P, Gould EA, de Lamballerie X (2000) Phylogeny of the genus flavivirus using complete coding sequences of arthropod-borne viruses and viruses with no known vector. J Gen Virol 81 Pt 3: 781–790
2. Briese T, Rambaut A, Pathmajeyan M, Bishara J, Weinberger M, Pitlik S, Lipkin WI (2002) Phylogenetic analysis of a human isolate from the 2000 Israel West Nile virus epidemic. Emerg Infect Dis 8: 528–531
3. Calisher CH, Karabatsos N, Dalrymple JM, Shope RE, Porterfield JS, Westaway EG, Brandt WE (1989) Antigenic relationships between flaviviruses as determined by cross-neutralization tests with polyclonal antisera. J Gen Virol 70: 37–43
4. Charrel RN, Zaki AM, Attoui H, Fakeeh M, Billoir F, Yousef AI, de Chesse R, De Micco P, Gould EA, de Lamballerie X (2001) Complete coding sequence of the Alkhurma virus, a tick-borne flavivirus causing severe hemorrhagic fever in humans in Saudi Arabia. Biochem Biophys Res Commun 287: 455–461
5. Chen WR, Tesh RB, Rico-Hesse R (1990) Genetic variation of Japanese encephalitis virus in nature. J Gen Virol 71: 2915–2922
6. Chen WR, Rico Hesse R, Tesh RB (1992) A new genotype of Japanese encephalitis virus from Indonesia. Am J Trop Med Hyg 47: 61–69
7. Ernek E (1960) Experimental Pathogenicity of the Tick-borne Encephalitis Virus for Domestic Ducks. In: Biology of viruses of the tick-borne encephalitis complex, Smolenice, Slovakia. Academic Press, New York, pp 387–390
8. Gao GF, Hussain MH, Reid HW, Gould EA (1993) Classification of a new member of the TBE flavivirus subgroup by its immunological, pathogenetic and molecular characteristics: identification of subgroup-specific pentapeptides. Virus Res 30: 129–144
9. Gao GF, Jiang WR, Hussain MH, Venugopal K, Gritsun TS, Reid HW, Gould EA (1993) Sequencing and antigenic studies of a Norwegian virus isolated from encephalomyelitic sheep confirm the existence of louping ill virus outside Great Britain and Ireland. J Gen Virol 74: 109–114
10. Gaunt MW, Sall AA, Lamballerie X, Falconar AK, Dzhivanian TI, Gould EA (2001) Phylogenetic relationships of flaviviruses correlate with their epidemiology, disease association and biogeography. J Gen Virol 82: 1867–1876
11. Gould EA, de Lamballerie X, Zanotto PMA, Holmes EC (2001) Evolution, epidemiology and dispersal of flaviviruses revealed by molecular phylogenies. Adv Virus Res 57: 71–103
12. Gritsun TS, Frolova TV, Pogodina VV, Lashkevich VA, Venugopal K, Gould EA (1993) Nucleotide and deduced amino acid sequence of the envelope gene of the Vasilchenko strain of TBE virus; comparison with other flaviviruses. Virus Res 27: 201–209
13. Gritsun TS, Venugopal K, Zanotto PM, Mikhailov MV, Sall AA, Holmes EC, Polkinghorne I, Frolova TV, Pogodina VV, Lashkevich VA, Gould EA (1997) Complete sequence of two tick-borne flaviviruses isolated from Siberia and the UK: analysis and significance of the 5′ and 3′-UTRs. Virus Res 49: 27–39
14. Gritsun TS, Lashkevich VA, Gould EA (2003) Tick-borne encephalitis. Antiviral Res 57: 129–146
15. Hayasaka D, Ivanov L, Leonova GN, Goto A, Yoshii K, Mizutani T, Kariwa H, Takashima I (2001) Distribution and characterization of tick-borne encephalitis viruses from Siberia and far-eastern Asia. J Gen Virol 82: 1319–1328

16. Heinz FX, Collett MS, Purcell RH, Gould EA, Howard CR, Houghton M, Moormann RJM, Rice CM, Thiel HJ (2000) Family *Flaviviridae*. In: Regenmortel MHV, Fauquet CM, Bishop DHL, Carstens E, Estes MK, Lemon S, Maniloff J, Mayo MA, McGeogch D, Pringle CR, Wickner RB (eds) Virus Taxonomy. 7th Report of the International Committee for the Taxonomy of Viruses. Academic Press, San Diego, pp 859–878
17. Holmes EC, Worobey M, Rambaut A (1999) Phylogenetic evidence for recombination in dengue virus. Mol Biol Evol 16: 405–409
18. Hubalek Z, Halouzka J (1999) West Nile fever – a reemerging mosquito-borne viral disease in Europe. Emerg Infect Dis 5: 643–650
19. Huong VT, Ha DQ, Deubel V (1993) Genetic study of Japanese encephalitis viruses from Vietnam. Am J Trop Med Hyg 49: 538–544
20. Jones LD, Gaunt M, Hails RS, Laurenson K, Hudson PJ, Reid H, Henbest P, Gould EA (1997) Transmission of louping ill virus between infected and uninfected ticks co-feeding on mountain hares. Med Vet Entomol 11: 172–176
21. Kharitonova NN, Leonov YA (1985) Omsk hemorrhagic fever. Ecology of the agent and epizootiology. Amerind Publishing Co Pvt Ltd, New Delhi
22. Kramer LD, Chandler L (2001) Phylogenetic analysis of the envelope gene of St. Louis encephalitis virus. Arch Virol 146: 2341–2355
23. Kuno G, Chang GJ, Tsuchiya KR, Karabatsos N, Cropp CB (1998) Phylogeny of the genus *Flavivirus*. J Virol 72: 73–83
24. Labuda M, Kozuch O, Zuffova E, Eleskova E, Hails RS, Nutall P (1997) Tick-borne encephalitis virus transmission between ticks cofeeding on specific immune natural hosts. Virology 235: 138–143
25. Lanciotti R, Ebel GD, Deubel V, Kerst AJ, Murri S, Meyer B, Bowen MD, McKinney N, Morril WE, Crabtree MB, Kramer LD, Roehrig JT (2002) Complete genome sequences and phylogenetic analysis of West Nile virus strains isolated from the United States, Europe and the Middle East. Virology 298: 96–105
26. Lanciotti RS, Roehrig JT, Deubel V, Smith J, Parker M, Steele K, Crise B, Volpe KE, Crabtree MB, Scherret JH, Hall RA, MacKenzie JS, Cropp CB, Panigrahy B, Ostlund E, Schmitt B, Malkinson M, Banet C, Weissman J, Komar N, Savage HM, Stone W, McNamara T, Gubler DJ (1999) Origin of the West Nile virus responsible for an outbreak of encephalitis in the northeastern United States. Science 286: 2333–2337
27. Malkinson M, Banet C, Weisman Y, Pokamunski S, King R, Drouet MT, Deubel V (2002) Introduction of West Nile virus in the Middle East by migrating white storks. Emerg Infect Dis 8: 392–397
28. Marin MS, McKenzie J, Gao GF, Reid HW, Antoniadis A, Gould EA (1995) The virus causing encephalomyelitis in sheep in Spain: a new member of the tick-borne encephalitis group. Res Vet Sci 58: 11–13
29. Marin MS, Zanotto PM, Gritsun TS, Gould EA (1995) Phylogeny of TYU, SRE, and CFA virus: different evolutionary rates in the genus *Flavivirus*. Virology 206: 1133–1139
30. McGuire K, Holmes EC, Gao GF, Reid HW, Gould EA (1998) Tracing the origins of louping ill virus by molecular phylogenetic analysis. J Gen Virol 79: 981–988
31. Monath TP, Heinz FX (1996) Flaviviruses. In: Fields BN (ed) Virology Vol 1. Lippincott-Raven, Philadelphia, pp 961–1034
32. Paranjpe S, Banerjee K (1996) Phylogenetic analysis of the envelope gene of Japanese encephalitis virus. Virus Res 42: 107–117
33. Platonov AE, Shipulin GA, Shipulina OY, Tyutyunnik EN, Frolochkina TI, Lanciotti RS, Yazyshina S, Platonova OV, Obukhov IL, Zhukov AN, Vengerov YY, Pokrovskii VI (2001) Outbreak of West Nile virus infection, Volgograd Region, Russia, 1999. Emerg Infect Dis 7: 128–132

34. Porterfield JS (1980) Antigenic characteristics and classification of *Togaviridae*. In: Schlesinger RW (ed) The Togaviruses. Academic Press, New York, pp 13–46
35. Proutski V, Gaunt MW, Gould EA, Holmes EC (1997) Secondary structure of the 3′-untranslated region of yellow fever virus: implications for virulence, attenuation and vaccine development. J Gen Virol 78: 1543–1549
36. Randolph SE, Green RM, Peacey MF, Rogers DJ (2000) Seasonal synchrony: the key to tick-borne encephalitis foci identified by satellite data. Parasitology 121: 15–23
37. Rappole JH, Derrickson SR, Hubalek Z (2000) Migratory birds and spread of West Nile virus in the western hemisphere. Emerg Infect Dis 6: 319–328
38. Rauscher S, Flamm C, Mandl CW, Heinz FX, Stadler PF (1997) Secondary structure of the 3′-noncoding region of flavivirus genomes: comparative analysis of base pairing probabilities. RNA 3: 779–791
39. Reid HW (1984) Epidemiology of louping ill. In: Mayo MA, Harrap KA (eds) Vectors in virus biology. Academic Press, London, pp 161–178
40. Reisen WK, Meyer RP, Milby MM, Presser SB, Emmons RW, Hardy JL, Reeves WC (1992) Ecological observations on the 1989 outbreak of St Louis encephalitis virus in the southern San Joaquin Valley of California. J Med Entomol 29: 472–482
41. Rice CM, Lenches EM, Eddy SR, Shin SJ, Sheets RL, Strauss JH (1985) Nucleotide sequence of yellow fever virus: implications for flavivirus gene expression and evolution. Science 229: 726–733
42. Rice CM (1996) *Flaviviridae*: The viruses and their replication. In: Fields BM (ed) Virology Vol 1. Lippincott-Raven, Philadelphia, pp 931–960
43. Roehrig JT, Layton M, Smith P, Campbell GL, Nasci R, Lanciotti R (2002) The emergence of West Nile virus in North America: ecology, epidemiology and surveillance. In: Mackenzie JS, Barrett ADT, Deubel V (eds) Japanese encephalitis and West Nile viruses. Curr Top Microbiol Immunol 267: 223–240
44. Salminen MO, Carr JK, Burke DS, McCutchan FE (1995) Identification of breakpoints in intergenotypic recombinants of HIV-1 by bootscanning. Aids Res Hum Retrovir 11: 1423–1425
45. Scherret HJ, Poidinger M, Mackenzie JS, Broom AK, Deubel V, Lipkin WI, Briese T, Gould EA, Hall RA (2001) Studies on the relationships between West Nile and Kunjin viruses. Emerg Infect Dis 7
46. Smorodintsev AA, Dubov AV (1986) The correlation between the pathogenesis and biological characteristics of tick-borne encephalitis virus. In: Drozdov SG, Safronov BN, Komandenko NI (eds) Tick-borne encephalitis and its vaccino-prophylaxis. Medicine, Moscow, pp 113–124
47. Solomon T, Dung NM, Kneen R, Thao LTT, Raengsakulrach B, Day NPJ, Farrar JJ, A. MKS, Nisalak A, White NJ (2000) Neurological manifestations of dengue infection. Lancet 355: 1053–1059
48. Solomon T, Ni H, Beasley DW, Ekkelenkamp M, Cardosa MJ, Barrett AD (2003) Origin and evolution of Japanese encephalitis virus in southeast Asia. J Virol 77: 3091–3098
49. St. George TD, Doherty RL, Carley JG, Filippich C, Brescia A, Casals J, Kemp DH, Brothers N (1985) The isolation of arboviruses including a new flavivirus and a new Bunyavirus from Ixodes (Ceratixodes) uriae (Ixodoidea: Ixodidae) collected at Macquarie Island, Australia, 1975–1979. Am J Trop Med Hyg 34: 406–412
50. Strode GK (1951) Yellow fever. McGraw-Hill, New York
51. Tolou HJG, Couissinier-Paris P, Durand J-P, Mercier V, de Pina J-J, de Micco P, Billoir F, Charrel RN, de Lamballerie X (2001) Evidence for recombination in natural populations of dengue virus type 1 based on the analysis of complete genome sequences. J Gen Virol 82: 1283–1290

52. Tsai TF, Canfield MA, Reed CM, Flannery VL, Sullivan KH, Reeve GR, Bailey RE, Poland JD (1988) Epidemiological aspects of a St Louis encephalitis outbreak in Harris County, Texas, 1986. J Infect Dis 157: 351–356
53. Twiddy SS, Holmes EC (2002) The extent of homologous recombination in the genus Flavivirus. J Gen Virol 84: 429–440
54. Uchil PD, Satchidanandam V (2001) Phylogenetic analysis of Japanese encephalitis virus: envelope gene based analysis reveals a fifth genotype, geographic clustering and multiple introductions of the virus into the Indian subcontinent. Am J Trop Med Hyg 65: 242–251
55. Uzcategui NY, Camacho D, Comach G, Cuello de Uzcategui R, Holmes EC, Gould EA (2001) Molecular epidemiology of dengue type 2 virus in Venezuela: evidence for in situ virus evolution and recombination. J Gen Virol 82: 2945–2953
56. Vesenjak-Hirjan J (1980) Arboviruses in Yugoslavia. In: Arboviruses in the Mediterranean countries. Gustav Fischer Verlag, Stuttgart, pp 165–177
57. Vrati S, Giri RK, Razdan A, Malik P (1999) Complete nucleotide sequence of an Indian strain of Japanese encephalitis virus: sequence comparison with other strains and phylogenetic analysis. Am J Trop Med Hyg 61: 677–680
58. Weissenbock H, Kolodziejek J, Url A, Lussy H, Rebel-Bauder B, Nowotny N (2002) Emergence of Usutu virus, an African mosquito-borne flavivirus of the Japanese encephalitis virus group, Central Europe. Emerg Infect Dis 8: 652–656
59. Work TH, Trapido H (1957) Kyasanur Forest disease, a new virus disease in India. Indian J Med Sci 11: 341–345
60. Work TH, Rodriguez FR, Bhatt PN (1959) Kyasanur forest disease: Virological epidemiology of the 1958 epidemic. Am J Publ Health 49: 869–874
61. Worobey M, Rambaut A, Holmes EC (1999) Widespread intra-serotype recombination in natural populations of dengue virus. Proc Natl Acad Sci USA 96: 7352–7357
62. Zaki AM (1997) Isolation of a flavivirus related to the tick-borne encephalitis complex from human cases in Saudi Arabia. Trans Roy Soc Trop Med Hyg 91: 179–181
63. Zanotto PM, Gao GF, Gritsun T, Marin MS, Jiang WR, Venugopal K, Reid HW, Gould EA (1995) An arbovirus cline across the northern hemisphere. Virology 210: 152–159
64. Zanotto PM, Gould EA, Gao GF, Harvey PH, Holmes EC (1996) Population dynamics of flaviviruses revealed by molecular phylogenies. Proc Natl Acad Sci 93: 548–553
65. Zlobin VI, Gorin OZ (1996) Tick-borne encephalitis: etiology, epidemiology and prophylactics in Siberia. Nauka, Novosibirsk

Author's address: Dr. E. A. Gould, Centre for Ecology and Hydrology, Mansfield Road, Oxford, OX1 3SR, United Kingdom; e-mail: eag@ceh.ac.uk

Arch Virol (2004) [Suppl] 18: 85–96

Archives of Virology

West Nile virus and other zoonotic viruses in Russia: examples of emerging-reemerging situations

D. K. Lvov[1,2], **A. M. Butenko**[1], **V. L. Gromashevsky**[1], **A. I. Kovtunov**[3], **A. G. Prilipov**[1], **R. Kinney**[4], **V. A. Aristova**[1], **A. F. Dzharkenov**[3], **E. I. Samokhvalov**[1], **H. M. Savage**[4], **M. Y. Shchelkanov**[1], **I. V. Galkina**[1], **P. G. Deryabin**[1,2], **D. J. Gubler**[4], **L. N. Kulikova**[3], **S. K. Alkhovsky**[1], **T. M. Moskvina**[1], **L. V. Zlobina**[3], **G. K. Sadykova**[1], **A. G. Shatalov**[1], **D. N. Lvov**[1,2], **V. E. Usachev**[1], and **A. G. Voronina**[1]

[1]D.I. Ivanovsky Institute of Virology RAMS, Moscow, Russia
[2]I.M. Sechenov Moscow Medical Academy, Moscow, Russia
[3]State Center of Sanitary-Epidemiological Inspection, Astrakhan, Russia
[4]Division of Vector-Borne Infection Diseases, National Center for Infectious Diseases, Centers for Disease Control and Prevention, Fort Collins, Colorado, U.S.A.

Summary. Studies of the interactions of vertebrates, viruses and arthropod vectors of these viruses were monitored in terms of different ecological groups of viruses transmitted by mosquitoes and ticks in Northern Eurasia in an area encompassing more than 15 million km^2. About 90 viruses were isolated, including 24 new to science. Newly recognized infections of vertebrates, including humans, were described. Many unusual epidemic situations were analysed. Permanent efforts were established to prevent bioterrorist activities and their consequences. Extensive epidemic outbreaks of West Nile fever (WNF; i.e., fever caused by West Nile virus) and Crimean-Congo hemorrhagic fever (CCHF) with unusual high mortality appeared in the last four years in southern Russia. We determined infection rates in humans, domestic and wild animals, mosquitoes and ticks from natural and synanthropic biocenoses [Editorial note: "synanthropic" means, roughly, all species living with (c.f. lice, fleas) or near people, such as in houses (c.f. house mice), parks (c.f. *Rattus* spp.), and the like, rather like "peridomestic", but not strictly so; "biocenosis" is the biome, the "totality of living populations in a particular habitat, which itself is only a part of the ecosystem".]. CCHF virus strains were phylogenetically similar to strains isolated in this area 35 years ago but different from Central-South-Asian and African strains. Before the outset of the current emergence of epidemic WNF, three genetic variants of this virus had been isolated in USSR, two African and one Indian. Phylogenetic analysis of complete genome sequences of epidemic strains demonstrated considerable similarity to strains from USA and Israel and differences from strains isolated in the same USSR areas 20–30 years before. In addition to strains of genotype 1, we isolated strains of second and third lineages and a strain of a fourth genetic variant. Nucleotide differences of these strains from all three genotypes was about 30%. The emerging WNF situation in Russia for the last 4 years probably has been the result of not only natural and social factors, but also to introduction of more virulent strains or by evolution of the virus.

Introduction

In the former Soviet Union there was a functioning and vigorous anti-terrorist system, similar in purpose to CDC's Epidemiological Intelligence Service [9, 14]. The purpose and methodological approach provided ecological evaluations of various regions, with collection of field materials in longitudinal sections about 2,000 km long throughout the entire territory of northern Eurasia. In this way a land mass of more than 15 million km^2 was studied. Twelve zones were distinguished, each passing through unique ecosystems, including arctic, tundra, taiga, deciduous forest, steppe, and deserts within 18 physico-geographical areas. About 90 different viruses were isolated, of which 24 were new to science (multiple publications). The etiological role of these viruses in human disease was evaluated, with four of them causing previously unrecognized illnesses. As well, the potential for emergence of epidemic situations in various landscape zones was determined. Predicting areas of emergence of new viruses also was done [17, 19–21] and all this was part of a permanent system to prevent bioterrorism events and consequences for the foreseeable future [16].

Within the past four years there are been examples of re-emerging epidemic situations in the south of Russia, including extensive outbreaks of Crimean-Congo hemorrhagic fever (CCHF), West Nile fever (WNF; i.e., fever caused by West Nile virus; WNV) and perhaps Dhori virus (DHOV) [16, 22–24]. WNF endemic territories in southern Russia have been recognized since the first isolation of this virus was made in the north of the Caspian Sea basin region 40 years ago. Sporadic morbidity and small outbreaks were observed in southern regions of the Soviet Union practically every year. Antibody prevalence within the population of the USSR was known, and highest indices were found in southern Russia [17–21]. Therefore the epidemic of 1999, originally diagnosed by local specialists as due to enteroviral infections, was not unexpected, although the epidemic had unique features, principally the magnitude of the epidemic. Laboratory-verified cases alone were more than 500 in 1999, and the number of infected patients exceeded 200,000, according to our calculations on the basis of the results of serological examination before and after the outbreak [15, 16, 23, 32]. Mortality was also unusually high, about 10% [16, 22, 23]. Interestingly, a simultaneous and similar epidemic outbreak occurred in the New York City area [6, 12, 26], one which has spread to essentially the entire continental United States and which continues [10, 11, 25]. The virus introduced into the U.S. is most closely related to virus strains from Israel and N Africa [13].

To the south of Europe, first of all through the deltas of large rivers, such as the Rhone, Danube, and Volga, constant introduction of viruses has taken place via migration routes of birds, with subsequent introduction of viruses into local populations of birds and mosquitoes [18, 19]. The main natural focus in our country is at the Volga River delta. Natural and social factors, as is case with CCHF, were of importance in the development of the epidemic, but we also had to consider the possibility that what had appeared was a viral population with unusual genetic characteristics. By the outset of the epidemic three genetic variants of WNV were known: two African and one Indian [3–5, 12, 13, 29, 31, 33]. Introduction of the agent into the Volga River delta takes place basically from Africa, where more than 90% of birds nesting or flying over the delta are members of overwintering species, with some proportion of birds having over-wintered in the Indian subcontinent [19]. We considered it useful to collect field materials in this region and to analyse them by virological, serological and genetic methods so that we might answer the question: What species of birds, mosquitoes, and domestic animals are involved in WNV circulation and what are their genetic properties? To answer these questions was the rationale for our studies of the genomes of epidemic strains of WNV isolated from different sources in various years in the south of Russia and other regions of the USSR; we then intended to compare our data with data regarding WNV strains from Romania, Israel, India, Africa, and USA.

Materials and methods

The delta of the Volga River consists of three basic areas with unique ecosystems features. The Low-delta is located at the edges of the Caspian Sea and is characterized by extensive, exposed water spaces. The depths of the sea in these areas usually do not exceed 1.0–1.5 meters, which promotes mass mosquito propagation and provides simultaneous nesting opportunity for waterfowl and shore birds. The Middle-delta is more distant from the sea and consists of powerful currents and shallow water with reeds and shrubs. Here natural ecosystems are combined with synanthropic areas around a number of settlements, where inhabitants keep cattle, sheep, camels, and horses. The High-delta adjoins the Volga-Akhtuba lowland as well as semi-desert and desert. Cities including Astrakhan are located here. Some species of birds that are usual to Middle-delta also occur in this belt, closely contacting domestic animals and synanthropic birds. Most morbidity caused by WNV occurs in the High-delta belt. The virus could be introduced from the delta to the Volga-Akhtuba area and probably further north to steppes of low Volga belt, with large cities and mostly synanthropic biocenosis.

Field materials (brain and spleen from birds and mammals; mosquitoes; ticks; blood from patients; sera from domestic animals) were collected in August of 2001–2002 (Tables 1, 2) and immediately put into liquid nitrogen. Mosquitoes were collected by hand-held, battery-powered aspirators and in light traps. For virus isolations we tested spleen and brain from birds and mammals and pools of mosquitoes, 100 per pool, which were tested for virus using VeroE6 and pig kidney cells. In some cases we also used 3–4 day-old suckling mice, which were inoculated intracranially with 0.01 ml of clarified tissue suspension; when necessary, one blind passage was made 6 to 7 days after inoculation. All manipulation with mice satisfied international requirements for occupational health and safety in the care and use of research animals. When mice began to show signs of illness, brain tissue was removed and used for identification by RT-PCR, enzyme-linked immunosorbent assay (ELISA), and

Table 1. Virus isolation * and WNV RNA detection, field materials collected 2001–2002

Object	Order	Family	Genera	Species	RT-PCR results								
					Volga-delta			Volga-Akhtuba			Total		
					Number	Positive		Number	Positive		Number	Positive	
						number	%		number	%		number	%
Birds	*Podicipidiformes*	*Podicipitidae*	*Podiceps*	2 species	7	0	0.0	12	0	0.0	19	0	0.0
	Anseriformes	*Anatidae*	*Anas*	5 species	22	0	0.0	19	0	0.0	41	0	0.0
	Gruiformes	*Rallidae*	*Fulica*	*F. atra*	158	18	11.4	3	0	0.0	161	18	11.2
	Pelecaniformes	*Phalacrocoracidae*	*Phalacrocorax*	2 species	112[1,*]	23	20.5	3	0	0.0	115	23	20.0
	Ciconiiformes	*Ardeidae*	4 genera	6 species	72	5	6.9	10	0	0.0	82	5	6.1
		Threskiornithidae	*Plegadis*	*P. falcinellus*	31	0	0.0	0	0	0.0	31	0	0
	Charadriiformes	*Charadriidae*	3 genera	3 species	43	1	2.3	1	0	0.0	44	1	2.3
		Laridae	2 genera	6 species	183	9	4.9	39	1	2.6	222	10	4.5
	Passeriformes	*Corvidae*	4 genera	5 species	139[2,*]	38	27.3	53	2	3.8	192	40	20.8
		4 families	4 genera	4 species	89	4	4.5	13	0	0.0	102	4	3.9
	Falconiformes	2 families	2 genera	2 species	11	0	0.0	2	0	0.0	13	0	0.0
	Cuculiformes	*Cuculidae*	*Cuculus*	*C. canoris*	5	1	20.0	0	0	0.0	5	1	20.0
	Caraciiformes	2 families	2 species	2 species	11	1	9.1	1	1	(100)	12	2	16.6
	Galliformes	*Phasianidae*	*Perdix*	*P. perdix*	7	1	14.3	1	0	0.0	8	1	12.5
	Columbiformes	*Columbidae*	*Columba*	*C. livia*	11[3,*]	2	18.2	5	0	0.0	16	2	12.5
Total	12 order	20 families	29 genera	42 species	901	103	11.4	162	4	2.5	1063	107	10.1
Mammals	*Lagomorpha*	*Leporidae*	*Lepus*	*L. europaeus*	8[4,*]	2	25.0	5	0	0.0	13	2	15.4
	Insectivora	*Erinaceidae*	*Hemiechinus*	*H. auritus*	3	0	0.0	2	0	0.0	5	0	0.0
	Rodentia	*Dipodidae*	*Allactaga*	*A. elater*	1	0	0.0	6	0	0.0	7	0	0.0
			Alactagulus	*A. acontion*	1	1	(100)	0	0	0.0	1	1	(100)
Total	3 order	3 families	4 genera		13	3	23.1	13	0	0.0	26	3	11.5

Ticks	*Parasitiformes*	*Ixodidae*	*Hyalomma*	*H. marginatum*		160[5,*]	nt	nt	0	0	0.0	160	nt	nt
Mosquitoes	*Diptera*	*Culicidae*	Culex	*C. pipiens*	(s)	11400	2	0.018	300	1	0.333	11700	3	0.026
				C. modestus	(s)	6700	2	0.030	2900	1	0.034	9600	3	0.031
					(n)	13000	0	0.0	0	0	0.0	13000	0	0
			Anopheles	*A. messeae*	(s)	15800[6,7,*]	10	0.063	2600	2	0.077	18400	12	0.065
				A. hyrcanus	(s)	6900	5	0.072	0	0	0.0	6900	5	0.072
					(n)	17000	2	0.012	0	0	0.0	17000	2	0.012
			Coquillettidia	*Cq. richiardii*	(n)	12300	6	0.049	0	0	0.0	12300	6	0.049
Total	1 order	1 family	3 genera	5 species	(s)	40800	19	0.047	5800	4	0.069	46600	23	0.049
					(n)	42300	8	0.019	0	0	0.0	42300	8	0.019
					$\sum$	83100	27	0.033	5800	4	0.069	88900	31	0.035

*Isolated strains:
[1] 3 WNV strains from *Phalacrocorax carbo*
[2] 4 WNV strains: 1 from *Corvus frugilegus*, 2 from *Corvus corone*, 1 from *Pica pica*
[3] 1 WNV strain from *Columba livia*
[4] 1 DHOV strain – from *Lepus europaeus*
[5] 7 strains from *H. marginatum*: 2 – WNV, 3 – DhoV, 1 – CCHFV, 1 – DhoV and CCHFV mix
[6] 19 Batai virus strains from Anopheles messeae (infection rate 0.039%)
[7] 2 WNV strains from *Anopheles messeae*
(s) – synanthropic biocenoses
(n) – natural biocenoses
Additionally one strain of CCHFV isolated from blood of patient in Volga-Akhtuba ecosystem

Table 2. Serological examination of domestic animals in Volga-delta and Volga-Akhtuba, 2001–2002

Animal	Ecosystem	NT			HIT		
		Number	Positive results		Number	Positive results	
			number	%		number	%
Cattle	Volga-delta	88	28	31.8	373	31	8.3
	Volga-Akhtuba	54	5	9.3	150	15	10.0
	Total:	142	33	23.2	523	46	8.8
Sheep	Volga-delta	91	15	16.5	332	10	3.0
	Volga-Akhtuba	28	9	32.1	188	2	1.1
	Total:	119	24	20.2	520	12	2.3
Camels	Volga-delta	24	2	8.3	76	9	11.8
	Volga-Akhtuba	38	6	15.8	83	2	2.4
	Total:	62	8	12.9	159	11	6.9
Horses	Volga-delta	nt*	nt	nt	68	15	22.1
	Volga-Akhtuba	nt	nt	nt	17	0	0.0
	Total:	nt	nt	nt	85	15	17.6
	Volga-delta	203	45	22.2	849	65	7.7
	Volga-Akhtuba	120	20	16.7	438	19	4.3
TOTAL:		323	65	20.1	1287	84	6.5

**nt* not tested

hemagglutination inhibition test (HIT) with immune ascite fluids (IAF) of mice to viruses previously isolated in this region: Tahyna and Batai (*Bunyaviridae*, *Orthobunyavirus*), CCHF (*Bunyaviridae*, *Nairovirus*), Bhanja (*Bunyaviridae*), Sindbis (*Togaviridae*, *Alphavirus*), WNV (*Flaviviridae*, *Flavivirus*), and Dhori (*Orthomyxoviridae*, *Thogotovirus*) [33]. Neutralization testing (NT) was done by the micro method using pig kidney cells, with a single dilution of sera or IAF and 10-fold dilutions of virus.

WNV RNA detection was extracted from 0.2–0.3 ml sample using a guanidine-iso-thiocyanate-phenol-chloroform method. cDNA synthesis was performed using WNV-specific antisense 23-mer primer GTGCACCAACAGTCGATGTCTTC, and AMV-derived reverse transcriptase (Promega). Amplification of WNV RNA-specific sequence was carried out by PCR using primers derived from a well-conserved 5′-region of the WNV genome. Briefly, the first-round PCR amplified a sequence with primers 1F/518R (25 cycles of 94 °C – 20 sec/55 °C – 20 sec/72 °C – 30 sec), and the second-round PCR amplified an inner sequence of 487 bp in length with primers 26F/518R (35 cycles of 94 °C – 20 sec/60 °C – 20 sec/72 °C – 30 sec). Primers: 23-mer 1F – AGTAGTTCGCCTGTGTGAGCTGA; 23-mer 518R – TCG GTAGCATTTACCGTCATCAT; 22-mer 26F – AACTTAGTAGTGTTTGTGAGGA. Both strands of the PCR product were then sequenced on a 377 automated sequencer (Applied Biosystems, USA). Sequences have been composed with Lasergene SeqMan computer program (DNAstar, USA). Consensus sequences then were aligned by using the program MegAlign (DNAstar, U.S.A.). WNV strains sequences tested in this study are presented in Table 3.

Table 3. West Nile virus strains used for phylogenetic study*

#	Strain	Year of isolation	Source of isolation	Place of isolation
1	Ast63-94-ticks	1963	Ticks *Hyalomma marginatum*	Russia, Volga delta
2	LEIV-Az67-1640-nathatch	1967	Bird *Sitta europaea*	Azerbaijan, Lencoran lowland
3	LEIV-Az67-1628-thrush	1967	Bird *Turdus merula*	Azerbaijan, Lencoran lowland
4	LEIV-Az70-72-ticks	1970	Ticks *Ornithodoros capensis*	Azerbaijan, Island in Caspian Sea
5	LEIV-Tur73-2914-ticks	1973	Ticks *Hyalomma detritum*	Turkmenistan, Uzboi valley
6	LEIV-Taj77-1304-tern	1977	Bird *Sterna albifrons*	Tajikistan, Pyanzh valley
7	LEIV-Ukr80-3266-rook	1980	Bird *Corvus frugilegus*	pre-Carpathian plain, Ukraine
8	LEIV-Krnd88-190-ticks	1988	Ticks *Dermacentor marginatus*	N-W Caucasus, mountain valley
9	LEIV-Vlg99-27889-human	1999	Human brain (section material)	Russia, Volgograd, low Volga
10	LEIV-Vlg00-27924-human	1999	Human blood	Russia, Volgograd, low Volga
11	Ast99-901-human	1999	Human blood	Russia, Astrakhan, Volga delta

*WNV strain sequences from GenBank used for phylogenetic study: Eg101 (Egypt, 1951), Ig2266 (India, 1955), NC-001563 (Nigeria), Italy 1998 (Italy, 1998), RO97-50 (Romania, 1997), IS-98 STD (Israel, 1998), Connecticut (U.S.A., 1999), NY-99 eqhs (U.S.A., 1999), HNY-1999 (U.S.A., 1999), NY-99 flamingo (U.S.A., 1999)

Results

Among 38 strains isolated from field materials collected in Volga-delta and Volga-Akhtuba ecosystems in 2001–2002, 12 were of WNV, 3 of CCHFV, 4 of DHOV, and 19 of Batai virus (Table 1). All strains of WNV were isolated in the Volga delta. All strains in natural habitats were obtained from cormorants and in synanthropic biocenoses – from *Anopheles messeae* mosquitoes, *Corvidae* birds (mostly crows) and, collected from them, pre-imago *Hyalomma marginatum* ticks. CCHFV strains were isolated from patients in Volga-Akhtuba ecosystem as well as from larva and nymphs of *H. marginatum* collected from a hare in synanthropic biocenoses of the Volga delta ecosystem. DHOV strains were also isolated from the same individual hare and ticks. One was a mixture of CCHFV and DHOV. Batai virus was isolated from *An. messeae* mosquitoes in synanthropic biocenoses (Table 1). WNV was detected by RT-PCR of tissues from birds, mostly those collected in the Volga delta and most often among cormorants, coots, herons, gulls, and terns in relatively undisturbed habitats. In these areas, positive results were obtained among ground-feeding birds, especially among corvids, and from mammals (Table 1).

Investigation by RT-PCR of mosquitoes showed that predominant species were involved in WNV circulation in synanthropic biocenoses, including *An. hyrcanus, Culex pipiens, Cx. modestus*, and *An. messeae*, and in natural biocenoses *Coquillettidia richiardii* and *An. hyrcanus* (Table 1).

All isolated strains and RT-PCR positive samples were shown to belong to genotype 1 of WNV, as determined by partial nucleotide sequencing of the E gene. Phylogenetic data of sequenced strains are presented on Fig. 1.

Serologic examination of domestic animals showed that all these species could be infected by WNV, more often in the Volga delta than in the Volga-Akhtuba area.

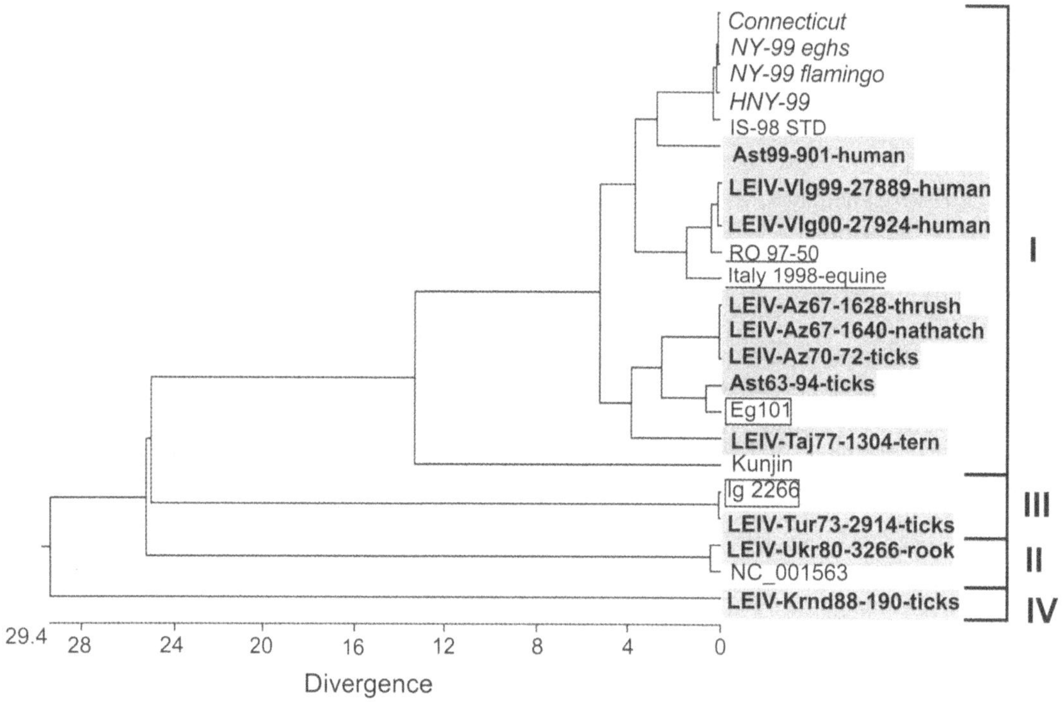

Fig. 1. Phylogenetic tree for historical and epidemic West Nile virus strains. Roman numbers designate genetic lineage. Font style designations for West Nile virus strains: grey scale – historic from Soviet Union; bold – Russian epidemic strain; frame – Topotype; underline – European epidemic strain; italic – American epidemic strain

Indices of infection rate were highest among horses, moderate among cattle and camels, and lowest among sheep (Table 2).

Discussion

Local complexes of arbovirus foci, including at least WNV, CCHFV, and DHOV, all recognized human pathogens, had been described in southern Russia [17–21]. In the course of laboratory manipulation with DHOV five cases of illness occurred with abrupt onset, marked general toxicity symptoms of subcortical pyramidal tract involvement and other lesions of the central nervous system [17]. Natural CCHF and DHOV foci in the world, as in our country, essentially coincide with the area of the main virus host and vector, *H. marginatum* ticks. In southern Russia this is the landscape belt of dry steppes, semi-deserts and deserts. In northern Russia the occurrence of these viruses is limited by isotherm of the sum of effective temperatures 3000 °C [17, 20, 21]. Recently, a number of circumstances have led to a considerable increase in ticks. This produced the emergence of an epidemic for CCHF [16]. Comparative studies were done with 310–529 fragments of S-segment genomes of epidemic strains isolated during the last four years from patients and ticks and strains isolated earlier in the south of Russia and Central Asia, as well as the strains from South Asia and Africa. These showed that in southern Russia, strains were genetically similar to those circulating 30 years previously, and that nucleotide sequence differences did not exceed 1–2%. Thus, the molecular evolution of CCHF virus is progressing very slowly [16, 35]. *H. marginatum* ticks are important for the long-term persistence of virus [19], for instance during winter in synanthropic foci of WNV in the Volga delta region.

We also isolated WNV, CCHF and DHOV from crow and hare and pre-imago stages of *H. marginatum* ticks collected from them [24], and developed diagnostic tests to detect evidence of human infection with these viruses.

In the U.S.A. WNV was detected by RT-PCR in New York State and elsewhere in 2000. Virus was detected among dead birds of 162 species, with highest indices in American crows (*Corvus brachyrhynchos*), which had 67% mortality in the epidemic epicenter; dead birds were found two weeks before the first human illness occurred [2, 11, 25]. ELISA for detection of antibodies to WNV was also used for the New York studies. Antibodies were found in birds of 23 species [7]. In studies done 30–40 years ago in the Volga delta region WNV was isolated from glossy ibis, bittern, blackbird, nuthatch, herring gull, and crow [17–19]. Our data showed that the greatest involvement in WNV circulation was among cormorants, twice the prevalence in coots, and five times that of herons, gulls, and terns in natural habitats. Positive results in synanthropic biocenoses were found among ground-feeding birds, especially among corvids. By RT-PCR, we determined that the most important mosquito vectors of WNV in synanthropic biocenoses are the ornithophilic species *An. hyrcanus, Cx. pipiens, Cx. modestus* and the cattle-feeding species *An. messae*. In natural habitats ornithophilic species *An. hyrcanus* in the middle belt and *Cq. richiardii* and *An. hyrcanus* in low Volga delta region are most important. Overall, the infection rate of mosquitoes in synanthropic biocenoses was twice that of the rate in natural habitats.

During the 1996 epidemic outbreak of WNV in Romania the principal vector was *Cx. pipiens* [31]. In the U.S.A. WNV has been isolated from *Cx. pipiens, Cx. restuans, Cx. salinarius, Culiseta melanura* [1, 2] and from *Ae. vexans, Ochlerotatus triseriatus, Psorophora ferox*, and *An. punctipennis*. It appears likely that *Cx. pipiens* and *Cx. restuans* are the primary enzootic and epizootic vectors among birds, *Cx. salinarius* for humans, and *Ae. vexans* and *Ochlerotatus* spp. for horses [11, 34]. Under experimental conditions the mosquitoes most susceptible to infection with WNV were *Cx. restuans* and *Cx. salinarius*; *Cx. quinquefasciatus*, and *Cx. nigripalpus* were of moderate sensitivity and *Coquillettidia perturbans* was of low sensitive [8, 30]. Our virological and serological data showed a high level of WNV activity in the Volga delta region, as compared with the Volga-Akhtuba ecosystem. Serologic investigations of domestic animals were useful as indicators of WNV activity.

Phylogenetic analysis of full-dimension genomes demonstrated the following: Astrakhan epidemic strains were similar to epidemic strains from New York and Israel (nucleotide differences 2.7%); Volgograd strains were similar to epidemic strains from Romania (nucleotide differences 0.4%). It is likely that the New York outbreak originated as a result of virus introduced via infected mosquitoes, humans or birds from the Mediterranean, but it is also possible that the virus was introduced from Black Sea ports, by infected mosquitoes within holds of ships. Recent epidemic strains are genetically different from strains isolated in Russia in the same region from various natural sources 20–30 years ago, during which interval there was an absence of epidemics (nucleotide differences about 6%). In addition to strains of genotype 1 WNV, we isolated a strain from a corvid (*Corvus frugilegus*) belonging to genotype 2, a strain from *Hyalomma detritum* ticks belonging to the Indian genetic lineage 3 (nucleotide distance of which from the first and second groups is about 25%), and a strain from *Dermacentor marginatus* ticks that is quite different from all recognized isolates (nucleotide differences from all three groups about 30%). Using 20% nucleotide differences as a criterion, we are able to differentiate four genetic lineages of WNV, one of which, is known thus far only from Russia [16, 27, 28]. It is of practical importance that the epidemic situation of the last three years probably was based not only by natural-social factors but also by alterations in the genetic characteristics of circulating viral populations. On the basis of these molecular genetic studies of epidemic and historic WNV strains, we developed ELISA-based systems to detect virus antigens and, as well, developed specific IgM antibody and IgG antibody assays, as well as PCR assays that are effective for detection of all four genotypes of WNV.

Acknowledgements

The work was partly supported by ISTC Grant 2087. The authors thank the following for their assistance – in Astrakhan: Guzhvin A.P. (Governor of Astrakhan region), Osipov A.A., Mikhailov G.M., Agoshkov V.M., Grishanova A.P., Zlobina L.B., Ustaev V.M., Zimin A.V., Stopkin N.V., Binkov S.V.; Ikryanoe: Odolevsky E.I., Kodyakov P.N., Kodyakov S.N., Shchetinkin P.F.; Liman: Ibragimov R.M., Pavly V.I., Seliverstov E.I.; Akhtubinsk: Shatilov V.P., Finogenov O.V., Lapenko V.V., Teplinskaya L.V.; Kharabaly: Chekrizov P.F., Moiseeva O.V., Inozemtseva I.G.; Krasny Yar: Volodin V.N., Turlaev Yu.V.; Moscow: Vakar E.I., Ivanova M.V., Belonozhkina G.M.; Fort Collins (U.S.A.): Tsuchiya R.

References

1. Andreadis TG, Anderson JF, Vossbrinck R (2001) Mosquito surveillance for West Nile virus in Connecticut, 2000: isolation from *Culex pipiens, Cx. restuans, Cx. salinarius,* and *Culiseta melanura*. Emerg Infect Dis 7: 670–674
2. Bernard KA, Maffei JG, Jones SA, Kauffman E, Ebel G, Dupuis A et al. (2001) West Nile virus infection in birds and mosquitoes. Emerg Infect Dis: 679–685
3. Berthet FX, Zeller HG, Drouet MT, Rauzier J, Digoutte JP, Deubel V (1997) Extensive nucleotide changes and deletions within the envelope glycoprotein gene of Euro-African West Nile viruses. J Gen Virol 78: 2293–2297
4. Briese T, Rambaut A, Pathmajeyan M, Bishara J, Weinberger M, Pitlik S, LipkinWI (2002) Phylogenetic analysis of human isolate from the 2000 Israel West Nile virus epidemic. Emerg Infect Dis 5: 528–531
5. Brinton MA (2002) The molecular biology of West Nile virus: A new invader of the Western Hemisphere. Annu Rev Microbiol 56: 371–402

6. Centers for Disease Control and Prevention (2002) West Nile virus activity – United States, September 5–11, and Texas, January 1–September 9, 2002, Morb Mortal Wkly Rep 51: 812–823
7. Ebel GD, Dupuis AP, Ngo K, Nickolas D, Kauffman E, Jones S et al. (2001) Partial genetic characterization of West Nile virus strains, New York state, 2000. Emerg Infect Dis 7: 650–653
8. Goddard LB, Roth AE, Reisen WK, Scott TW (2002) Vector competence of California mosquitoes for West Nile virus. Emerg Infect Dis 8: 1385–1391
9. Goodman RA, Bauman CF, Gregg MB, Videtto JF, Stroup DF, Chalmers NP (1990) Epidemiologic field investigations by the Centers for Disease Control and Epidemic Intelligence Service, 1996–1987. Public Health Reports 105: 604–610
10. Huang C, Slater B, Rudd R, Parchuri N, Hull R, Dupuis M, Hindenburg A (2002) First isolation of West Nile virus from a patient with encephalitis in the United States. Emerg Infect Dis 8: 1367–1371
11. Kulasekera VL, Kramer L, Nasci RS, Mostashari F, Cherry B, Trock SC, Glaser C, Miller JR (2001) West Nile virus infection in mosquitoes, birds, horses, and humans, Staten Island, New York, 2000 (2001) Emerg Infect Dis 7: 722–725
12. Lanciotti RS, Ebel GD, Deubel V, Kerst AJ, Murri S, Meyer R et al. (2002) Complete genome sequences and phylogenetic analysis of West Nile virus strains isolated from the United States, Europe, and the Middle East. Virology 298: 96–105
13. Lanciotti RS, Roehrig JT, Deubel V, Smith J, Parker M, Steele K et al. (1999) Origin of the West Nile virus responsible for an outbreak of encephalitis in the northeastern United States. Science 286: 2333–2337
14. Langmuir AD (1980) The Epidemic Intelligence Service of the Center for Disease Control. Public Health Reports Sept-Oct 95: 470–477
15. Larichev VF, Butenko AM, Rusakova NV, Shishkina LV, Zhukov AN, Shishkina EO, Lvov DK (2002) Indexes of specific antibodies to West Nile virus among reconvalescents. Vopr Virusol 47: 11–13 (in Russian)
16. Lvov DK (2002) Importance of Emerging-ReEmerging Infections for biosecurity. Vopr Virusol 47: 4–7 (in Russian)
17. Lvov DK (1994) Arboviral Zoonoses on Northern Eurasia (Eastern Europe and the Commonwealth of Independent States). In: Beran GW, Steel JH (eds), Handbook of zoonoses, 2nd edn. CRC Press, Boca Raton, pp 237–260
18. Lvov DK (2000) West Nile fever. Vopr. Virusol 45: 4–9 (in Russian)
19. Lvov DK, Ilichev VD (1979) Migrations of birds and transfer of infections agents. "Science" Moscow: 1–270 (in Russian)
20. Lvov DK, Klimenko SM, Gaidamovich SY (1989) Arboviruses and arboviral infections. "Meditsina" Moscow: 1–335 (in Russian)
21. Lvov DK, Deryabin PG, Aristova VA, Butenko AM, Galkina IV, Gromashevsky VL, Davydova AA, Kolobukhina LV, Lvov SD, Scchelkanov MY (2001) Atlas of distribution of zoonotic viral infections on territory of Russia. Ministry of Public Health of Russian Federation, Moscow, 193 pp (in Russian)
22. Lvov DK, Butenko AM, Gromashevsky VL. Larichev VP, Salko VN, Kovtunov AI, Galimzyanov KM, Platonov AE, Morozova TN, Khutoretskaya NV, Shishkina EO, Skvortsova TM (2000) Isolation of two strains of West Nile virus during an outbreak in Southern Russia, 1999. Emerg Infect Dis 6: 373–376
23. Lvov DK, Butenko AM, Gaidamovich SY, Larichev VF, Leschinskaya EV, Lazarenko VV et al. (2000) Epidemic outbreak of meningitis and meningo-encephalitis in Krasnodar territory and Volgograd region provoked by West Nile virus. Vopr Virusol 45: 37–38 (in Russian)

24. Lvov DN, Jarkenov AF, Aristova VA, Kovtinov AT, Gromashevsky VL, Galkina IV et al. (2002) Isolation of Dhori (*Orthomyxoviridae, Thogotovirus*) and Crimean-Congo hemorrhagic fever (*Bunyaviridae, Nairovirus*) from hare (*Lepus europaeus*) and collected from ticks *Hyalomma marginatum* in the middle part of Volga-delta, Astrakhan region, 2001. (2002) Vopr Virusol 47: 32–36 (in Russian)
25. Marfin AA, Petersen LR, Eidson M, Miller J, Hadler J, Farello C et al. (2001) Widespread West Nile activity, Eastern United States, 2000. Emerg Infect Dis 7: 730–735
26. Nasci RS, White DJ, Stirling H, Oliver J-A, Daniels TJ, Falco RC et al. (2001) West Nile virus isolates from mosquitoes in New York and New Jersey, 1999. Emerg Infect Dis 7: 626–630
27. Prilipov AG, Kinney R, Samokhvalov EI, Savage HM, Alkhovsky SV, Tsychia R, Gromashevsky VL, Sadykova GK, Shatalov AG, Vyshemirsky OI, Usachev EV, Mokhonov VV, Voronina AG, Butenko AM, Larichev VF, Gubler DJ, Lvov DK (2002) Phylogenetic analysis of a new variants of West Nile virus. Vopr Virusol 47: 36–41 (in Russian)
28. Prilipov AG, Samokhvalov EI, Lvov DK, Gromashevsky VL, Butenko AM, Larichev VF et al. (2001) Genetic analysis of West Nile viruses isolated on the south of Russian Plain (Volgograd and Astrakhan regions) in 1999. Vopr Virusol 46: 8–12 (in Russian)
29. Scherret JH, Poidinger M, Mackenzie JS, Broom AK, Deubel V et al. (2001) The relationships between West Nile and Kunjin viruses. Emerg Infect Dis 7: 697–705
30. Sardelis MR, Turell MJ, Dohm DJ, O'Guinn ML (2001) Vector competence of selected North American *Culex* and *Coquilletidia* mosquitoes for West Nile virus. Emerg Infect Dis 7: 1018–1022
31. Savage HM, Ceianu C, Nicolescu G, Karabatsos N, Lanciotti R, Vladimiresku A et al. (1999) Entomologic and avian investigations of an epidemic of West Nile fever in Romania in 1996, with serological and molecular characterization of a virus isolated from mosquitoes. Am J Trop Med Hyg 61: 600–611
32. Shishkina EO, Butenko AM, Gaidamovich SY, Larichev VF, Zhuravlev VI, Shcherbakova SA et al. (2002) Immune structure of population to West Nile virus in Astrakhan region in 1998 and 1999. Vopr Virusol 47: 13–17 (in Russian)
33. Van Regenmortel MHV, Fauquet CM, Bishop DHL, Carstens EB, Estes MK, Lemon SM, Maniloff J, Mayo MA, McGeoch DJ, Pringle CR, Wicner RB (eds) (2000) Virus Taxonomy. 7th Report of the ICTV. Academic Press, San Diego
34. White DJ, Kramer LD, Backenson PB, Lukacik G, Johnson G, Oliver JA et al. (2001) Mosquito surveillance and polymerase chain reaction detection of West Nile virus. New York state. Emerg Infect Dis 7: 643–649
35. Yashina L, Petrov VC, Vyshemirsky OI, Aristova VA, Moskvina TM, Lvov DK et al. (2002) Characteristics of Crimean-Congo hemorrhagic fever virus circulating in the Russia and Central Asian Republics. Vopr Virusol 47: 11–15 (in Russian)

Author's address: D. K. Lvov, D.I. Ivanovsky Institute of Virology RAMS, Gamaleya 16, Moscow 123098, Russia; e-mail: dk_lvov@mail.ru

Arch Virol (2004) [Suppl] 18: 97–111

Archives of
Virology

Emerging encephalitogenic viruses: lyssaviruses and henipaviruses transmitted by frugivorous bats

J. S. Mackenzie[1] and **H. E. Field**[2]

[1]Department of Microbiology and Parasitology, School of Molecular and Microbial Sciences, University of Queensland, Brisbane, Australia
[2]Department of Primary Industries, Animal Research Institute, Yeerongpilly, Queensland, Australia

Summary. Three newly recognized encephalitogenic zoonotic viruses spread from fruit bats of the genus *Pteropus* (order *Chiroptera*, suborder *Megachiroptera*) have been recognised over the past decade. These are: Hendra virus, formerly named equine morbillivirus, which was responsible for an outbreak of disease in horses and humans in Brisbane, Australia, in 1994; Australian bat lyssavirus, the cause of a severe acute encephalitis, in 1996; and Nipah virus, the cause of a major outbreak of encephalitis and pulmonary disease in domestic pigs and people in peninsula Malaysia in 1999. Hendra and Nipah viruses have been shown to be the first two members of a new genus, *Henipavirus*, in the family *Paramyxoviridae*, subfamily *Paramyxovirinae*, whereas Australian bat lyssavirus is closely related antigenically to classical rabies virus in the genus *Lyssavirus*, family *Rhabdoviridae*, although it can be distinguished on genetic grounds. Hendra and Nipah viruses have neurological and pneumonic tropisms. The first humans and equids with Hendra virus infections died from acute respiratory disease, whereas the second human patient died from an encephalitis. With Nipah virus, the predominant clinical syndrome in humans was encephalitic rather than respiratory, whereas in pigs, the infection was characterised by acute fever with respiratory involvement with or without neurological signs. Two human infections with Australian bat lyssavirus have been reported, the clinical signs of which were consistent with classical rabies infection and included a diffuse, non-suppurative encephalitis. Many important questions remain to be answered regarding modes of transmission, pathogenesis, and geographic range of these viruses.

Introduction

Three novel emergent viruses that can cause encephalitis or meningo-encephalomyelitis have been described over the past decade from frugivorous bats (flying foxes) in the genus *Pteropus* from Australasia and South-East Asia [45]. Two of these, Hendra and Nipah viruses, are members of a new genus, *Henipavirus*, in the family *Paramyxoviridae*, and the third, Australian bat lyssavirus, is a recently recognized member of the genus *Lyssavirus* in the family *Rhabdoviridae*. Additionally, two non-encephalitogenic viruses have also been isolated from pteropid bats, Menangle and Tioman viruses, which are members of the genus *Rubulavirus* in the family *Paramyxoviridae*. This paper describes the emergence and biological characteristics of these viruses, and addresses some of the questions that remain to be answered about their ecology and pathogenesis.

Henipaviruses

Hendra virus

The first of the novel bat-borne viruses to be recognized was Hendra virus, formerly known as equine morbillivirus, which was the cause of a 1994 outbreak of acute respiratory disease in two humans and 21 thoroughbred race horses in Hendra, a suburb of Brisbane, Australia [52, 61]. In that outbreak, one human, the racehorse trainer, and 14 horses died, and a further 7 horses with subclinical infection were later euthanased. A number of recent reviews have described the isolation, ecology and molecular biology of Hendra virus [13, 16, 29, 34, 43, 66, 68]. The clinical features and human and equine autopsy findings of the original outbreak were consistent with interstitial pneumonia [51, 52, 61]. A virus was isolated from equine lung tissue and from human kidney tissue taken at autopsy, and was initially named equine morbillivirus on the basis of its structure and a weak one-way cross reaction with rinderpest virus.

A second small outbreak came to light about 12 months later, although it actually pre-dated the Hendra outbreak by more than a month. Two horses died (one of acute respiratory distress and the other with rapid-onset of neurological symptoms) of unknown cause in August, 1994, on a farm near Mackay, about 1000 km north of Brisbane. The farmer became ill with a mild meningitic illness after assisting in the necropsy of the dead horses, but subsequently recovered. About 13 months later, he became ill and died from a severe encephalitis, which was shown to have been caused by Hendra virus [54]. Retrospective investigations showed that the horses had died of Hendra virus infection [30, 59] and that the farmer had also been infected at the same time, indicating that the virus had presumably remained latent and reactivated 13 months later [54]. These were the first indications that Hendra virus could cause neurological disease in humans and horses. A third incident of equine infection occurred in Cairns in January 1999, when a single infected horse was identified [15, 31].

In efforts to find the source of the virus, an extensive serological investigation was undertaken of wild and domestic animals, and over 5000 animals were tested [51, 59, 70, 79]. Only fruit bats (flying foxes) of the genus *Pteropus* in the suborder *Megachiroptera* were found to be seropositive [78, 79]. At least some of all four members of the genus found in Australia (the black flying fox, *Pteropus alecto*; the grey-headed flying fox, *P. poliocephalus*; the little red flying fox, *P. scapulatus*; and the spectacled flying fox, *P. conspicillatus*) were shown to be seropositive.

Indeed approximately 47% of flying foxes sampled over their full geographic range have been found to have antibodies to Hendra virus, although variation in prevalence was found between species [16]. Hendra virus isolated from the uterine fluid and from foetal tissues from a euthanased flying fox [23, 24] was found to be indistinguishable from the equine and human isolates.

Molecular studies have clearly shown that Hendra virus is a member of the sub-family *Paramyxovirinae* within the family *Paramyxoviridae*, but that it differs significantly from members of other genera in the sub-family in a number of molecular and biological properties [66, 68]. Thus, sequence studies have shown that although the gene arrangement is very similar to viruses in the genera *Respirovirus* and *Morbillivirus*, the genome is much larger (18.2 kb), due partly to a larger P gene and partly to longer untranslated regions at the 3′ ends of the six transcription units [19, 23, 67, 69, 80, 81]. It was therefore suggested that Hendra virus should be the type species of a new genus in the family and that the new genus should be called *Henipavirus* [67]. Biological and other properties also distinguish Hendra virus from other paramyxoviruses. It was shown to have an unusual double fringe morphology comprising 15 nm and 18 nm projections [33, 34], a wide *in vitro* host range, and a predilection for endothelial cells [29].

A number of major questions about the ecology of Hendra virus are still to be addressed. The most important questions relate to fruit bats, the reservoir hosts of the virus, and the mode of transmission of the virus from bats to horses, as the basis for the genesis of further epidemic activity, and thus of possible management strategies. There is also an apparent predilection of Hendra virus for pregnancy, which may also be an important epidemiological guide in providing an indication of seasons of potential transmission. Natural infection in fruit bats appears to be asymptomatic, and virus has been isolated from the kidney [16] and is present in uterine fluids of pregnant animals [24]. These findings are borne out by experimental infection; most bats seroconverted, but no gross lesions were observed at necropsy, although systemic vascular lesions were observed, and there was positive staining for virus in the endothelial cells [76] and in the placentas of two bats, with virus isolated from the foetus of one [74]. All three occurrences of Hendra virus activity have been associated with infection in horses, and in two instances, transmission from horses to humans. Interestingly, no evidence of direct transmission from fruit bats to humans has been found, despite the close relationships often over many years between bat carers and sick, injured or orphan fruit bats [61] nor with the exception of the three human cases, among people who had variable levels of exposure to infected horses (such as veterinarians with necropsy contact) [49]. Furthermore, no cases of Hendra virus infection have been found in archival human tissue specimens from autopsies of people who had died of pneumonia or acute lung infection of unknown aetiology (C Allen, LA Selvey, JS Mackenzie, unpublished results).

A characteristic pathology has emerged for Hendra virus, regardless of differences between hosts, which is primarily vasotropic and/or neurotropic, generating interstitial pneumonia or encephalitis, with syncytial cells in vascular tissues and with a pronounced endothelial tropism. Investigations of experimental infections in a range of animal species provide some indication of modes of transmission. Thus horses, cats and guinea pigs have been investigated. Horses generally develop severe pulmonary oedema and congestion. A characteristic change was the appearance of endothelial syncytial cells in the capillaries and arterioles in lung, lymph node, spleen, brain, heart and kidney. Interestingly, experimental infections differ from field infections in that the latter often display a frothy nasal discharge, sometimes blood-tinged, which fills the airways, whereas this has not been observed in experimental infections. This discharge has been suspected as a means of horse-to-horse transmission during the outbreak in Mackay [16]. Non-suppurative encephalitis was also observed in some horses [28, 76]. Virus has been isolated from various horse tissues, and also from urine and in one case, from the oral cavity. No transmission was observed between infected bats and in-contact horses and bats, or between infected horses and in-contact horses or cats [76]. The disease in experimentally-infected cats resembled that in horses and was characterised by generalised vascular disease with the most severe effects in the lung [27, 73]. Encephalitis has not been observed in cats [29]. Virus was isolated from various tissues and from blood and urine of infected cats. Transmission was observed from an infected cat to a horse [76]. Guinea pigs also developed a generalised vascular disease, but there was little pulmonary oedema, and congestion and oedema were seen more commonly in the gastrointestinal tract [27]. Encephalitis was seen in some guinea pigs after inoculation with high doses of virus [75], and thus they may provide a model system for exploring an encephalitic tropism. Virus can also be found in lesions in the bladder by immunostaining, and virus detected in the urine [75]. Infection of pregnant guinea pigs resulted in vascular disease in the placenta and virus infection of the foetuses [74].

The importance of pregnancy in providing a potential time of increased risk of transmission has not been adequately considered in management plans. Thus the virus was first isolated from fruit bat uterine fluid and subsequently from aborted foetuses; the index or initial horse in each of the outbreaks was pregnant; the season of each outbreak corresponded to the birthing period of the local fruit bat population [16]; the virus has been shown experimentally to infect the placenta in fruit bats and guinea pigs, and to invade the foetus. As pregnancy leads to immunosuppression, it may be that the virus is reactivated in pregnant bats, excreted in the urine, and transmitted to other bats during mutual grooming during which they use urine in the grooming process. Transmission from bats to horses must be a rare event; it may be due to infected urine contaminating pasture, or possibly through placental material and afterbirth dropping to the ground from 'birthing trees' in fields where horses graze.

Are other wild or domestic animal species at risk? No information is yet available about the susceptibility of sheep, goats, deer or pigs, all of which are potentially at risk of exposure. Similarly, there is no information on the susceptibility of indigenous species. Information of this nature is important in formulating management plans.

Nipah virus

Nipah virus first came to light as the causative agent of a major outbreak of disease in pigs and humans in Peninsula Malaysia between September 1998 and April 1999, which resulted in 265 human cases with 106 fatalities, and the eventual culling of about 1.1 million pigs [11, 12]. The emergence, ecology, clinical course, molecular biology and pathogenesis of Nipah virus have been reviewed recently [5, 16, 29, 34, 38, 64, 68, 77], and so will only be described briefly here. The outbreak began as sporadic cases of encephalitis among pig farmers in the Ipoh area in the state of Perak in late 1998. The outbreak was initially ascribed to Japanese encephalitis virus (JEV), and wide scale JEV immunisation was therefore undertaken. However, as pigs were also dying, which is not a characteristic of JEV infection, and cases of encephalitis were occurring in JEV-immunised individuals, it became apparent that the cause of the outbreak had to be another pathogen. The epidemic then spread through movement of infected animals to a second epicentre, about 160 miles south, in the state of Negri Sembilan in an intensive pig farming area, where the outbreak increased significantly in late February 1999, reaching a peak in late March and early April. The outbreak was contained by the slaughter of more than 1.1 million pigs. Almost all cases were in people in close contact with pigs, especially those associated with pig farming or involved in the transport or slaughter of pigs. There were 106 deaths among 265 reported cases of encephalitis, a mortality rate of nearly 40% [55]. The outbreak later spread to abattoir workers in Singapore, where pigs from Negri Sembilan had been sent for slaughter [56, 63]. A virus was first isolated from brain tissue of a fatal human case and found to be a novel agent closely related to Hendra virus, and thus the second member of the *Henipavirus* genus [11, 12]. The virus was called Nipah virus after Kampung Sungai Nipah (Nipah River Village) from where this first virus isolate had come [12, 39].

Most patients presented with a severe rapidly progressive encephalitis syndrome, although some also had significant pulmonary symptoms [18, 39, 56]. The main presenting features were fever, headache, myalgia, dizziness, drowsiness and vomiting [11, 18]. About half of the patients had a reduced level of consciousness and prominent brain-stem dysfunction. Indeed death was probably due to severe brain-stem involvement [18]. There was a significant association between the presence of virus in the cerebral spinal fluid and mortality [9]. The key pathological findings included systemic endothelial infection and cell damage accompanied by vasculitis with syncytia in affected vessels [29]. In a follow up 24 months after the outbreak, 7.5% of patients who survived the acute encephalitis had recurrent neurological disease or relapsed encephalitis, 3.4% of patients who initially had non-encephalitic disease or asymptomatic infections subsequently developed late onset encephalitis after an average interval of about 8.4 months, and 4 of the 22 relapsed or late onset encephalitis patients (18%) died, suggesting that both conditions were due to complications caused by persistent Nipah virus infection of the central nervous system [65]. There was no evidence of human-to-human transmission, even between close family contacts. Nevertheless, secretion of Nipah virus was reported from the upper respiratory tract and urine of some patients [7].

Epidemiological evidence suggested that the primary means of spread between farms and between regions was the movement of pigs. The primary mode of transmission on pig farms was believed to be via the respiratory route, which was subsequently confirmed by laboratory evidence [50]. Retrospective investigations suggest that Nipah virus has been responsible for disease in pigs in Peninsular Malaysia since late 1996, but was not recognised as a new syndrome because the clinical signs were not markedly different from those of several endemic diseases, and because morbidity and mortality were not remarkable [16]. The clinical disease in pigs could result in lesions in both the brain and lungs of individual pigs or in either organ independently. Importantly, the lesions were those of hyperplasia of the columnar epithelium, peribronchiolar and peribronchial infiltration of lymphocytes, and exudation of live and dead cells to the bronchial and bronchiolar lumens. Also, there were numerous alveolar macrophages, and often neutrophils, in a proteinaceous oedema fluid in the alveoli, bronchioles and bronchi [29]. Syncytial cells were seen in the small blood and lymphatic vessels, and possibly as multinucleated macrophages in the alveolar spaces. Thus, unlike Hendra or Nipah virus infection of humans, the infection in pigs targeted epithelial cells rather than endothelial cells. Of other species, a serological survey showed that many dogs from infected pig farms had antibodies to Nipah virus, but only two cases of disease in canids were detected, one in a moribund animal and the other in a dead animal. The former had clinical symptoms resembling those of canine distemper, histologically characterized as severe pulmonary oedema and interstitial pneumonia with numerous alveolar macrophages [29]. A single infected cat was found to have a generalised vasculitis, and experimentally, infection was similar to that observed in cats with Hendra virus, except that there was virus antigen in the respiratory epithelium and virus shedding in the oropharynx [50].

As with Hendra virus, a number of questions must be addressed, the most important of which relates to the transmissibility of the virus from pigs. Thus the most significant difference observed between Nipah virus infection of pigs and possibly cats compared to either Hendra or Nipah virus infections of most other species, was the epithelial tropism, resulting in virus excretion in the respiratory tract. This undoubtedly explains the contagious nature of the pig infection both to other pigs and to humans, and why there has been little evidence of transmission from or between other species in which the infection was largely endothelial [29]. Exceptions to this are the vascular transmission of Hendra virus through the placenta to the foetus in bats and guinea pigs, and the excretion of the two viruses in urine in bats and cats, and occasionally in horses infected with Hendra virus.

As is obvious for two viruses in the same genus, Nipah and Hendra viruses share many of the same properties, although with small and subtle differences, including morphological similarities [34], genomic sequence and organisational similarities [26, 68], and ecological similarities [16].

Questions still remain concerning the genesis of outbreaks, and the transmission from the wildlife reservoir to pigs. The close genetic and antigenic similarities between Nipah and Hendra viruses has led to the suggestion that bats, particularly pteropid bats (flying foxes), might also be the wildlife hosts of Nipah virus. This was confirmed initially by a seroepidemiological study of bats, which found seropositive individuals among four species of fruit bats (*P. hypomelanus*, the island flying fox; *P. vampyrus*, the Malaysian flying fox; *Cynopterus brachyotis*, the lesser dog-faced fruit bat; and *Eonycteris spelaea*, the cave bat) and one species of insectivorous bat (*Scotophilus kuhli*, the house bat) [35]. Subsequently Nipah virus was isolated from urine samples collected from island fruit bats on Tioman Island and from partially eaten fruit [8]. The mechanism by which the virus passed from bats to pigs to initiate the outbreak still remains to be determined, but it is of major importance with respect to formulating management plans [44] (Field, Mackenzie and Daszak, this volume).

Are there similar or related viruses waiting to emerge elsewhere? There are about 60 species within the genus *Pteropus*, with a distribution that extends from the west Indian Ocean islands of Mauritius, Madagascar, Comoro, and Pembe, along the sub-Himalayan region of Pakistan and India, through southeast Asia, Philippines, Indonesia, New Guinea, southwest Pacific Islands as far east as the Cook Islands, and Australia, excluding Tasmania. They are not found on mainland Africa, Europe, Asia, North America or South America [13]. With the exception of the populations in the western Indian Ocean, the various pteropid species occur as a series of overlapping populations, and thus it might be suspected that other related henipaviruses would occur throughout the range [13, 16, 43]. Indeed, bats have been reported with antibodies to Hendra-like or Nipah-like viruses in Papua New Guinea [16] and Cambodia [53], and suspected outbreaks of disease associated with a Nipah-like virus have been recently reported twice from Bangladesh [58] (RF Breiman, ICDDR,B Centre for Health and Population Research, Dhaka, Bangladesh, personal communication) and Uttar Pradesh [37] and North Bengal, India. Interestingly, the outbreaks in Bangladesh and India had no apparent association with an animal source, although it is probable that epidemiological investigations may have been inadequate to demonstrate this, and in India may have involved human-to-human transmission. The major observations and findings from the latter outbreak still remain to be shared with the international community, and given its importance to our understanding of the biology, geographic range and risk factors associated with this novel group of viruses, it is hoped that this information will be published in the near future. Perhaps some of the most interesting future results might come from investigations of the pteropid bat populations of the western Indian Ocean, which are believed to have been separated from other pteropid species for many thousands of years; they might provide a valuable insight into the evolution of this new genus in the family *Paramyxoviridae*.

Two additional paramyxoviruses have been described recently from pteropid bats, Menangle and Tioman viruses [10, 57]. Menangle was the first of the two to be recognised, and shown to be the cause of an outbreak of reproductive disease and congenital malformations in a large piggery in New South Wales in 1997 [57]. There was a decrease in the farrowing percentages and a decline in the number of live piglets per litter. Mummified foetuses, stillborn piglets with arthrogryposis, craniofacial deformities and degeneration of the brain and spinal cord, were observed along with occasional abortions.[40, 57]. Menangle virus also appeared able to infect humans, with two piggery workers becoming ill with an influenza-like illness and rash [4, 57]. Sero-epidemiological studies indicated that the virus came from pteropid bats, with neutralising antibodies found in sera sampled from animals at two colonies in the vicinity of the piggery [36, 57]. There was no evidence of infection in cattle, sheep, birds, rodents or feral cats, nor in a dog at the affected piggery [36]. Over 96% of the pigs in the affected piggery were found to have neutralising antibodies to the virus, and 88% of pigs at two other piggeries which had received animals from the affected piggery were also found to be sero-positive for the virus [36]. No serological cross-reactivity was detected between Menangle virus and other known paramyxoviruses, Molecular studies, however, showed that Menangle virus was a new member of the genus *Rubulavirus* [3], with only limited amino acid sequence homology to other members of the genus. A second virus, Tioman virus, was isolated in 1999 from urine collected from island flying foxes on Tioman Island, and shown to be closely related to Menangle virus serologically and genetically [6, 10]. It has not been associated with disease in animals or humans.

Australian bat lyssavirus

Australian bat lyssavirus (ABLV) was first reported in July 1996 from a black flying fox (*P. alecto*) from Ballina in northern New South Wales; the bat had displayed abnormal behaviour and had been euthanased [17]. The virus subsequently was shown to be closely related to classical rabies virus antigenically but distinct from it genetically [20]. Subsequent studies demonstrated that ABLV occurred in all four Australian species of flying-fox (black flying-fox, grey-headed flying-fox, little red flying-fox, and spectacled flying-fox) throughout their Australian range, and in at least one species of insectivorous bat in the suborder *Microchiroptera*, the yellow-bellied sheath-tailed bat (*Saccolaimus flaviventris*) [21, 32, 46] (HE Field, unpublished observations). The prevalence of ABLV in these species, and its possible occurrence in other microchiropterans, are not yet known. However, the prevalence in 366 sick, injured, or orphaned bats in southern Queensland, or in bats submitted for testing after biting or scratching individuals, or for other public health reasons, was 6–9% [46, 72]. The virus was found to be antigenically similar to classical rabies virus and therefore a member of Lyssavirus serotype 1, but was distinguishable on genetic sequence and was therefore ascribed to a new genotype 7 [20]. Results from the U.S. Centers for Disease Control and Prevention have indicated that rabies vaccine may elicit a protective immune response to ABLV [32, 41].

The first recognised human infection with ABLV was in November 1996 in a patient who had been scratched and possibly bitten 5 weeks earlier by a yellow-bellied sheathtail bat [1, 32, 60]. The patient developed progressive weakness in all limbs and eventually a depressed conscious state, and required ventilation. An electroencephalogram was consistent with a diffuse encephalitis. The patient exhibited progressive evidence of cerebral damage, and died [1]. A second case occurred in December 1998 after a 27-month incubation period following a bite from a flying fox [25]. After a history of fever, vomiting, anorexia, pain about the left shoulder girdle and a sore throat, the patient's condition deteriorated with muscle spasms, and then paralysis requiring ventilation, and she died 17 days after onset. Thus both cases died following an encephalitis similar to that described for classical rabies.

The complete sequence of the ABLV isolate from the second human infection has been reported and, at 11,918 nucleotides, shown to be the smallest lyssavirus yet sequenced [71]. Further sequences have been reported for various parts of the genome [20–22], and a comparison of the sequence data from various isolates has shown that the isolates can be separated into two clades, based on the host of origin. Thus flying fox isolates can be distinguished from yellow-bellied sheathtail bat isolates, with nucleotide sequence divergence of up to about 20% and amino acid sequence divergence of 4–12% [21, 22, 71]. These analyses also provided additional evidence to distinguish ABLV from rabies on genetic grounds [22].

The disease caused by ABLV has been described in a captive juvenile black flying fox which was probably infected by its mother after birth [14], and between a wild black flying fox and a captive grey-headed flying fox housed in a captive colony of 23 animals [71]. The incubation period for the former was 6–9 weeks, and for the latter, about 4 weeks. In both cases, the disease began initially as an abnormal pattern of behaviour. The juvenile animal then exhibited dysphagia, and both animals exhibited progressive neurological symptoms to euthanasia or death. Experimental intramuscular infection of flying foxes with either ABLV or bat-associated rabies virus also has given some insight into the progression and outcomes of infection. Three of ten ABLV-infected animals and two of four rabies-infected animals developed abnormal clinical signs 15–24 days post-infection, and all five were found to have histological lesions and antigen in the central nervous system. None of the five animals developed anti-lyssavirus antibodies, although five of the seven ABLV-inoculated survivors and one of two rabies-inoculated survivors had seroconverted by 3 months after infection [48].

Murine studies have indicated that rabies vaccine offers protection against infection with ABLV [32, 41, 42], a finding that has had major positive public health implications. Pre-exposure vaccination with rabies vaccine is recommended for high-risk occupations such as wildlife workers and veterinarians. Post-exposure treatment with combined vaccine and rabies immune globulin is offered to anyone exposed to bats through bites or scratches unless ABLV infection can be ruled out in the bat. These management issues are discussed in an accompanying manuscript (Field, Mackenzie and Daszak, this volume).

The major questions still to be fully addressed are the incidence of ABLV in wild populations of flying foxes and yellow-bellied sheathtail bats, the possibility of transmission and adaptation of ABLV to terrestrial species, and the geographic range of ABLV outside Australia. Despite retrospective and progressive case investigation and passive surveillance, no evidence of ABLV infection has been found in domestic vertebrate species [47] or targeted wild mammalian species (HE Field, unpublished data). However, limited susceptibility studies to date have revealed that dogs and cats experimentally infected with ABLV do seroconvert and, in some cases, show transient clinical signs of disease (KA McColl, CSIRO Australian Animal Health Laboratory, Geelong, personal communication). As with henipaviruses in pteropid bats, it would be expected that similar viruses might occur elsewhere in the geographic range of other members of this bat genus, but there has been no evidence of ABLV-like viruses in either Malaysia or Papua New Guinea. However, neutralising antibodies to an ABLV-like virus have been reported from bats collected in the Philippines [2].

Conclusions

Some important questions have yet to be addressed about the ecology, pathogenesis, and host range of the two henipaviruses and ABLV, and which undoubtedly will have significant implications for management practices if we are to minimise the risk of future outbreaks. Some of this information would not be difficult to generate, but could be essential to our understanding of transmission cycles and of species susceptibilities. Of particular importance is the knowledge of the domestic animal host ranges of the two henipaviruses, and of the potential for terrestrial transmission of ABLV. In addition, a better understanding of tissue tropisms and its genetic basis is crucial to determining potential contagiousness of the henipaviruses and the possibility of selection for increased transmissibility.

The emergence of three novel bat-borne viruses in Australia and Malaysia over the past 8 years and their associations with encephalitic and pneumonic diseases demonstrates the importance of wildlife as a source of disease emergence, and the need for continued vigilance through international surveillance programmes for the emergence of related viruses throughout the range of the pteropid bats. From recent experiences in Bangladesh and India, it appears that there might not always be an obvious zoonotic association with novel henipaviruses, and human-to-human transmission might be a major feature, which was not seen in either of the well-documented outbreaks described here. Thus, it is possible that the biological properties of presently unrecognized henipaviruses may show a number of differences from those described above.

Acknowledgements

Some of the unpublished material described in this manuscript was undertaken with financial support from the National Health and Medical Research Council of Australia.

References

1. Allworth A, Murray K, Morgan J (1996) A case of encephalitis due to a lyssavirus recently identified in fruit bats. Commun Dis Intell 20: 504
2. Arguin PM, Murray-Lillibridge K, Miranda MEG, Smith JS, Calaor AB, Rupprecht CE (2002) Serologic evidence of Lyssavirus infection among bats, the Philippines. Emerg Infect Dis 8: 258–262
3. Bowden TR, Westenberg M, Wang LF, Eaton BT, Boyle DB (2001) Molecular characterization of Menangle virus, a novel paramyxovirus which infects pigs, fruit bats, and humans. Virology 283: 358–373

4. Chant K, Chan R, Smith M, Dwyer DE, Kirkland P (1998) Probable human infection with a newly described virus in the family *Paramyxoviridae*. The NSW Expert Group. Emerg Infect Dis 4: 273–275
5. Chua KB (2003) Nipah virus outbreak in Malaysia. J Clin Microbiol 26: 265–275
6. Chua KB, Wang LF, Lam SK, Eaton BT (2002) Full length genome sequence of Tioman virus, a novel paramyxovirus in the genus *Rubulavirus* isolated from fruit bats in Malaysia. Arch Virol 147: 1323–1348
7. Chua KB, Lam SK, Goh KJ, Hooi PS, Ksiazek TG, Kamaeulzaman A, Olson J, Tan CT (2001) The presence of Nipah virus in respiratory secretions and urine of patients during an outbreak of Nipah virus encephalitis in Malaysia. J Infect 42: 40–43
8. Chua KB, Koh CK, Hooi PS, Wee KF, Khong JH, Chua BH, Chan YP, Lim ME, Lam SK (2002) Isolation of Nipah virus from Malaysian Island flying-foxes. Microbes Infect 4: 145–151
9. Chua KB, Lam SK, Tan CT, Hooi PS, Goh KJ, Chew NK, Tan KS, Kamarulzaman A, Wong KT (2000) High mortality in Nipah encephalitis is associated with presence of virus in cerebrospinal fluid. Ann Neurol 48: 802–805
10. Chua KB, Wang LF, Lam SK, Crameri G, Yu M, Wise T, Boyle D, Hyatt AD, Eaton BT (2001) Tioman virus, a novel paramyxovirus isolated from fruit bats in Malaysia. Virology 283: 215–229
11. Chua KB, Goh KJ, Wong KT, Kamarulzaman A, Tan PS, Ksiazek TG, Zaki SR, Paul G, Lam SK, Tan CT (1999) Fatal encephalitis due to Nipah virus among pig-farmers in Malaysia. Lancet 354: 1256–1259
12. Chua KB, Bellini WJ, Rota PA, Harcourt BH, Tamin A, Lam SK, Ksiazek TG, Rollin PE, Zaki SR, Shieh WJ, Goldsmith CS, Gubler DJ, Roehrig JT, Eaton B, Gould AR, Olson J, Field H, Daniels P, Ling AE, Peters CJ, Anderson LJ, Mahy BWJ (2000) Nipah virus: a recently emergent deadly paramyxovirus. Science 288: 1432–1435
13. Field HE, Mackenzie JS, Hall LS (2001) Emerging zoonotic paramyxoviruses – the role of pteropid bats. In: Dodet B, Viscari M (eds) Emergence and control of zoonotic ortho- and paramyxovirus diseases. John Libby Eurotext, Montrouge, France, pp 205–209
14. Field H, McCall B, Barrett J (1999) Australian bat lyssavirus infection in a captive juvenile black flying fox. Emerg Infect Dis 5: 438–440
15. Field HE, Barratt PC, Hughes RJ, Shield J, Sullivan ND (2000) A fatal case of Hendra virus infection in a horse in north Queensland: clinical and epidemiological features. Aust Vet J 78: 279–280
16. Field H, Young P, Yob JM, Mills J, Hall L, Mackenzie J (2001) The natural history of Hendra and Nipah viruses. Microbes Infect 3: 307–314
17. Fraser GC, Hooper PT, Lunt RA, Gould AR, Gleeson LJ, Hyatt AD, Russell GM, Kattenbelt JA (1996) Encephalitiss caused by a lyssavirus in fruit bats in Australia. Emerg Infect Dis 2: 327–331
18. Goh KJ, Tan TC, Chew NK, Tan PS, Kamarulzaman A, Sarji SA, Wong KT, Abdullah BJJ, Chua KB, Lam SK (2000) Clinical features of Nipah virus encephalitis among pig farmers in Malaysia. N Engl J Med 342: 1229–1235
19. Gould AR (1996) Comparison of the deduced matrix and fusion protein sequences of equine morbillivirus with cognate genes of the *Paramyxoviridae*. Virus Res 43: 17–31
20. Gould AR, Hyatt AD, Lunt R, Kattenbelt A, Hengstberger S, Blacksell SD (1998). Characterisation of a novel lyssavirus isolated from pteropid bats in Australia. Virus Res 54: 165–187
21. Gould AR, Kattenbelt JA, Gumley SG, Lunt RA (2002) Characterisation of an Australian bat lyssavirus variant isolated from an insectivorous bat. Virus Res 89: 1–28

22. Guyatt KJ, Twin J, Davis P, Holmes EC, Smith GA, Smith IL, Mackenzie JS, Young PL (2003) A molecular epidemiological study of Australian bat lyssavirus. J Gen Virol 84: 485–496
23. Halpin K (2000) Genetic studies of Hendra virus and other novel paramyxoviruses. PhD Thesis, University of Queensland
24. Halpin K, Young PL, Field HE, Mackenzie JS (2000) Isolation of Hendra virus from pteropid bats: a natural reservoir of Hendra virus. J Gen Virol 81: 1927–1932
25. Hanna JN, Carney IK, Smith GA, Tannenberg EG, Deverill JE, Botha JA, Serafin IL, Harrower BJ, Fitzpatrick PF, Searle JW (2000) Australian bat lyssavirus infection: a second human case, with a long incubation period. Med J Aust 172: 597–599
26. Harcourt BH, Tamin A, Ksiazek TG, Rollin PE, Anderson LJ, Bellini WJ, Rota PA (2000) Molecular characterisation of Nipah virus, a newly emergent paramyxovirus. Virology 271: 334–349
27. Hooper PT, Westbury HA, Russell GM (1997) The lesions of experimental equine morbillivirus disease in cats and guinea pigs. Vet Pathol 34: 323–329
28. Hooper PT, Ketterer PJ, Hyatt AD, Russell GM (1997) Lesions of experimental equine morbillivirus pneumonia in horses. Vet Pathol 34: 312–322
29. Hooper P, Zaki S, Daniels P, Middleton D (2001) Comparative pathology of the diseases caused by Hendra and Nipah viruses. Microbes Infect 3: 315–322
30. Hooper PT, Gould AR, Russell GM, Kattenbelt JA, Mitchell G (1996) The retrospective diagnosis of a second outbreak of equine morbillivirus infection. Aust Vet J 74: 244–245
31. Hooper PT, Gould AR, Hyatt AD, Braun MA, Kattenbelt JA, Hengstberger SG, Westbury HA (2000) Identification and molecular characterisation of Hendra virus in a horse in Queensland. Aust Vet J 78: 281–282
32. Hooper P, Lunt R, Gould A, Samaratunga H, Hyatt AD, Gleeson LF, Rodwell BJ, Rupprecht CE, Smith JS, Murray PK (1997) A new lyssavirus – the first endemic rabies-related virus recognized in Australia. Bull Inst Pasteur 95: 209–218
33. Hyatt AD, Selleck PW (1996) Ultrastructure of equine morbillivirus. Virus Res 43: 1–15
34. Hyatt AD, Zaki SR, Goldsmith CS, Wise TG, Hengstberger SA (2001) Ultrastructure of Hendra virus and Nipah virus within cultured cells and host animals. Microbes Infect 3: 297–306
35. Johara MY, Field H, Rashdi AM, Morrissy C, van der Heide B, Rota P, bin Adzhar A, White J, Daniels P, Jamaluddin A, Ksiazek T (2001) Nipah virus infection in bats (Order Chiroptera) in Peninsular Malaysia. Emerg Infect Dis 7: 439–441
36. Kirkland PD, Love RJ, Philbey AW, Ross AD, Davis RJ, Hart KG (2001) Epidemiology and control of Menangle virus in pigs. Aust Vet J 79: 199–206
37. Kumar S (2003) Inadequate research facilities fail to tackle mystery disease. Brit Med J 326: 12
38. Lam SK, Chua KB (2002) Nipah virus encephalitis outbreak in Malaysia. Clin Infect Dis 34 Suppl 2: S48–S51
39. Lee KE, Umapathi T, Tan CB, Tjia HT, Chua TS, Oh HM, Fock KM, Kurup A, Das A, Tan AK, Lee WL (1999) The neurological manifestations of Nipah virus encephalitis, a novel paramyxovirus. Ann Neurol 46: 428–432
40. Love RJ, Philbey AW, Kirkland PD, Ross AD, Davis RJ, Morrissey C, Daniels PW (2001) Reproductive disease and congenital malformations caused by Menangle virus in pigs. Aust Vet J 79: 192–198
41. Lyssavirus Expert Group (1996) Prevention of human lyssavirus infection. Commun Dis Intell 20: 505–507
42. Lyssavirus Expert Group (1996). Update on bat lyssavirus. Commun Dis Intell 20: 535

43. Mackenzie JS, Field HE (2001) Hendra virus: a new zoonotic Paramyxovirus from fruit bats in Australia. In: Dodet B, Viscari M (eds) Emergence and control of zoonotic ortho- and paramyxovirus diseases. John Libby Eurotext, Montrouge, France, pp 177–184
44. Mackenzie JS, Field HE, Guyatt KJ (2003) Managing emerging diseases borne by fruit bats (flying foxes), with particular reference to henipaviruses and Australian bat lyssavirus. J Appl Microbiol 94 Suppl 1: 59–69
45. Mackenzie JS, Chua KB, Daniels PW, Eaton BT, Field HE, Hall RA, Halpin K, Johansen CA, Kirkland PD, Lam SK, McMinn P, Nisbet DJ, Paru R, Pyke AT, Ritchie SA, Siba P, Smith DW, Smith GA, van den Hurk AF, Wang LF, Williams DT (2001) Emerging viral diseases of South-East Asia and the Western Pacific: a brief review. Emerg Infect Dis 7 (Suppl 3): 497–504
46. McCall BJ, Epstein JH, Neill AS, Heel K, Field H, Barrett J, Smith GA, Selvey LA, Rodwell B, Lunt R (2000) Potential exposure to Australian bat lyssavirus, Queensland, 1996–1999. Emerg Infect Dis 6: 259–264
47. McColl KA, Tordo N, Aguilar Setién A (2000) Bat lyssavirus infections. Rev Sci Tech Off Int Epizoot 19: 177–196
48. McColl KA, Chamberlain T, Lunt RA, Newberry KM, Middleton D, Westbury HA (2002) Pathogenesis studies with Australian bat lyssavirus in grey-headed flying foxes (Pteropus poliocephalus). Aust Vet J 80: 636–641
49. McCormack JG, Allworth AM, Selvey LA, Selleck PW (1999) Transmissibility from horses to humans of a novel paramyxovirus, equine morbillivirus (EMV). J Infect 38: 22–23
50. Middleton DJ, Westbury HA, Morrissy CJ, van der Heide BM, Russell GM, Braun MA, Hyatt AD (2002) Experimental Nipah virus infection in pigs and cats. J Comp Path 126: 124–136
51. Murray K, Rogers R, Selvey L, Selleck P, Hyatt A, Gould A, Gleeson L, Hooper P, Westbury H (1995) A novel morbillivirus pneumonia of horses and its transmission to humans. Emerg Infect Dis 1: 31–33
52. Murray K, Selleck P, Hooper P, Hyatt A, Gould A, Gleeson L, Westbury H, Hiley L, Selvey L, Rodwell B (1995) A morbillivirus that caused fatal disease in horses and humans. Science 268: 94–97
53. Olson JG, Rupprecht C, Rollin P, An US, Niezgoda M, Clemins T, Watson J, Ksiazek T (2002) Antibodies to Nipah-like virus in bats (Pteropus lylei), Cambodia. Emerg Infect Dis 8: 987–988
54. O'Sullivan JD, Allworth AM, Paterson DL, Snow TM, Boots R, Gleeson LJ, Gould AR, Hyatt AD, Bradfield J (1997) Fatal encephalitis due to novel paramyxovirus transmitted from horses. Lancet 349: 93–95
55. Parashar UD, Sunn LM, Ong F, Mounts AW, Arif MT, Ksiazek TG, Kamaluddin MA, Mustafa AN, Kaur H, Ding LM, Othman G, Radzi HM, Kitsutani PT, Stockton PC, Arokiasamy J, Gary HE, Anderson LJ (2000) Case-contol study of risk factors for human infection with a new zoonotic paramyxovirus , Nipah virus, during a 1998–1999 outbreak of severe encephalitis in Malaysia. J Infect Dis 181: 1755–1759
56. Paton NI, Leo YS, Zaki SR, Auchus AP, Lee KE, Ling AE, Chew SK, Ang B, Rollin PE, Umapathi T, Sng I, Lee CC, Lim E, Ksiazek TG (1999) Outbreak of Nipah virus infection among abattoir workers in Singapore. Lancet 354: 1253–1256
57. Philbey AW, Kirkland PD, Ross AD, Davis RJ, Gleeson AB, Love RJ, Daniels PW, Gould AR, Hyatt AD (1998) An apparently new virus (family *Paramyxoviridae*) infections for pigs, humans, and fruit bats. Emerg Infect Dis 4: 269–271
58. ProMED (2002) Bangladesh: A Nipah-like virus linked to neurological disease outbreak in 2001. <http://www.promedmail.org> Archive number 20020830.5187

59. Rogers RJ, Douglas IC, Baldock FC, Glanville RJ, Seppanen KT, Gleeson LJ, Selleck PN, Dunn KJ (1996) Investigation of a second focus of equine morbillivirus infection in coastal Queensland. Aust Vet J 74: 243–244
60. Samaratunga H, Searle JW, Hudson N (1998) Non-rabies lyssavirus human encephalitis from fruit bats. Australian bat lyssavirus (peropid lyssavirus) infection. Neuropathol Appl Neurobiol 24: 331–335
61. Selvey L, Taylor R, Arklay A, Gerrard J (1996) Screening of bat carers for antibodies to equine morbillivirus. Comm Dis Intell 20: 477–478
62. Selvey LA, Wells RM, McCormack JG, Ansford AJ, Murray K, Rogers RJ, Lavercombe PS, Selleck P, Sheridan JW (1995) Infection of humans and horses by a newly described morbillivirus. Med J Aust 162: 642–645
63. Tambyah PA, Tan JH, Ong BK, Ho KH, Chan KP (2001) First case of Nipah virus in Singapore. Intern Med J 31: 132–133
64. Tan CT, Wong KT (2003) Nipah encephalitis outbreak in Malaysia. Ann Acad Med Singapore 32: 112–117
65. Tan CT, Goh KJ, Wong KT, Sarji SA, Chua KB, Chew NK, Murugasu P, Loh YL, Chong HT, Tan KS, Thayaparan T, Kumar S, Jusoh MR (2002). Relapsed and late-onset Nipah encephalitis. Ann Neurol 51: 703–708
66. Wang LF, Yu M, Eaton BT (2001) Molecular biology of Hendra virus. In: Dodet B, Viscari M (eds) Emergence and control of zoonotic ortho- and paramyxovirus diseases. John Libby Eurotext, Montrouge, France, pp 185–197
67. Wang LF, Yu M, Hansson E, Pritchard LI, Michalski WP, Eaton BP (2000) The exceptionally large genome of Hendra virus: support for creation of a new genus within the family *Paramyxoviridae*. J Virol 74: 9972–9979
68. Wang LF, Harcourt BH, Yu M, Tamin A, Rota PA, Bellini WJ, Eaton BT (2001) Molecular biology of Hendra and Nipah viruses. Microbes Infect 3: 279–287
69. Wang LF, Michalski WP, Yu M, Pritchard LI, Crameri G, Shiell B, Eaton BT (1998) A novel P/V/C gene in a new member of the *Paramyxoviridae* family, which causes lethal infection in humans, horses, and other animals. J Virol 72: 1482–1490
70. Ward MP, Black PF, Childs AJ, Baldock FC, Webster WR, Rodwell BJ, Brouwer SL (1996) Negative findings from serological studies of equine morbillivirus in the Queensland horse population. Aust Vet J 74: 241–242
71. Warrilow D, Smith IL, Harrower B, Smith GA (2002) Sequence analysis of an isolate from a fatal human infection of Australian bat lyssavirus. Virology 297: 109–110
72. Warrilow D, Harrower B, Smith IL, Field H, Taylor R, Walker C, Smith GA (2003) Public health surveillance for Australian bat lyssavirus in Queensland, Australia 2000–2001. Emerg Infect Dis 9: 262–264
73. Westbury HA, Hooper PT, Brouwer SL, Selleck PW (1996) Susceptibility of cats to equine morbillivirus. Aust Vet J 74: 132–134
74. Williamson M, Hooper P, Selleck P, Westbury H, Slocombe R (1999) The effect of Hendra virus on pregnancy in fruit bats and guinea pigs. J Comp Pathol 122: 201–207
75. Williamson M, Hooper P, Selleck P, Westbury H, Slocombe R (2001) Guinea pigs as an experimental model of Hendra virus encephalitis. J Comp Pathol 124: 273–279
76. Williamson M, Hooper P, Selleck P, Gleeson LJ, Daniels PW, Westbury HA, Murray PK (1998) Transmission studies of Hendra virus (equine morbillivirus) in fruit bats, horses and cats. Aust Vet J 76: 813–818
77. Wong KT, Shieh WJ, Kumar S, Norain K, Abdullah W, Guarner J, Goldsmith CS, Chua KB, Lam SK, Tan CT, Goh KJ, Chong HT, Jusoh R, Rollin PE, Ksiazek TG, Zaki SR; Nipah Virus Pathology Working Group (2002) Nipah virus infection: pathology and pathogenesis of an emerging paramyxoviral zoonosis. Am J Pathol 161: 2153–2167

78. Young P, Halpin K, Field H, Mackenzie J (1997). Finding the wildlife reservoir of equine morbillivirus. In: Asche V (ed) Recent advances in microbiology, Vol 5. Australian Society for Microbiology Inc, Melbourne, pp 1–12
79. Young PL, Halpin K, Selleck PW, Field H, Gravel JL, Kelly MA, Mackenzie JS (1996) Serologic evidence for the presence in pteropus bats of a paramyxovirus related to equine morbillivirus. Emerg Infect Dis 2: 239–240
80. Yu M, Hansson E, Langedijk JP, Eaton BP, Wang LF (1998) The attachment protein of Hendra virus has high structural similarity but limited primary sequence homology compared with viruses in the genus *Paramyxovirus*. Virology 251: 227–233
81. Yu M, Hansson E, Shiell B, Michalski W, Eaton BT, Wang LF (1998) Sequence analysis of the Hendra virus nucleoprotein gene: comparison with other members of the subfamily *Paramyxovirinae*. J Gen Virol 79: 1775–1780

Author's address: Dr. John S. Mackenzie, Department of Microbiology and Parasitology, School of Molecular and Microbial Sciences, University of Queensland, Brisbane, Queensland 4072, Australia; e-mail: john.mackenzie@uq.edu.au

Arch Virol (2004) [Suppl] 18: 113–121

Archives of Virology

Novel viral encephalitides associated with bats (*Chiroptera*) –host management strategies

H. Field[1], **J. Mackenzie**[2], and **P. Daszak**[3]

[1]Animal Research Institute, Department of Primary Industries, Yeerongpilly, Queensland, Australia
[2]Department of Microbiology and Parasitology, School of Molecular and Microbial Sciences, University of Queensland, Australia
[3]Consortium for Conservation Medicine, Palisades, New York, U.S.A.

Summary. Several novel viruses recently described in bats of the genus *Pteropus* (sub-order *Megachiroptera*) in Australia and southeast Asia cause encephalitic disease in animals and humans. These viruses include Hendra virus and Nipah virus (genus *Henipavirus*, family *Paramyxoviridae*) and Australian bat lyssavirus (ABLV; genus *Lyssavirus*, family *Rhabdoviridae*). Broadly, strategies for disease prevention and control in the spillover host are directed at minimising direct or indirect contact with the natural host, improving farm-gate and on-farm biosecurity, and better disease recognition and diagnosis. Additional strategies for ABLV include the use of rabies vaccine for effective pre- and post-exposure prophylaxis in humans. Effective management strategies in the natural host are predicated on an understanding of the ecology of the disease in the natural host, and the identification and avoidance of factors putatively associated with emergence, such as habitat loss, land use change and demographic shifts. A possible future management strategy for ABLV in reservoir populations is immunisation using bait or plant-derived vaccination.

Introduction

Several novel viruses recently described in bats of the genus *Pteropus* (sub-order *Megachiroptera*) in Australia and southeast Asia cause encephalitic disease in animals and humans. These are Hendra virus (HENV) and Nipah virus (NIPV), of the genus *Henipavirus*, family *Paramyxoviridae*, and Australian bat lyssavirus (ABLV), of the genus *Lyssavirus*, family *Rhabdoviridae*. This paper builds on a previous discussion of management strategies for these and other bat-associated viruses [31], presenting strategies for the natural, intermediate (spillover) and human hosts. Consideration of factors associated with the (apparent temporally clustered) emergence of these agents from pteropid bats, and an understanding of agent and host ecology, forms an integral part of the paper.

The temporal emergence of HENV, NIPV, and ABLV

Hendra virus, a novel paramyxovirus (initially called equine morbillivirus) was first described in September 1994, in an outbreak of disease in horses in Australia. Twenty-one horses and two humans were infected, with the resultant deaths of 14 of the horses and one of the humans [4, 14, 35]. In a further two foci (the most recent in January 1999), another three horses and one human were infected, all fatally [15, 25, 26, 40]. The predominant clinical syndrome in horses and humans was an acute respiratory syndrome, although neurological symptoms were seen in both, most notably in the third human case, which suffered a relapsing encephalitic disease [1]. Human infection was attributed to exposure to infected horses [38, 41]. Extensive wildlife surveillance has identified bats of the genus *Pteropus* (flying foxes) as a natural host of HENV, with infection endemic in the four mainland Australian species (*Pteropus alecto*, *P. poliocephalus*, *P. scapulatus* and *P. conspicillatus*) across their Australian range [22, 42] (HE Field: in preparation).

Nipah virus was first identified in 1999 as the primary aetiological agent in a major outbreak of disease in pigs and humans in Peninsular Malaysia [5, 6]. Respiratory and neurological syndromes were seen in pigs [36]. The predominant clinical syndrome in humans was encephalitic, with clinical signs including fever, headache, myalgia, drowsiness, and disorientation sometimes proceeding to coma within 48 hours [10, 19]. Approximately 1.1 million pigs were culled to contain the outbreak. Of 265 reported human cases, 105 were fatal. Direct contact with infected pigs was identified as the predominant mode of human infection [7, 19]. Preliminary wildlife surveillance found serological evidence of infection in two species of flying foxes, *Pteropus vampyrus* and *P. hypomelanus* [27]. The recent isolation of virus from *P. hypomelanus* strengthens the contention that flying foxes are a natural host of NIPV [8].

In October 1996, a rabies-like disease caused the death of a wildlife carer in Australia. The causative agent, ABLV was first described earlier that same year when it was identified in an Australian flying fox [18]. Both flying foxes (Field, HE: unpublished data) and humans [2, 23] exhibit a progressive neurological syndrome similar to that associated with rabies. Australia was previously believed free of viruses of the genus *Lyssavirus*, although other rhabdoviruses were known to occur there. While phylogenetically distinct from classical rabies virus, ABLV has strong serotypic, antigenic and sequence analysis similarities with rabies virus [21]. ABLV infection has also been identified in insectivorous bats [20] (Field, HE: in preparation). A second fatal human case occurred in December 1998 [23].

A more detailed review of the emergence of HENV, NIPV and ABLV is presented elsewhere in this volume (JS Mackenzie and HE Field) and previously [16, 30].

Putative risk factors for emergence

Why is the emergence of these diseases apparently clustered in time? In answering this question, it is appropriate to distinguish between emergence and detection. The identification of flying foxes as a probable natural host of ABLV and NIPV can largely be attributed to increased surveillance after the initial discovery of antibodies to HENV in Australian flying foxes. However, improved surveillance efforts or diagnostic capabilities cannot explain the emergence *per se* of either. We contend that emergence of all three agents from flying foxes has been driven by environmental factors that have facilitated movement of the pathogens beyond their ecological niche. Morse [33], Krause [28] and Lederberg et al. [29] propose that a series of commonly occurring anthropogenic environmental changes drive disease emergence by pushing pathogens outside their normal population parameters. Thus, introduction of pathogens via global air travel and trade, the encroachment of human activities into wilderness regions, urbanization, climatic changes and agricultural intensification are common drivers of emergence. For zoonotic diseases with wildlife reservoirs, anthropogenic factors that alter wildlife population structure, migration patterns and behaviour may also drive emergence of disease in human populations [12, 13]. For example, human population encroachment into wildlife habitat may increase the risk of Lyme disease and other tick-borne diseases by driving the loss of less competent reservoir hosts [37]. Likewise, the introduction of a "new" infection into a human or domestic animal population may follow the incursion of humans (accompanied by their domestic animals) into previously remote natural habitats where unknown disease agents exist in harmony with wild reservoir hosts. Upon contact with new species, an agent may jump species barriers, thereby spilling over into humans or livestock. Unlike the natural host, the new host may have no natural immunity or evolved resistance. Additionally, high population densities and management practices may facilitate the rapid spread of pathogens throughout livestock populations. Infection may be transmitted to humans directly from the natural host or via an intermediate (livestock) host [17].

The emergence of NIPV disease illustrates the two step process described by Morse [34]. The establishment of pig farms within the range of the natural host supported the initial introduction into the pig population; the maintenance of high densities of pigs and the transport of pigs led to the establishment and rapid dissemination of infection within the pig population in Peninsular Malaysia. Amplification of virus within pig populations then facilitated transmission to humans. We propose that an environmental factor or combination of factors increased the opportunity for effective contact between flying foxes and pigs, and thus the initial introduction of infection into the pig population. Plausible hypotheses include:

- The unsustainable hunting of *Pteropus* bat species has caused localized niche vacuums ('sinks') with relative resource abundancy, creating regional gradients along which neighbouring bat populations move, resulting in a net movement of virus into human-inhabited areas and so an increased probability of effective contact and spillover.
- Regional deforestation has changed the seasonal foraging movements of *Pteropus* bats and lead to an increased reliance on horticultural crops, resulting in a relative increased density of bats proximate to human and livestock populations. (Climatic changes, forest fires and associated haze events have similarly been hypothesized to influence flying fox movement patterns [9].

- The marked increase in the number, density and distribution of the Malaysian pig population in the last 10 years has led to an increased probability of contact between fruit bats and pigs. This probability has been further increased by the practice of planting fruit orchards immediately adjacent to piggeries.

The increasing urban presence of flying foxes in many Australian cities and towns (thought due to more reliable and abundant food resources) [39] and the associated changes in flying fox population dynamics, represents a similar emergence-promoting scenario (for HENV and ABLV) to those described above for NIPV.

Host management strategies

Effective disease management requires an understanding of the epidemiology of the disease (knowledge of its cause, maintenance and transmission, host range of the aetiologic agent, and the nature of the host-agent relationship), an ability to detect disease (surveillance and diagnostic capabilities) and political/public/industry support. Broadly, current strategies for the management of the recently emerged bat-associated agents are directed at minimising direct or indirect contact with the natural host, monitoring intermediate hosts, improving biosecurity on farms, and better disease recognition and diagnosis [31].

Hendra virus

The sporadic (and rare) nature of spillover events from flying foxes to horses, the low infectivity for horses (and consequently the limited economic impact), and the apparent absence of direct transmission from flying foxes to people have resulted in more emphasis on management strategies for horses than for flying foxes. Quarantine of infected premises, movement controls on stock, and disinfection have proved effective strategies [4]. A HENV vaccine is not currently available and development of one is not foreseen. Australian veterinarians have a high awareness of HENV, and HENV exclusion is routinely undertaken for horses exhibiting an acute respiratory syndrome. Veterinarians involved in these disease investigations wear appropriate protective equipment and use a limited necropsy approach, as horses have been the source of infection for all three human cases. Field et al. [15] observed that breed (thoroughbred), sex (female), age (>8 yo) and housing (paddocked) have been consistent putative horse-level risk factors for the index case in each of the three spillover events to date. Reproductive status may be an additional risk factor, as two of the three index cases were heavily pregnant. A considerable research focus on the ecology of HENV is yet to establish the route of virus excretion or any temporal pattern of infection in flying foxes. This information, and knowledge of the actual mode of flying fox-to-horse transmission would facilitate a risk management approach to spillover infection in horses.

Nipah virus

In strong contrast to HENV, the NIPV outbreak in Peninsular Malaysia in 1999 had an enormous economic and social impact. Nipah virus was highly infectious for pigs, with all classes of pigs susceptible. The pattern of on-farm infection was consistent with respiratory transmission; between-farm spread was generally associated with the movement of pigs. The extensive post-outbreak surveillance program in Malaysia showed that farms that did not receive pigs generally remained uninfected even when neighbouring farms were infected. Human infections were predominantly attributed to contact with live pigs; none was attributed to contact with bats. Horizontal transmission was not a feature of infection in humans. Recommended host management strategies primarily target pig-to-pig transmission, secondarily the flying fox-pig interface (that is, the natural host-spillover host interface). The central strategy is the implementation of sound farm management practices, such as monitoring herd health and early recognition of disease syndromes. The latter includes maintaining the high level of farmer and veterinarian awareness of the disease generated by the outbreak. A second key strategy is the strict application of farm-gate biosecurity [11], with clearly defined protocols for the introduction of new stock. These may include quarantine and/or exclusion testing. A NIPV vaccine is not currently available and development of one is unlikely in the near future. Overarching the above is a strategy of advanced planning for emergency management of disease outbreaks. This involves established surveillance, detection, and emergency response capabilities. The pre-existence of the latter in Malaysia enabled the implementation of effective quarantine, movement controls, and culling to bring the outbreak under control. The Malaysian pig population is now free of NIPV infection.

While strategies directed at the flying fox-pig interface are limited by our incomplete knowledge of the ecology of NIPV, several simple on-farm measures can be taken to reduce the likelihood of spillover events occurring. The removal of fruit orchards and other favoured flying fox food trees from the immediate vicinity of pig sheds greatly reduces the probability of flying fox-pig contact. Similarly, the wire screening of open-sided pig sheds is a simple and inexpensive strategy to prevent direct contact between flying foxes and pigs. Indirect contact (with flying fox urine or faeces or partially eaten fruit) can be avoided by ensuring roof run-off does not enter pig pens [9].

Australian bat lyssavirus

Epidemiological evidence suggests that ABLV is endemic in bat populations in Australia, both in bats of the sub-order *Megachiroptera* (which includes the *Pteropus* species) and in bats of the sub-order *Microchiroptera*. No evidence of infection has been found in any species other than bats and humans. The two known (fatal) human cases of ABLV infection have been attributed to direct contact with bats. This contrasts to the situation with HENV and NIPV, in which human infection resulted from an intermediate (spillover) host. Thus host management strategies seek to minimize bat-human interactions, and extensive public awareness campaigns have sought to educate members of the general public to avoid any direct contact with bats [31]. This means not picking up sick or injured bats and preventing (small microchiropteran) bats from gaining roost access to houses. The former is heavily emphasised, as sick and injured bats have been found to have a significantly higher prevalence of infection than do wild-caught bats (Field, HE: unpublished data). The risk posed by sick and injured bats is highlighted by a study of 'occupations' associated with potential exposure, which found that volunteer animal handlers accounted for 39% of potential exposures, their family members for 12%, professional animal handlers for 14%, community members who intentionally handled bats for 31%, and community members where contact was initiated by the bat for 4% [32].

The strategies of public awareness and contact minimisation are supported by vaccination. Murine vaccine protection studies showed that rabies vaccine offers protection against infection with ABLV [3, 24]. This provides a management strategy not available for HENV or NIPV, and pre-exposure administration of rabies vaccine is recommended for high risk 'occupations', such as those described above. Post-exposure treatment with vaccine and with rabies immune globulin is offered to anyone exposed to bats through bites or scratches, unless ABLV infection can be ruled out in the bat. Bats having at-risk contact with humans or domestic animals are routinely tested for ABLV infection. Vaccination strategies are restricted to humans, although limited research has been undertaken to explore the efficacy of parenteral rabies vaccination in flying foxes. Certainly oral rabies vaccines have dramatically reduced the incidence and spread of rabies in wildlife hosts in Europe and North America. Preliminary research in Australia has also investigated bait or plant-derived vaccination of flying foxes as a future management strategy [31]. Oral sub-unit vaccines appear to offer a less expensive, more stable, more easily administered and safer (incapable of causing infection) alternative to conventional rabies vaccine [31]. Nonetheless, the development of successful management strategies based on vaccination requires careful consideration of the management methods available as well as an understanding of both the agent and target species behaviour and ecology.

Conclusion

Host management strategies for natural hosts seek to minimize the opportunity for effective contact between the natural host and the spillover host. Incomplete knowledge of the ecology of the agents (the maintenance and transmission of infection, the natural and domestic host range, and the nature of the host-agent relationship) and of the factors associated with emergence (habitat loss, land use change and demographic shifts) precludes a more targeted approach at this time. Strategies for intermediate hosts (particularly for NIPV) emphasize surveillance, detection and emergency response capabilities. Strategies for human hosts promote awareness, minimization of contact, use of personal protective equipment and (for ABLV) vaccination.

It is notable that neither the indiscriminate nor targeted killing of bats is contemplated as an effective host management strategy. Certainly such a strategy with nomadic species such as flying foxes is biologically flawed, and as argued earlier, may cause a net influx of bats to the resultant niche vacuum. The forced movement of roosting flying foxes (flying foxes communally roost in trees, frequently in groups of thousands) by sustained noise, smoke, and flags has been attempted with mixed success. Permanent removal of flying fox roosts can only be guaranteed by the removal of the roost trees.

The evidence indicates that HENV, NIPV, and ABLV are 'old' viruses [21, 35] that have likely evolved with flying foxes, remaining in this niche until ecological changes precipitated their emergence.

References

1. Allworth A, O'Sullivan J, Selvey L, Sheridan J (1995) Equine morbillivirus in Queensland. Comm Dis Intell 19: 575
2. Allworth A, Murray K, Morgan J (1996) A human case of encephalitis due to a lyssavirus recently identified in fruit bats. Comm Dis Intell 20: 504
3. Anon (1996) Prevention of human lyssavirus infection. Comm Dis Intell 20: 505–507
4. Baldock FC, Douglas IC, Halpin K, Field H, Young PL, Black PF (1996) Epidemiological investigations into the 1994 equine morbillivirus outbreaks in Queensland, Australia. Sing Vet J 20: 57–61
5. CDC (1999) Outbreak of Hendra-like virus – Malaysia and Singapore, 1998–1999. Morbid Mortal Wkly Rep 48: 265–269
6. CDC (1999) Update: Outbreak of Nipah virus – Malaysia and Singapore, 1999. MMWR 48: 335–337
7. Chua K, Bellini W, Rota P, Harcourt B, Tamin A, Lam S, Ksiazek T, Rollin P, Zaki S, Shieh W-J, Goldsmith C, Gubler D, Roehrig J, Eaton B, Gould A, Olson J, Field H, Daniels P, Ling A, Peters C, Anderson L, Mahy B (2000) Nipah virus: A recently emergent deadly paramyxovirus. Science 288: 1432–1435
8. Chua K, Koh C, Hooi P, Wee K, Khong J, Chua B, Chan Y, Lim M, Lam S (2002) Isolation of Nipah virus from Malaysian Island flying foxes. Microb Infect 4: 145–151
9. Chua K (2003) Nipah virus outbreak in Malaysia. J Clin Virol 26: 265–275
10. Chua KB, Goh KJ, Wong KT, Kamarulzaman A, Tan PSK, Ksiazek TG, Zaki SR, Paul G, Lam SK, Tan CT (1999) Fatal encephalitis due to Nipah virus among pig-farmers in Malaysia. Lancet 354: 1256–1259
11. Daniels P (2000) The Nipah virus outbreak in Malaysia: Overview of the outbreak investigation and the issues that remain. In: 16th International Pig Veterinary Society Congress, Ocean Grove, USA, 2000. International Pig Veterinary Society, Ocean Grove, USA, pp 553–554
12. Daszak P, Cunningham AA, Hyatt AD (2000) Emerging infectious diseases of wildlife-threats to biodiversity and human health. Science 287: 443–448
13. Daszak P, Cunningham A, Hyatt A (2001) Anthropogenic environmental change and the emergence of infectious diseases in wildlife. Acta Tropica 78: 103–116
14. Douglas I, Baldock C, Black P (1997) Outbreak investigation of an emerging disease (equine morbillivirus). In: Epidemiology Sante Animal. In: Proc 8th Int Sympos Veterinary Epidemiology and Economics, Paris, France 1997. International Society for Veterinary Epidemiology and Economics, Paris, France, pp 81–83
15. Field H, Barratt P, Hughes R, Shield J, Sullivan N (2000) A fatal case of Hendra virus infection in a horse in north Queensland: clinical and epidemiological features. Aust Vet J 78: 279–280

16. Field H, Mackenzie J, Hall L (2001) Emerging zoonotic paramyxoviruses – the role of pteropid bats. In: Dodet B VM (ed) Emergence and control of zoonotic ortho- and paramyxovirus diseases. John Libby Eurotext, Montrouge, France, pp 205–209
17. Field H, Young P, Johara M, Mills J, Hall L, Mackenzie J (2001) The natural history of Hendra and Nipah viruses. Microbes Infect 3: 315–322
18. Fraser G, Hooper P, Lunt R, Gould A, Gleeson L, Hyatt A, Russell G, Kattenbelt J (1996) Encephalitis caused by a lyssavirus in flying-foxes in Australia. Emerg Infect Dis 2: 327–331
19. Goh KJ, Tan TC, Chew NK, Tan PSK, Karmarulzaman A, Sarji SA, Wong KT, Abdullah BJJ, Chua KB, Lam SK (2000) Clinical features of Nipah virus encephalitis among pig farmers in Malaysia. N Engl J Med 342: 1229–1235
20. Gould A, Kattenbelt J, Gumley S, Lunt R (2002) Characterisation of an Australian bat lyssavirus variant isolated from an insectivorous bat. Virus Res 89: 1–28
21. Gould AR (1996) Comparison of the deduced matrix and fusion protein sequences of equine morbillivirus with cognate genes of the Paramyxoviridae. Virus Res 43: 17–31
22. Halpin K, Young P, Field H, Mackenzie J (2000) Isolation of Hendra virus from pteropid bats: a natural reservoir of Hendra virus. J Gen Virol 81: 1927–1932
23. Hanna J, Carney I, Smith G, Tannenberg A, Deverill J, Botha J, Serafin I, Harrower B, Fitzpatrick P, Searle I (2000) Australian bat lyssavirus infection: a second human case, with a long incubation period. Med J Aust 172: 597–599
24. Hooper P, Lunt R, Gould A, Samaratunga H, Hyatt A, Gleeson L, Rodwell B, Rupprecht C, Smith J, Murray P (1997) A new lyssavirus-the first endemic rabies-related virus recognised in Australia. Bull Inst Pasteur 95: 209–218
25. Hooper P, Gould A, Hyatt A, Braun M, Kattenbelt J, Hengstberger S, Westbury H (2000) Identification and molecular characterisation of Hendra virus in a horse in Queensland. Aust Vet J 78: 281–282
26. Hooper PT, Gould AR, Russell GM, Kattenbelt JA, Mitchell G (1996) The retrospective diagnosis of a second outbreak of equine morbillivirus infection. Aust Vet J 74: 244–245
27. Johara M, Field H, Rashdi A, Morrissy C, vanderHeide B, Rota P, Azri A, White J, Daniels P, Jamaluddin A, Ksiazek T (2001) Serological evidence of infection with Nipah virus in bats (order Chiroptera) in Peninsular Malaysia. Emerg Infect Dis 7: 439–441
28. Krause R (1992) The origins of plagues: old and new. Science 257: 1073–1078
29. Lederberg J, Shope RE, Oaks SC (1992) Emerging infections: microbial threats to health in the United States. National Academy Press, Washington, DC, USA
30. Mackenzie J, Field H (2001) Hendra virus: a new zoonotic paramyxovirus from fruit bats in Australia. In: Dodet B VM (ed) Emergence and control of zoonotic ortho- and paramyxovirus diseases. John Libby Eurotext, Montrouge, France, pp 177–184
31. Mackenzie J, Field H, Guyatt K (2003) Managing emerging diseases borne by fruit bats (flying foxes) with particular reference to henipaviruses and Australian bat lyssavirus. J Appl Microbiol 94: 59S–69S
32. McCall B, Epstein J, Neill A, Heel K, Field H, Barrett J, Smith G, Selvey L, Rodwell B, Lunt R (2000) Potential human exposure to Australian bat lyssavirus: Brisbane South and South Coast, Queensland, 1996–1999. Emerging Inf Dis 6: 259–264
33. Morse SS (1993) Examining the origins of emerging viruses. In: Morse SS (ed) Emerging viruses. Oxford University Press, New York, pp 10–28
34. Morse SS (1995) Factors in the emergence of infectious diseases. Emerg Infect Dis 1: 7–15
35. Murray K, Selleck P, Hooper P, Hyatt A, Gould A, Gleeson L, Westbury H, Hiley L, Selvey L, Rodwell B, Ketterer P (1995) A morbillivirus that caused fatal disease in horses and humans. Science 268: 94–97

36. Nor M, Gan C, Ong B (2000) Nipah virus infection of pigs in Peninsular Malaysia. Rev Sci Off Int Epiz 19: 160–165
37. Ostfeld R, Keesing F (2000) Biodiversity and disease risk: The case of Lyme disease. Conserv Biol 14: 722–728
38. O'Sullivan JD, Allworth AM, Paterson DL, Snow TM, Boots R, Gleeson LJ, Gould AR, Hyatt AD, Bradfield J (1997) Fatal encephalitis due to novel paramyxovirus transmitted from horses. Lancet 349: 93–95
39. Parry-Jones K, Augee M (2001) Factors affecting the occupation of a colony site in Sydney, New South Wales by the Grey-headed Flying-fox *Pteropus poliocephalus* (Pteropodidae). Austral Ecol 26: 47–55
40. Rogers RJ, Douglas IC, Baldock FC, Glanville RJ, Seppanen KT, Gleeson LJ, Selleck PN, Dunn KJ (1996) Investigation of a second focus of equine morbillivirus infection in coastal Queensland. Aust Vet J 74: 243–244
41. Selvey L, Wells RM, McCormack JG, Ansford AJ, Murray PK, Rogers RJ, Lavercombe PS, Selleck P, Sheridan JW (1995) Infection of humans and horses by a newly described morbillivirus. Med J Aust 162: 642–645
42. Young PL, Halpin K, Selleck PW, Field H, Gravel JL, Kelly MA, Mackenzie JS (1996) Serologic evidence for the presence in *Pteropus* bats of a paramyxovirus related to equine morbillivirus. Emerg Infect Dis 2: 239–240

Author's address: Dr. Hume Field, Animal Research Institute, Department of Primary Industries, 665 Fairfield Road, Yeerongpilly, Queensland 4105, Australia; e-mail: hume.field@dpi.qld.gov.au

Arch Virol (2004) [Suppl] 18: 123–131

Archives of Virology

Henipaviruses: recent observations on regulation of transcription and the nature of the cell receptor

B. T. Eaton[1], P. J. Wright[2], L.-F. Wang[1], O. Sergeyev[1], W. P. Michalski[1], K. N. Bossart[3], and C. C. Broder[3]

[1]Australian Animal Health Laboratory, CSIRO Livestock Industries, Geelong, Australia
[2]Department of Microbiology, Monash University, Clayton, Australia
[3]Department of Microbiology, Uniformed Services University, Bethesda, Maryland, U.S.A.

Summary. Hendra virus (HENV) and Nipah virus (NIPV) are classified in the new genus *Henipavirus*, within the subfamily *Paramyxovirinae*, family *Paramyxoviridae*. The genetic and biological characteristics that differentiate henipaviruses from other members of the subfamily are summarized. Although they do not display neuraminidase and hemagglutination activities and in that regard resemble viruses in the genus *Morbillivirus*, several recent observations highlight similarities between henipaviruses and respiroviruses (genus *Respirovirus*) in structure and replication strategy. First, three-dimensional modeling studies suggest that the external globular head domain of the HENV G protein resembles that of respiroviruses rather than morbilliviruses. Second, the pattern of transcriptional attenuation in HENV-infected cells resembles that observed with Sendai virus, a respirovirus, and differs from that found in cells infected with measles virus, a morbillivirus.

Henipaviruses have a broad host range *in vitro* and *in vivo*, indicating wide distribution of cellular receptor molecules. The extensive host range has been confirmed in a quantitative *in vitro* cell-fusion assay using recombinant vaccinia viruses expressing the attachment and fusion proteins of HENV and NIPV. Cell lines of diverse origin and which are permissive in the *in vitro* cell fusion assay have been identified and the pattern of relative susceptibilities is the same for both HENV and NIPV, implying that both viruses use the same cell receptor. Protease treatment of permissive cells destroys their ability to fuse with cells expressing viral envelope glycoproteins. Virus overlay protein binding assay (VOPBA) and radio-immune precipitation assays confirm that both HENV and NIPV bind to membrane proteins in the 35–50 kD range. Treatment of cell membrane proteins with N-glycosidase eliminates HeV binding activity in VOPBA whereas treatment with neuraminidase has no effect on binding. Thus preliminary evidence suggests that NIPV and HENV bind to the same glycoprotein receptor via a non-sialic acid-dependant mechanism.

Introduction

Members of the family *Paramyxoviridae* contain a non-segmented negative strand RNA genome and are divided into two subfamilies, *Paramyxovirinae* and *Pneumovirinae*. In 2001 the number of genera in the subfamily *Paramyxovirinae* was raised from three (*Respirovirus*, *Morbillivirus* and *Rubulavirus*) to five by the addition of two new genera *Avulavirus* and *Henipavirus*. Whereas the *Avulavirus* genus contains avian paramyxoviruses previously classified in the *Rubulavirus* genus, the *Henipavirus* genus was created to accommodate Hendra virus (HENV) and Nipah virus (NIPV), highly pathogenic, zoonotic viruses that emerged in Australasia in the last decade of the twentieth century [28]. The appearance of new paramyxoviruses is not an unusual event. Since 1979 almost a dozen new paramyxoviruses have been discovered, predominantly as a result of investigations into outbreaks of disease in human, livestock and wild animal populations. A number of these viruses were readily classified as members of the genus *Morbillivirus* (Phocine distemper virus and Cetacean morbillivirus) and the genus *Rubulavirus* (Mapuera virus, Porcine rubulavirus, Menangle virus and Tioman virus) but several (Salem virus, Tupaia virus) remain unclassified (for review see [28]).

One of the most interesting aspects associated with the finding of new paramyxoviruses is the fact that four of them (HENV, NIPV, Menangle virus and Tioman virus) appear to have flying foxes (fruit bats in the genus *Pteropus*) as a natural host. It is indeed quite remarkable that in just a 5 year period starting in 1994 HENV, Menangle virus and NIPV "jumped", in ways that remain unclear, from flying foxes to horses (HENV) and pigs (NIPV and Menangle virus) and from thence to humans. In three unrelated incidents HENV caused fatal respiratory infection of 16 horses and the death of two humans in Queensland [31]. Menangle virus reduced the farrowing rate and caused fetal abnormalities in a piggery in New South Wales and an influenza-like illness in two piggery workers [7]. NIPV spread undetected through the Malaysian pig population before manifesting itself as the causative agent of encephalitis that killed 105 of 265 infected humans who came into close contact with infected pigs or pig carcasses [8]. Genetic analysis revealed that HENV and NIPV were so distant from morbilliviruses, rubulaviruses and respiroviruses that they required classification in a new genus, *Henipavirus*, in the subfamily *Paramyxovirinae* [29]. This paper will review the differences that justified the creation of a new genus to accommodate HENV and NIPV and summarize some recent observations made in our laboratories on the nature of the cell receptor and the attenuation of transcription in HENV-infected cells.

Distinguishing characteristics of henipaviruses

A summary of the key features that serve to differentiate henipaviruses from other members of the family *Paramyxoviridae* is given in Table 1. Features include genetic criteria such as genome organization, the coding pattern of the P gene, the sequence and size of intragenic regions (IGR) and genome terminal sequences

Table 1. Differential genetic and biological features in genera of the subfamily *Paramyxovirinae*

Feature	Genus				
	Henipavirus	*Respirovirus*	*Morbillivirus*	*Rubulavirus*	*Avulavirus*
Hemagglutination	−	+	+/−	+	+
Neuraminidase	−	+	+/−	+	+
Nuclear inclusions	−	−	+	−	−
Sialic acid receptor	−	+	−	+	+
C protein expression	+	+	+	−	−
Product of unedited P/V mRNA	P	P	P	V	P
Trimeric IGS	+	+	+	−	−
UTR (200–568 nt)	+	−	−	−	−
Relative size of P protein (number of amino acids)	large >700	medium ~500–600	medium ~500–600	small <400	small <400
Genome length (kb)	>18	15–16	15–16	15–16	15–16

and biological criteria related to the structure of the cell attachment and fusion proteins.

Genetic criteria

Members of the *Paramyxoviridae* contain between 6 and 10 transcription units and their genome organization resembles that of other nonsegmented negative stranded RNA viruses. Like the majority of viruses in the subfamily *Paramyxovirinae*, henipaviruses have six transcription units encoding six major structural proteins designated NP (nucleocapsid protein), P (phophoprotein), M (matrix protein), F (fusion protein), G (glycoprotein or attachment protein) and L (large protein or RNA polymerase) in the order 3′-NP-P-M-F-G-L-5′. Overall, the arrangement of genes in henipaviruses most closely resembles that of respiroviruses and morbilliviruses [27, 30].

All genomes in the *Paramyxoviridae* contain two short extragenic regions at the 3′ (leader) and 5′ (trailer) termini. Approximately the first 12 nt of the leader and the last 12 nt of the trailer are highly conserved and complementary, features consistent with their function as promoters for transcription and replication, respectively [14]. The leader and trailer nucleotide sequences are unique and conserved within each of the five genera and the discovery of henipaviruses has increased the known diversity of genome terminal sequence patterns in the subfamily *Paramyxovirinae* [10, 30]. It is worth noting that the terminal sequences provide the only unique molecular feature not found in other viruses in the family. Other features, such as IGR variability and/or P gene coding strategy are generally shared by viruses of more than one genus [27].

The nature of the IGR between the transcription units is a significant feature that differentiates members of the *Rubulavirus* and *Avulavirus* genera from others in the subfamily *Paramyxovirinae*. Rubulaviruses and avulaviruses have IGRs that vary significantly in both size and sequence whereas other members of the subfamily, including henipaviruses, have fixed-size (3-nt) IGRs whose sequence is highly conserved, with the majority being 3′-GAA-5′ [27, 30].

Prior to the discovery of the henipaviruses and Tupaia virus, the genome size of paramyxoviruses appeared quite uniform, with lengths varying between 15.2 and 15.9 kb, that is by less than 5%. This changed significantly with the discovery that the genome lengths of HENV and NIPV were 18,234 and 18,246 nt, respectively [27, 30]. The size of the genomes was not reflected in the size of encoded proteins, with the possible exception of the P proteins, which were more than 100 amino acids longer than other P proteins in the subfamily. Thus the percentage of the henipavirus genomes coding for proteins fell to 82%, compared with an average of 92% for other paramyxoviruses. The extra genome length is due to the presence in all but the L gene, of a long non-translated region located at the 3′ end of the mRNA. In this respect, the mRNA of HENV and NIPV resembles that of the Ebola virus and Marburg virus in the family *Filoviridae*. Comparison of the nucleotide sequence of HENV and NIPV revealed significant homology in the coding but not the non-coding regions (74.2% versus 41.8% for N, P, M, F and G genes) [10].

All members of the subfamily *Paramyxovirinae* produce multiple proteins from the P gene using mechanisms such as RNA editing and internal translational initiation [14, 26]. The majority of members of the *Paramyxovirinae* utilize either of two P gene-coding strategies. In that used by respiroviruses, morbilliviruses, and henipaviruses, the P gene encodes not only the P protein, but also the V and C proteins. An unedited gene transcript encodes the P protein. A separate transcript in which a 1G-insertion is made at a conserved editing site encodes the V protein. The third protein, C, is produced via internal translation initiation, and is encoded at the 5′ end of both P and V mRNAs. In contrast, rubulaviruses use a different P gene coding strategy in which the unedited transcript encodes the V protein and a separate mRNA with a 2G-insertion at an editing site encodes the P protein. Neither the V nor P mRNA encodes a C protein. Avulaviruses use the same P/V editing strategy as those of respiroviruses, morbilliviruses and henipaviruses, but they lack the coding capacity to specify a C protein.

Biological criteria

The cell attachment proteins of members of the *Paramyxovirinae* display haemagglutination (H) and neuraminidase (N) activities in a predominantly genus-specific manner. Members of the *Respirovirus, Avulavirus* and *Rubulavirus* genera (with the exception of Menangle virus and Tioman virus in the latter genus) display both activities [5, 9, 14]. Morbilliviruses do not behave uniformly [16] and henipaviruses fail to display both haemagglutination and neuraminidase activities [21]. This failure is consistent with the observation that, among the seven active site residues known to be important for neuraminidase activity (Table 2), only one is present in the henipavirus G protein whereas six are conserved in the neuraminidase-positive respiroviruses and rubulaviruses and four in morbilliviruses, which are predominantly neuraminidase-negative [16, 30]. In addi-

Table 2. Presence of neuraminidase active site and sialic acid binding residues in genera of the subfamily *Paramyxovirinae*

	Neuraminidase catalytic site residues in the attachment protein[1]							Sialic acid binding site
	R(1)	D(2)	E(3)	R(4)	R(5)	Y(6)	E(7)	NRKSCS
Respirovirus genus	√	√	√	√	√	√	√	NRKSCS
Rubulavirus genus	√	√	√	√	√	√	√	NRKSCS
Morbillivirus genus	√				√	√	√	- - - - C -
Menangle virus	√						√	- - - - CS
Tioman virus								
Henipavirus genus							√	- - - - CS
Tupaia paramyxovirus								N - - - CS

[1]See [16]

tion, the failure of henipaviruses to haemagglutinate is consistent with the presence of only 2 of the 6 amino acids in the hexapeptide NRKSCS which is believed to play a role in sialic acid binding [12, 20, 32].

On the basis of preliminary sequence data and the failure to detect neuraminidase and hemagglutinating activities in purified virus preparations, HENV was originally thought to more closely resemble morbilliviruses than viruses in other *Paramyxovirinae* genera [21]. However, several recent observations highlight similarities between respiroviruses and henipaviruses in structure and replication strategy. First, three-dimensional modeling studies of the external globular head domain of the HENV G protein indicated that, although the protein displays limited primary sequence homology with other members of the *Paramyxovirinae*, including respiroviruses, the locations of proposed disulfide bonds and a large number of glycine, proline and aromatic amino acids are absolutely conserved between HENV and respiroviruses. More significantly, the predicted folding pattern of the extracellular globular head of the G protein of HENV resembled that of respiroviruses rather than of morbilliviruses [32]. The second similarity between henipaviruses and respiroviruses has been observed as a result of studies examining transcription attenuation in HENV-infected cells and is described below.

Attenuation of transcription

The genome of non-segmented negative stranded RNA viruses is tightly associated with NP proteins within a complex that also contains P and L proteins. The ratio of NP, P and L proteins in Sendai virus (a respirovirus) is 100:30:2 [15]. Genomic RNA in RNP is transcribed by the viral polymerase, which associates with the ribonucleoprotein complex at a single 3′ proximal site and sequentially generates discrete, predominantly monocistronic mRNAs from each of the viral genes. The mRNAs are not produced in equimolar amounts and there is a transcription gradient from the NP to the L gene. The levels of gene expression are thus regulated according to their position on the genome. However, the attenuation is not linear because subtle differences in gene start or stop/polyadenylation signals may influence the attenuation that is observed at gene boundaries [13].

The extent of attenuation at gene junctions has been determined for Sendai virus and measles virus by quantifying the steady state levels of mRNA species using a method in which RNA from virus-infected cells was mixed with known concentrations of RNA complementary to defined regions of each gene and the mixture electrophoresed, blotted onto nylon membranes and hybridized with radiolabelled RNA probes for each gene. Comparison of the radioactivity in the mono-cistronic mRNA bands compared with that in the faster-migrating standard bands permitted quantification of each mRNA species. With Sendai virus attenuation was prominent at the M–F gene and HN–L gene junctions. There was little decrease in the relative amounts of NP, P and M mRNAs and only a small difference was observed between the F and HN mRNAs [11]. The situation observed in measles virus-infected cells differed, with attenuation observed at all junctions but with maximum attenuation detected at the NP–P junction [6]. The reason for the different attenuation patterns of Sendai virus, a respirovirus, and measles virus, a morbillivirus is not understood and whether the difference is genus specific remains unknown.

Steady state levels of mRNA in HENV-infected cells have been determined using gene-specific, real time PCR assays. RNA was extracted from HENV-infected cells at various times post infection and random hexamer-primed cDNA synthesized on poly(A) + RNA and cDNA was quantified by real-time PCR using gene-specific primers, targeting amplicons of identical length near the 3′ end of each gene. The results demonstrated a gene-specific mRNA gradient with major attenuation at the M–F and G–L junctions, as was observed for Sendai virus. The relative levels of mRNA remained unchanged from early in infection, before the release of detectable virus, through 8 to 10 h post infection, when released virus yield increased approximately 1000 fold.

Nature of the henipavirus cell receptor

For viruses within the *Respirovirus*, *Rubulavirus* and *Avulavirus* genera, hemagglutination and neuraminidase activities are correlated with capacity of the virus to bind to sialic acid-containing receptor molecules. Newcastle disease virus and Sendai virus, members of the *Respirovirus* and *Avulavirus* genera respectively, have the receptors identified as glycolipids [19, 24]. In the *Morbillivirus* genus, vaccine strains of measles virus bind to the glycoprotein CD46 [22] but not via a sialic acid-dependent process [18]. A number of morbilliviruses including field isolates and vaccine strains of measles virus, and strains of canine distemper virus and rinderpest virus, have been shown to use signaling lymphocyte activation molecule (SLAM; CD150) as a receptor, thus providing an explanation for the lymphotropism and immunosuppressive nature of morbilliviruses [25].

A virus overlay protein binding assay (VOPBA) was utilized to determine whether HENV and NIPV bound to Vero cell proteins that had been separated by polyacrylamide gel electrophoresis and transferred to nitrocellulose membrane [17, 18]. Bound virus was detected using a monoclonal antibody to HENV G protein and alkaline phosphatase-conjugated anti-mouse antibody. Both HENV and NIPV bound to an identical set of discrete proteins in the 35 to 50 kDa range. Treatment of cell proteins with N-glycosidase F, which removes N-linked oligosaccharide chains, strongly reduced HENV binding activity in VOPBA, whereas neuraminidase treatment had no effect on binding. Attempts were made to block HENV binding in VOPBA and infection of Vero cells using lectins with different specificities. Wheat germ agglutinin, with specificity for N-acetylglucosaminyl oligomers, proved effective at a concentration of 10 μg/ml for 1 h at 37 °C. Putative receptor molecules have been partially purified using wheat germ agglutinin-based affinity column chromatography. VOPBA on Vero cell proteins separated by 2 dimensional gel electrophoresis has identified a number of discrete HENV-binding proteins in the 35 to 50 kDa range. These preliminary observations indicate that HENV and NIPV bind via a sialic acid-independent mechanism to glycoprotein molecules. Failure to abrogate virus infection by removal of sialic acid from susceptible cells by prior neuraminidase treatment is consistent with a sialic acid-independent infection process.

Further evidence implicating a protein as the cell receptor for henipaviruses has come from studies using recombinant vaccinia viruses expressing the F and G proteins in a quantitative assay for henipavirus-mediated fusion [3, 4]. This system relies on the vaccinia virus-mediated co-expression in an effector cell population of viral F and G proteins and either bacteriophage T7 RNA polymerase or a T7 promoter-driven *Escherichia coli LacZ* cassette. When mixed with target cells infected with recombinant vaccinia virus expressing either bacteriophage T7 RNA polymerase or a T7 promoter-driven *Escherichia coli LacZ* cassette as appropriate, cell fusion can be quantified by the amount of β-galactosidase produced. This assay has proven especially useful in the study of envelope glycoproteins derived from viruses that employ a pH-independent mechanism of membrane fusion for virion entry [1, 2, 23].

Henipaviruses cause systemic infections in a number of animal species including flying foxes (HENV and NIPV), horses (HENV), pigs (NIPV) and cats. Unlike other members of the *Paramyxoviridae*, henipaviruses replicate *in vitro* in cells derived from a wide variety of species. This broad species tropism is reflected in the identification of permissive target cells from cat, bat, horse, human, simian, mouse and rabbit using the quantitative fusion assay. Several cell lines and some primary cell types have been found to be likely receptor-negative for henipaviruses. HENV and NIPV display the same receptor recognition pattern in the cell lines tested and yield similar fusion rates with receptor-positive cell line. These observations support the suggestion made above, that HENV and NIPV may use the same cellular receptors for virus entry [3, 4]. Perhaps this is not surprising given that the NIPV and HENV glycoproteins are capable of highly efficient heterotypic functional activity with each other [4]. Protease treatment of target cells completely abolished HENV-mediated fusion, thus providing further evidence that the virus was employing a cell surface protein as its receptor [3].

References

1. Alkhatib G, Broder CC, Berger EA (1996) Cell type-specific fusion cofactors determine human immunodeficiency virus type 1 tropism for T-cell lines versus primary macrophages. J Virol 70: 5487–5494
2. Bagai SL, Lamb RA (1995) Quantitative measurement of paramyxovirus fusion: differences in requirements of glycoproteins between simian virus 5 and human parainfluenza virus 3 or Newcastle disease virus. J Virol 69: 6712–6719
3. Bossart KN, Wang LF, Eaton BT, Broder CC (2001) Functional expression and membrane fusion tropism of the envelope glycoproteins of Hendra virus. Virology 290: 121–135
4. Bossart KN, Wang LF, Flora MN, Chua KB, Lam SK, Eaton BT, Broder CC (2002) Membrane fusion tropism and heterotypic functional activities of the Nipah virus and Hendra virus envelope glycoproteins. J Virol 76: 11186–11198
5. Bowden TR, Westenberg M, Wang LF, Eaton BT, Boyle DB (2001) Molecular characterization of Menangle virus, a novel paramyxovirus which infects pigs, fruit bats, and humans. Virology 283: 358–373
6. Cattaneo R, Rebmann G, Baczko K, ter Meulen V, Billeter MA (1987) Altered ratios of measles virus transcripts in diseased human brains. Virology 160: 523–526
7. Chant K, Chan R, Smith M, Dwyer DE, Kirkland P (1998) Probable human infection with a newly described virus in the family Paramyxoviridae. Emerg Infect Dis 4: 273–275
8. Chua KB, Bellini WJ, Rota PA, Harcourt BH, Tamin A, Lam SK, Ksiazek TG, Rollin PE, Zaki SR, Shieh WJ, Goldsmith CS, Gubler DJ, Roehrig JT, Eaton BT, Gould AR, Olson J, Field H, Daniels P, Ling AE, Peters CJ, Anderson LJ, Mahy BWJ (2000) Nipah virus: A recently emergent deadly paramyxovirus. Science 288: 1432–1435
9. Chua KB, Wang LF, Lam SK, Eaton BT (2002) Full length genome sequence of Tioman virus, a novel paramyxovirus in the genus *Rubulavirus* isolated from fruit bats in Malaysia. Arch Virol 147: 1323–1348
10. Harcourt BH, Tamin A, Ksiazek TG, Rollin PE, Anderson LJ, Bellini WJ, Rota PA (2000) Molecular characterization of Nipah virus, a newly emergent paramyxovirus. Virology 271: 334–349
11. Homann HE, Hofschneider PH, Neubert WJ (1990) Sendai virus gene expression in lytically and persistently infected cells. Virology 177: 131–140
12. Jorgensen ED, Collins PL, Lomedico PT (1987) Cloning and nucleotide sequence of Newcastle disease virus hemagglutinin-neuraminidase mRNA: identification of a putative sialic acid binding site. Virology 156: 12–24
13. Kato A, Kiyotani K, Hasan MK, Shioda T, Sakai Y, Yoshida T, Nagai Y (1999) Sendai virus gene start signals are not equivalent in reinitiation capacity: moderation at the fusion protein gene. J Virol 73: 9237–9246
14. Lamb RA, Kolakofsky D (2001) *Paramyxoviridae*: The viruses and their replication. In: Knipe DM, Howley PM, Griffin DE, Lamb RA, Martin MA, Roizman B, Straus SE (eds) Fields virology vol 1. Lippincott, Williams & Wilkins, Philadelphia, pp 1305–1340
15. Lamb RA, Mahy BW, Choppin PW (1976) The synthesis of Sendai virus polypeptides in infected cells. Virology 69: 116–131
16. Langedijk JP, Daus FJ, van Oirschot JT (1997) Sequence and structure alignment of Paramyxoviridae attachment proteins and discovery of enzymatic activity for a morbillivirus hemagglutinin. J Virol 71: 6155–6167

17. Loffler S, Lottspeich F, Lanza F, Azorsa DO, ter Meulen V, Schneider-Schaulies J (1997) CD9, a tetraspan transmembrane protein, renders cells susceptible to canine distemper virus. J Virol 71: 42–49
18. Maisner A, Schneider-Schaulies J, Liszewski MK, Atkinson JP, Herrler G (1994) Binding of measles virus to membrane cofactor protein (CD46): importance of disulfide bonds and N-glycans for the receptor function. J Virol 68: 6299–6304
19. Markwell MA, Paulson JC (1980) Sendai virus utilizes specific sialyloligosaccharides as host cell receptor determinants. PNAS 77: 5693–5697
20. Mirza AM, Deng R, Iorio RM (1994) Site-directed mutagenesis of a conserved hexapeptide in the paramyxovirus hemagglutinin-neuraminidase glycoprotein: effects on antigenic structure and function. J Virol 68: 5093–5099
21. Murray K, Selleck P, Hooper P, Selleck P, Hyatt A, Gould A, Gleeson L, Westbury H, Hiley L, Selvey L, Rodwell B (1995) A morbillivirus that caused fatal disease in horses and humans. Science 268: 94–97
22. Naniche D, Varior-Krishnan G, Cervoni F, Wild F, Rossi B, Rabourdin-Combe C, Gerlier D (1993) Human membrane cofactor protein (CD46) acts as a cellular receptor for measles virus. J Virol 67: 6025–6032
23. Nussbaum O, Broder CC, Moss B, Stern LB, Rozenblatt S, Berger EA (1995) Functional and structural interactions between measles virus hemagglutinin and CD46. J Virol 69: 3341–3349
24. Reyes-Leyva J, Hernández-Jáuregui P, Montaño LF, Zenteno E (1993) The porcine paramyxovirus LPM specifically recognizes sialyl (alpha 2,3) lactose-containing structures. Arch Virol 133: 195–200
25. Tatsuo H, Ono N, Yanagi Y (2001) Morbilliviruses use signaling lymphocyte activation molecules (CD150) as cellular receptors. J Virol 75: 5842–5850
26. Thomas SM, Lamb RA, Paterson RG (1988) Two mRNAs that differ by two nontemplated nucleotides encode the amino coterminal proteins P and V of the paramyxovirus SV5. Cell 54: 891–902
27. Wang L-F, Chua KB, Yu M, Eaton BT (2003) Genome diversity of emerging paramyxoviruses. Curr Genom 4: 263–273
28. Wang LF, Eaton BT (2001) Emerging paramyxoviruses. Infect Dis Rev 3: 52–69
29. Wang LF, Eaton BT (2002) Henipavirus (*Paramyxoviridae*, *Paramyxovirinae*). In: Tidona CA, Darai G (eds) The Springer index of viruses. Springer, Berlin Heidelberg New York Tokyo, pp 641–644
30. Wang LF, Harcourt BH, Yu M, Tamin A, Rota PA, Bellini WJ, Eaton BT (2001) Molecular biology of Hendra and Nipah viruses. Microb Infect 3: 279–287
31. Westbury HA (2000) Hendra virus disease in horses. Rev Sci Tech Office Int Epizoot 19: 151–159
32. Yu M, Hansson E, Langedijk JP, Eaton BT, Wang LF (1998) The attachment protein of Hendra virus has high structural similarity but limited primary sequence homology compared with viruses in the genus *Paramyxovirus*. Virology 251: 227–233

Author's address: Bryan Eaton, Australian Animal Health Laboratory, CSIRO Livestock Industries, 5 Portarlington Road, Greenlong, Vic. 3220, Australia; e-mail: Bryan.eaton@csiro.au

Arch Virol (2004) [Suppl] 18: 133–137

Archives of Virology

The entry machinery of flaviviruses

F. X. Heinz, K. Stiasny, and **S. L. Allison**

Institute of Virology, University of Vienna, Austria

Summary. We have been using the flavivirus tick-borne encephalitis virus (TBEV) as a model system for investigating the molecular mechanisms underlying the membrane fusion process mediated by a class II viral fusion protein, the flavivirus envelope protein E. In the mature virion this protein exists as a metastable dimer that dissociates at the acidic pH in endosomes and is converted into a more stable trimeric conformation. The dimer dissociation step liberates an internal fusion peptide that interacts with the target endosomal membrane, and then further conformational changes are believed to drive membrane fusion. Although flavivirus fusion appears to be a more facile and efficient process than that of alphaviruses, which also possess a class II viral fusion protein, the fusion mechanism in both viral systems involves structurally related interactions with lipids, specifically the 3ß-hydroxyl group at C3 of cholesterol. The class II viral fusion machineries are structurally different from those involving class I viral fusion proteins, such as those found in orthomyxoviruses, paramyxoviruses, retroviruses, and filoviruses, but have certain similarities in common with bacterial pore-forming proteins.

Introduction

Flaviviruses are small enveloped viruses that belong to the genus *Flavivirus* in the family *Flaviviridae* and which possess three structural proteins [9] (Fig. 1). The isometric nucleocapsid is composed of a single capsid protein (C) and contains the plus-stranded RNA genome of approximately 11,000 nucleotides. The assembly of enveloped virus particles occurs in the endoplasmic reticulum and first leads to the formation of immature virions that have two membrane-associated proteins, E and prM, and are transported through the cellular exocytotic pathway. Virus maturation requires proteolyic cleavage of the prM protein by a cellular protease (presumably furin) in the trans-Golgi network [13], which generates infectious virions that are released by exocytosis. The E protein carries both receptor-binding and fusion activity and the atomic structure of this protein isolated from tick-borne encephalitis (TBEV) has been determined by X-ray crystallography (Fig. 2) [11]. It is an elongated dimeric molecule that is oriented in parallel to the viral membrane and composed of three distinct domains, designated I, II, and III. A very

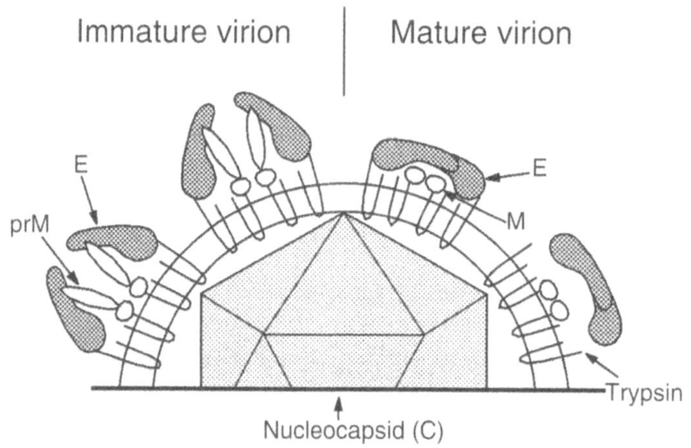

Fig. 1. Schematic of the composition of immature and mature flaviviruses. Limited trypsin digestion of mature TBE virions yielded a soluble dimeric ectodomain fragment of the E protein, which was used for structure determination [11]. Reproduced from Heinz and Allison [5]

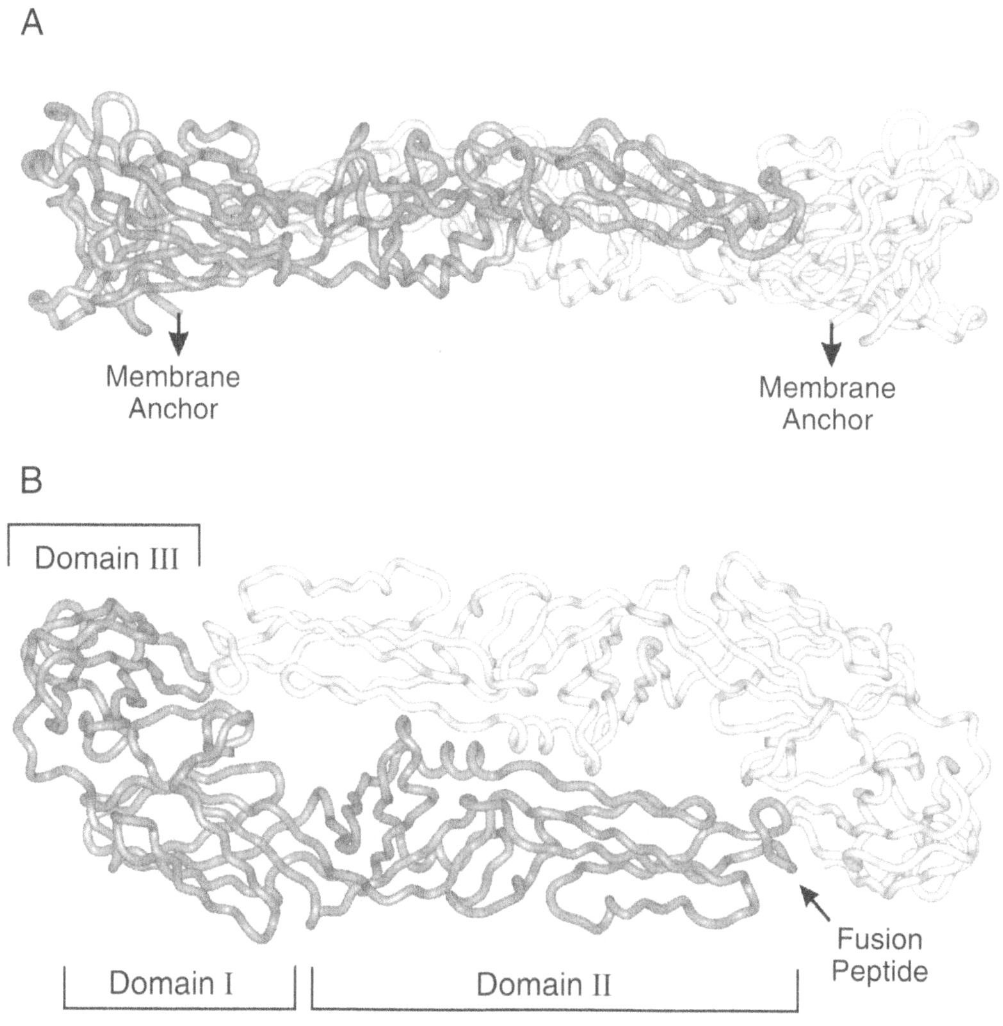

Fig. 2. Ribbon diagram of the TBEV E protein dimer [11]. The monomeric subunits are shown in different shades of grey. **A** side view **B** top view

similar molecular structure has more recently been found for the fusion protein of an alphavirus (family *Togaviridae*), the E1 protein of Semliki Forest virus [8]. These proteins are now referred to as class II viral fusion proteins because of their striking structural differences from the class I viral fusion proteins found in orthomyxoviruses, paramyxoviruses, retroviruses, and filoviruses [12, 18]. We have been using TBEV as a model flavivirus to study the molecular mechanisms of membrane fusion in a class II viral fusion protein system.

Structures and structural changes involved in TBEV fusion

Flaviviruses enter cells by receptor-mediated endocytosis, and the acidic pH in the endosome triggers the fusion of the viral membrane with the endosomal membrane [5, 6]. In vitro fusion experiments with pyrene phospholipid-labeled TBEV and artificial liposomes have revealed that the fusion reaction of this virus is a very facile process, occurring faster than that of any other enveloped virus analyzed to date [2]. Exposure of the virus to acidic pH causes dramatic structural changes in the viral fusion protein E that also lead to a quantitative oligomeric switch from dimers to trimers [1]. These changes are irreversible and apparently drive the fusion process. Thermal denaturation experiments have shown that the trimeric form is more stable than the dimeric form, suggesting that the E protein in mature virions exists in a metastable state that requires the acidic pH-trigger to be converted into its final stable conformation [14].

Experiments with a C-terminally truncated form of the E protein dimer that lacks both the membrane anchor and a stem structure of about 40 amino acids have revealed that acidic pH causes dimer dissociation and that the monomeric form is required to make the first contact with the target membrane [15]. This contact is mediated by an internal fusion peptide that has been shown by mutagenesis experiments with recombinant subviral particles to reside in a loop at the tip of the molecule (Fig. 2) [1]. This structure is buried by specific interactions between the monomers in the E protein dimer and only becomes exposed upon low-pH-induced dissociation of the two subunits. In contrast to the solubilized full-length E protein, which trimerizes irreversibly, the truncated dimeric form of E dissociates at acidic pH, and is not converted into trimers in solution [16]. Low-pH-induced trimerization of the truncated E protein, however, can be achieved in the presence of membranes, suggesting that the interactions of the fusion protein with lipids in the target membrane might contribute to the structural rearrangements necessary for fusion [15].

Role of lipids in TBEV fusion

Despite the striking structural similarities between the flavivirus and alphavirus fusion proteins, a significant difference appears to exist with respect to the dependence of fusion on certain lipids in the target membrane. Alphavirus fusion has been shown to be absolutely dependent on the presence of both cholesterol and sphingomyelin, which appear to be required at different stages of the fusion process [7]. In contrast, flavivirus fusion can proceed in the absence of both lipids, although fusion is more efficient when cholesterol is present [3]. In a recent study, we have investigated details of the TBEV fusion mechanism in experimental systems that allowed us to study the effect of specific lipids not only on overall fusion, but also on the early step of membrane binding and the oligomeric switch from dimers to trimers [17]. For these studies we used artificial liposomes composed of different lipids, including the cholesterol analogs coprostanol, androstanol, cholesteryl methylether, and cholestanone. Surprisingly, the effect of cholesterol on liposome binding and lipid-induced E protein trimerization was quite significant and, as with alphaviruses, the interactions observed involved the 3ß-hydroxyl group. In the case of TBEV, however, overall fusion in vitro did not exhibit the same absolute requirement for these fusion protein-lipid interactions as had been observed for alphaviruses. It was shown to be completely independent of sphingomyelin and less dependent on the specific interaction with the 3ß-hydroxyl group at C3 of cholesterol. Despite these quantitative differences, it appears that the fusion process mediated by the class II viral fusion proteins of both alphaviruses and flaviviruses involves structurally related interactions with certain lipids in the target membrane, which is also a characteristic feature of certain bacterial pore-forming proteins. These bacterial proteins exhibit the same structural specificity for cholesterol as do the class II viral fusion proteins, and these interactions have also been shown to be involved in conformational changes and allosteric transitions [4, 10]. The structural elements in the class II viral fusion and bacterial pore-forming proteins that specifically interact with lipids have yet to be determined.

References

1. Allison SL, Schalich J, Stiasny K, Mandl CW, Kunz C, Heinz FX (1995) Oligomeric rearrangement of tick-borne encephalitis virus envelope proteins induced by an acidic pH. J Virol 69: 695–700
2. Corver J, Moesby L, Erukulla RK, Reddy KC, Bittman R, Wilschut J (1995) Sphingolipid-dependent fusion of Semliki Forest virus with cholesterol-containing liposomes requires both the 3-hydroxyl group and the double bond of the sphingolipid backbone. J Virol 69: 3220–3223
3. Corver J, Ortiz A, Allison SL, Schalich J, Heinz FX, Wilschut J (2000) Membrane fusion activity of tick-borne encephalitis virus and recombinant subviral particles in a liposomal model system. Virology 269: 37–46
4. Gilbert RJ (2002) Pore-forming toxins. Cell Mol Life Sci 59: 832–844
5. Heinz FX, Allison SL (2000) Structures and mechanisms in flavivirus fusion. Adv Virus Res 55: 231–269
6. Heinz FX, Allison SL (2001) The machinery for flavivirus fusion with host cell membranes. Curr Opin Microbiol 4: 450-455
7. Kielian M, Chatterjee PK, Gibbons DL, Lu YE (2000) Specific roles for lipids in virus fusion and exit. Examples from the alphaviruses. Subcell Biochem 34: 409–455
8. Lescar J, Roussel A, Wien MW, Navaza J, Fuller SD, Wengler G, Rey FA (2001) The fusion glycoprotein shell of Semliki Forest virus. An icosahedral assembly primed for fusogenic activation at endosomal pH. Cell 105: 137–148

9. Lindenbach BD, Rice MC (2001) Flaviviridae: the viruses and their replication. In: Knipe DM, Howley PM, Griffin DE, Lamb RA, Martin MA, Roizman B, Strauss SE (eds.), Fields virology, 4th edn., Lippincott-Raeven, Philadelphia, pp 991–1041
10. Palmer M, Vulicevic I, Saweljew P, Valeva A, Kehoe M, Bhakdi S (1998) Streptolysin O: a proposed model of allosteric interaction between a pore-forming protein and its target lipid bilayer. Biochemistry 37: 2378–2383
11. Rey FA, Heinz FX, Mandl C, Kunz C, Harrison SC (1995) The envelope glycoprotein from tick-borne encephalitis virus at 2 Å resolution. Nature 375: 291–298
12. Skehel JJ, Wiley DC (1998) Coiled coils in both intracellular vesicle and viral membrane fusion. Cell 95: 871–874
13. Stadler K, Allison SL, Schalich J, Heinz FX (1997) Proteolytic activation of tick-borne encephalitis virus by furin. J Virol 71: 8475–8481
14. Stiasny K, Allison SL, Mandl CW, Heinz FX (2001) Role of metastability and acidic pH in membrane fusion by tick-borne encephalitis virus. J Virol 75: 7392–7398
15. Stiasny K, Allison SL, Schalich J, Heinz FX (2002) Membrane interactions of the tick-borne encephalitis virus fusion protein E at low pH. J Virol 76: 3784–3790
16. Stiasny K, Allison SL, Marchler-Bauer A, Kunz C, Heinz FX (1996) Structural requirements for low-pH-induced rearrangements in the envelope glycoprotein of tick-borne encephalitis virus. J Virol 70: 8142–8147
17. Stiasny K, Koessl C, Heinz FX (2003) Involvement of lipids in different steps of the flavivirus fusion mechanism. J Virol 77: 7856–7862
18. Weissenhorn W, Dessen A, Calder LJ, Harrison SC, Skehel JJ, Wiley DC (1999) Structural basis for membrane fusion by enveloped viruses. Mol Membr Biol 16: 3–9

Author's address: Franz X. Heinz, Institute of Virology, Kinderspitalgasse 15, 1095 Vienna, Austria; e-mail: franz.x.heinz@univie.ac.at

Arch Virol (2004) [Suppl] 18: 139–147

Archives of Virology

Persistent infection and suppression of host response by alphaviruses

I. Frolov

Department of Microbiology and Immunology, University of Texas Medical Branch, Galveston, Texas, U.S.A.

Summary. Alphaviruses cause chronic noncytopathic infection in mosquito cells and develop a highly cytopathic infection in a wide variety of cells of vertebrate origin. Upon infection, alphaviruses modify cellular processes to meet the virus needs for propagation. Downregulation of translation and transcription caused by viral infection appears to reduce interferon (IFN) and cytokine gene expression and allows more efficient dissemination of infection. Alphaviruses with mutations in nonstructural protein nsP2 can become less cytopathic and capable of persisting in some vertebrate cell lines for a number of passages. nsP2 likely functions as an important regulator of virus-host cell interactions and plays a significant role in suppressing the antiviral response. Mammalian cells having no defects in type I IFN system react to replication of the nsP2 viral mutants by more efficient activation of IFN and IFN-dependent genes and are capable of eliminating established alphavirus infection. Blocking of IFN-α/β signaling makes mouse fibroblasts unable to stop replication of Sindbis virus (SINV) with mutated nsP2 and leads to persistent infection. Downregulation of transcription and translation during alphavirus infection are quite independent events, and both probably are involved in inhibition of the antiviral response.

Introduction

The genus *Alphavirus* in the family *Togaviridae* contains a variety of significant human and animal pathogens [27]. Some of them, including Venezuelan equine encephalitis, eastern equine encephalitis and western equine encephalitis viruses, can cause fatal diseases in humans and vertebrates. Other alphaviruses produce serious, but not life-threatening, illnesses or clinically inapparent disease with mild symptoms, including headache, fever, skin rash and arthritis [22]. The nearly 30 known alphaviruses are believed to have similar life cycles, viral particle structure, and replication strategies. The majority of the alphaviruses are arthropod-borne viruses (arboviruses), which are maintained in nature through biological transmission between vertebrate hosts by arthropod vectors, predominantly mosquitoes

[29]. Their survival implies continuous infection of susceptible hosts under conditions whereby the density of vertebrates is too low for virus transmission by aerosol or direct contact. Therefore, in vertebrates, alphaviruses must cause a short-term acute infection with viremia levels sufficient for transmission before infection can be cleared by the developing immune response. In insect vectors, they cause a long-term infection that may be either chronic or persistent. The geographic distribution of individual alphaviruses principally is determined by the presence of competent mosquito vectors and migration of susceptible avian hosts [28]. At the same time, other vertebrates may be important for virus maintenance in certain geographical areas. Infection in tissue culture generally correlates with the natural circulation of alphaviruses. They cause a chronic or persistent noncytopathic infection in mosquito cells in vitro and develop a highly cytopathic infection in a wide variety of cells of vertebrate origin [22].

This overview briefly discusses the strategy that alphaviruses use to develop acute infection and high-titer viremia in the vertebrate host, including the approaches that they use to escape the antiviral response.

Alphavirus replication

The alphavirus genome is packaged into an icosahedral nucleocapsid formed by 240 molecules of a basic capsid protein (C), surrounded by a lipid envelope with 240 heterodimers of E1 and E2 glycoproteins organised into 80 trimeric spikes. The regularity of the spike distribution is determined by interaction of glycoproteins with C subunits [3].

The alphavirus genome is a single-stranded RNA molecule of positive polarity, almost 12 kb in size. Like cellular messenger RNAs, it contains both a 5′ methylguanylate cap and a 3′ polyadenylate tail, and translation of viral proteins by host cell machinery starts immediately after release of the genome RNAs from the nucleocapsids [26]. The 5′ two-thirds of the genome is translated into the nonstructural proteins (nsPs) that represent the viral components of the RNA-dependent RNA polymerase (RdRp) required for replication of the viral genome and transcription of the subgenomic RNA (Fig. 1). For the majority of alphaviruses, nonstructural proteins are translated as two polyproteins. Synthesis of polyprotein P123, a precursor of nsP1, nsP2 and nsP3, is terminated at an opal codon. Polyprotein P1234 is synthesized upon readthrough of the P123 termination codon with 5–10% efficiency [16]. P1234 is ultimately cleaved into nsP1, nsP2, nsP3 and nsP4, forming mature replicative complexes. The subgenomic RNA corresponds to the 3′ third of the genome. It is synthesized from the subgenomic promoter and translated into the viral structural proteins. Early in infection, the viral genome serves as a template for synthesis of a full-length minus-strand intermediate. Later in infection, large quantities of plus-strand genomic and subgenomic RNAs are produced. Synthesis of these three alphavirus RNA species is asymmetric and highly regulated by the differential cleavage of the nonstructural polyproteins, P123 and P1234, by protease activity of nsP2. Processing of the P123 changes the composition of the replicative enzyme complex and its ability to recognize

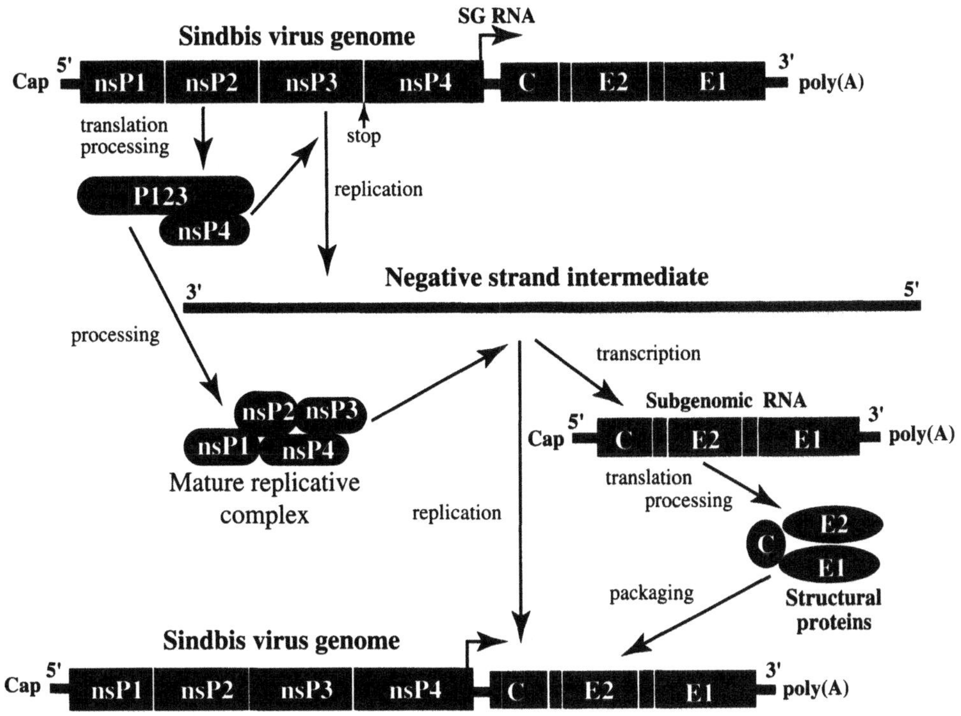

Fig. 1. Organization of SIN genome and schematic representation of SIN replication. Processing of P1234 makes it initially capable of minus-strand RNA synthesis. Further processing to nsP1, nsP2, nsP3 and nsP4 leads to formation of a mature replicative complex capable of synthesis of plus-strand genomes and transcription of the subgenomic RNA

different RNA promoters in the viral genome and in its minus-strand intermediate [13, 14, 23]. In addition, nsP1, nsP2 and nsP3 accumulate in infected cells in a 5–10 fold excess to nsP4, the catalytic subunit of viral RdRp, and this fact suggests that the nonstructural proteins may have other functions in addition to their roles in RNA replication.

Translational shutoff

Alphavirus replication profoundly affects metabolism in vertebrate cells, which lose integrity and die within 24–48 hours post infection. Like many other viral infections, the alphaviruses cause a cytopathic effect (CPE) that often is accompanied by apoptotic changes [15]. Translational shutoff is one of the central events during alphavirus infection. Studies of SINV and Semliki Forest virus (SFV) replication demonstrated that viral structural proteins are dispensable for inhibition of translation. Replication of the viral genome or accumulation of viral nonstructural proteins is sufficient to downregulate translation and to cause cell death [17]. However, the expression of envelope glycoproteins can significantly accelerate CPE development [6].

It is generally believed that cells respond to viral infection by inducing the antiviral response that downregulates virus growth and prevents dissemination of the infection. Activation of PKR (*p*rotein *k*inase, *R*NA-activated) by dsRNA is a cellular reaction to many viral infections [4, 12]. Binding to dsRNA, formed during replication of viral genomes, leads to PKR autophosphorylation and dimerization. The activated form of this protein is capable of phosphorylation of the alpha subunit of the eukaryotic initiation factor 2 (eIF2α), inactivating its ability to initiate translation, thus affecting protein synthesis in the infected cells. However, SINV and SFV not only successfully escape the translational shutoff, but also appear to employ inhibition of translation to promote virus propagation. The subgenomic RNAs of these viruses contain the structural elements that upregulate the translation of structural genes in the presence of phosphorylated eIF2α, observed during virus infection [7, 8, 24, 25]. Viral subgenomic RNAs become the only templates translated in the conditions of the translational shutoff in the late stages of the infection, when large quantities of viral structural proteins are required for assembly and release of high titers of infectious viral particles. In addition, some data suggest that alphaviruses may directly inhibit translation of cellular mRNA: i) a number of SINV and SFV mutants are incapable of causing translational shutoff in spite of efficient replication [9, 19]; ii) mouse embryonic fibroblasts lacking PKR ($PKR^{-/-}$ MEFs) infected with wild-type (wt) SINV or SFV demonstrate the same translational shutoff as do PKR-competent cells; iii) expression of high concentrations of PKR lacking kinase activity does not affect development of dramatic cessation of cellular protein synthesis during SINV infection. However, this aspect of alphavirus-host cell interactions has not yet been well-studied.

Inhibition of transcription in alphavirus-infected cells

Downregulation of transcription is a second critical element of alphavirus infection of vertebrate cells. Both messenger and ribosomal RNAs are synthesized significantly less efficiently, if at all, following infection of many permissive cell lines [10]. This can reduce interferon (IFN) and cytokine gene transcription in infected cells and allow more efficient dissemination of infection. In uninfected NIH 3T3 cells, ~4,700 mRNAs can be detected by microarray analysis on Affymetrix gene chips. However, in the same cells infected with wt SINV, fewer than 2,000 mRNAs are present and, for most of them these, concentration drops 3- to 150-fold after infection [10].

There is no reason to expect that in vertebrate hosts all cells are as susceptible to alphavirus replication as are the cell lines normally used in the laboratory practice. The infected animals do respond to infection with production of IFN-α/β. However, its level appears to be not sufficient to prevent viremia development. Mutations in nonstructural protein nsP2 can attenuate alphaviruses and, in the infected animals, they induce a high level of IFN-α/β that does not correlate with lower virus titer [10].

Alphavirus nsP2 mutants

Alphavirus genomes encode only four mature nonstructural and three structural proteins. These proteins are required for replication of the viral genetic material and for its release from the cells in the form of virions. In addition, viral proteins may interact with host machinery to minimize activation of antiviral responses in infected cells. The interesting observation that nsP1, nsP2 and nsP3 are synthesized in 10-fold excess to nsP4 (the core subunit of RdRp) suggests that the former three proteins could serve to counter host cell defense [27]. Recently, new types of mutant alphaviruses and alphavirus replicons (self-replicating RNAs lacking the genes of structural proteins) have been selected for further studies by a number of research groups [5, 9, 19, 30]. These SINV and SFV mutants are less cytopathic, but capable of persisting in some vertebrate cell lines for many passages. All of the SINV mutants detected alterations at the same position of the nsP2 protein, in which Pro_{726} was replaced by other amino acids. This remarkable convergence of mutations suggested that the noncytopathic phenotype might be a result of changes in nsP2 functioning in virus-host cell interactions, rather than in the general downregulation of virus-specific RNA replication that can occur in a wide variety of mutants. Further analyses demonstrated that in contrast to the wt virus infection leading to transcriptional shutoff, cells infected with a SINV nsP2-$P_{726} \rightarrow G$ mutant activated transcription of 170 genes, of which 113 were undetectable in uninfected cells. For the remaining 57 genes, mRNA levels increased between 3- and 200-fold [10]. A large number of the activated genes was previously described as lipopolysaccharide- or IFN-α/β-activated, while the products of most of the remainder were cytokines or proteins involved in antigen presentation. Cells infected with this nsP2 mutant, also released significantly higher levels of IFN-α/β than those infected with the wt virus. Additional analysis indicated that activation of the genes was mediated by two related factors: by virus replication itself and by autocrine action of IFN-α/β. The nsP2 mutants of SINV were not

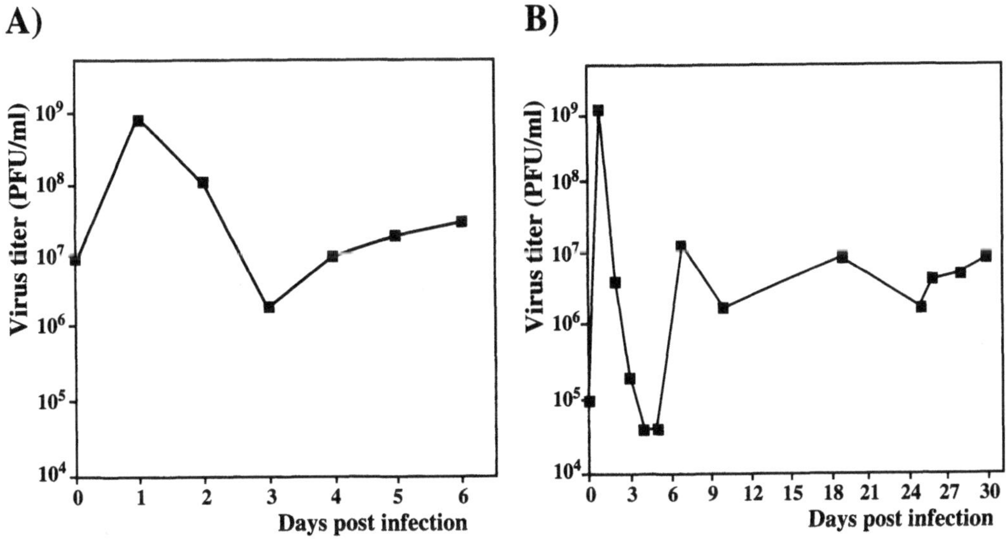

Fig. 2. Growth of SIN with $P_{726} \rightarrow G$ mutation in nsP2 in fibroblasts derived from IFN-$\alpha/\beta R^{-/-}$ mice (**A**) and $PKR^{-/-}$ mice (**B**). Cells were infected at an MOI of 20 PFU/cell, and virus titers were determined in the media at the indicated time points

only less cytopathic, but infected cells were also capable of suppressing and terminating viral replication. Activation of type I IFN signaling was a critical factor in virus clearance, and cells incapable of responding to IFN [mouse embryo fibroblasts lacking IFN-α/β receptor (IFN-$\alpha/\beta R^{-/-}$MEFs)] developed persistent infection (Fig. 2A). Long-term persistence also was achieved by infecting cells with an effective type I IFN system with nsP2 mutants in the presence of anti-IFN-α/β antibodies. PKR was also a critical component in clearance of SINV mutants. Specifically, $PKR^{-/-}$ cells were incapable of eliminating replication of noncytopathic SINV nsP2 mutants and remained persistently infected (Fig. 2B), as was observed in the experiments with IFN-$\alpha/\beta R^{-/-}$ MEFs.

In cells infected with the wt SINV, a significant fraction of the nsP2 protein is normally transported to the nuclei [20, 21]. Interestingly, in the cells infected with above-described nsP2 mutant, nsP2 accumulated in the nuclei at levels higher than those found for the wt SINV.

Another SINV with a mutation in the nsP2/nsP3 cleavage site [23], abrogating processing of the precursor protein, displayed a very different phenotype. The genomic RNA of this mutant replicated noticeably better than the genome of the wt virus, and the nsP2 protein was localized only in the cytoplasm, but not in the nuclei, of the infected cells. The distinguishing feature of this mutant was its ability to cause only translational, but not transcriptional shutoff in mouse cells. Consequently, these cells were able to react by activation of the same genes as those described for infections with SINV containing mutation of P_{726} in nsP2. Detailed investigation of the nsP2/nsP3 cleavage mutant has confirmed that the wt form of the nsP2 protein is required for suppression of host cell defense, and also demonstrated that inhibition of translation and transcription caused by SINV infection are quite independent events, and probably both are involved in downregulation of antiviral response.

Concluding remarks

Within the past few years, a great deal of progress has been made in understanding the molecular mechanisms of alphavirus replication, infectious particle formation and viral pathogenesis on an organismal level. At the same time, the main events of virus-host cell interaction still remain obscure. Upon infection, alphaviruses modify cellular processes to meet the needs of virus propagation. At the same time, cells respond to virus replication by downregulating virus growth and activating cytokine release, preventing dissemination of the infection. The balance between these competing mechanisms can determine the outcome of the infection at the cellular and organismal levels. Viral nonstructural protein nsP2, a component of the RdRp, appears to function as an important regulator of virus-host cell interactions and plays a significant role in suppressing the antiviral response in infected cells (Fig. 3). Alphaviruses thus may be added to a growing number of infectious agents that developed mechanisms to interfere with IFN-α/β production [1, 2, 11, 18, 31]. Mutations in the nsP2 protein have a significant impact on development of both translational and transcriptional shutoff, an efficient way to downregulate

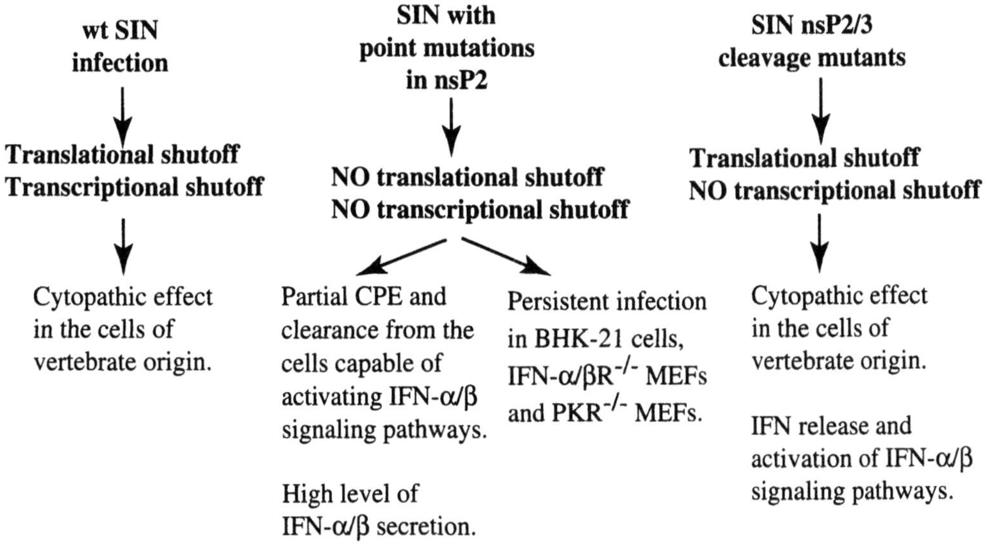

Fig. 3. Schematic representation of infections caused by wt SIN and SIN nsP2 mutants in different cell types

the synthesis of IFN and other cytokines. At the same time, mutations in the nsP2 cause alphaviruses to be less cytopathic, thus capable of persisting in cell lines having defects in IFN-α/β production or activating type I IFN pathways. Mammalian cells having no defects in the type I IFN system react to replication of the nsP2 mutants by more efficient activation of IFN and IFN-dependent genes and can eliminate established alphavirus infections. Preliminary data indicate that PKR plays a very significant role in shutoff of alphavirus replication, and this role may be different from phosphorylation of the eIF2α. However, this hypothesis needs additional experimental support.

References

1. Basler CF, Wang X, Muhlberger E, Volchkov V, Paragas J, Klenk HD, Garcia-Sastre A, Palese P (2000) The Ebola virus VP35 protein functions as a type I IFN antagonist. Proc Natl Acad Sci U S A 97: 12289–12294
2. Bouloy M, Janzen C, Vialat P, Khun H, Pavlovic J, Huerre M, Haller O (2001) Genetic evidence for an interferon-antagonistic function of rift valley fever virus nonstructural protein NSs. J Virol 75: 1371–1377
3. Cheng RH, Kuhn RJ, Olson NH, Rossmann MG, Choi HK, Smith TJ, Baker TS (1995) Nucleocapsid and glycoprotein organization in an enveloped virus. Cell 80: 621–630
4. Davies MV, Chang HW, Jacobs BL, Kaufman RJ (1993) The E3L and K3L vaccinia virus gene products stimulate translation through inhibition of the double-stranded RNA-dependent protein kinase by different mechanisms. J Virol 67: 1688–1692
5. Dryga SA, Dryga OA, Schlesinger S (1997) Identification of mutations in a Sindbis virus variant able to establish persistent infection in BHK cells: The importance of a mutation in the nsP2 gene. Virology 228: 72–83

6. Frolov I, Schlesinger S (1994) Comparison of the effects of Sindbis virus and Sindbis virus replicons on host cell protein synthesis and cytopathogenicity in BHK cells. J Virol 68: 1721–1727
7. Frolov I, Schlesinger S (1994) Translation of Sindbis virus mRNA: effects of sequences downstream of the initiating codon. J Virol 68: 8111–8117
8. Frolov I, Schlesinger S (1996) Translation of Sindbis virus mRNA: analysis of sequences downstream of the initiating AUG codon that enhance translation. J Virol 70: 1182–1190
9. Frolov I, Agapov E, Hoffman TA Jr, Prágai BM, Lippa M, Schlesinger S, Rice CM (1999) Selection of RNA replicons capable of persistent noncytopathic replication in mammalian cells. J Virol 73: 3854–3865
10. Frolova EI, Fayzulin RZ, Cook SH, Griffin DE, Rice CM, Frolov I (2002) Roles of nonstructural protein nsP2 and alpha/beta interferons in determining the outcome of Sindbis virus infection. J Virol 76: 11254–11264
11. Garcia-Sastre A, Egorov A, Matassov D, Brandt S, Levy DE, Durbin JE, Palese P, Muster T (1998) Influenza A virus lacking the NS1 gene replicates in interferon-deficient systems. Virology 252: 324–330
12. Lee TG, Tang N, Thompson S, Miller J, Katze MG (1994) The 58,000-dalton cellular inhibitor of the interferon-induced double-stranded RNA-activated protein kinase (PKR) is a member of the tetratricopeptide repeat family of proteins. Mol Cell Biol 14: 2331–2342
13. Lemm JA, Rice CM (1993) Roles of nonstructural polyproteins and cleavage products in regulating Sindbis virus RNA replication and transcription. J Virol 67: 1916–1926
14. Lemm JA, Rümenapf T, Strauss EG, Strauss JH, Rice CM (1994) Polypeptide requirements for assembly of functional Sindbis virus replication complexes: A model for the temporal regulation of minus and plus-strand RNA synthesis. EMBO J 13: 2925–2934
15. Levine B, Huang Q, Isaacs JT, Reed JC, Griffin DE, Hardwick JM (1993) Conversion of lytic to persistent alphavirus infection by the *bcl*-2 cellular oncogene. Nature 361: 739–742
16. Li G, Rice CM (1989) Mutagenesis of the in-frame opal termination codon preceding nsP4 of Sindbis virus: studies of translational readthrough and its effect on virus replication. J Virol 63: 1326–1337
17. Liljeström P, Garoff H (1991) A new generation of animal cell expression vectors based on the Semliki Forest virus replicon. BioTechnology 9: 1356–1361
18. Park MS, Shaw ML, Munoz-Jordan J, Cros JF, Nakaya T, Bouvier N, Palese P, Garcia-Sastre A, Basler CF (2003) Newcastle disease virus (NDV)-based assay demonstrates interferon-antagonist activity for the NDV V protein and the Nipah virus V, W, and C proteins. J Virol 77: 1501–1511
19. Perri S, Driver DA, Gardner JP, Sherrill S, Belli BA, Dubensky TW Jr, Polo JM (2000) Replicon vectors derived from Sindbis virus and Semliki forest virus that establish persistent replication in host cells. J Virol 74: 9802–9807
20. Rikkonen M, Peranen J, Kaariainen L (1992) Nuclear and nucleolar targeting signals of Semliki Forest virus nonstructural protein nsP2. Virology 189: 462–473
21. Rikkonen M, Peranen J, Kaariainen L (1994) Nuclear targeting of Semliki Forest virus nsP2. Arch Virol (Suppl 9): 369–377
22. Schlesinger S, Schlesinger MJ (1996) *Togaviridae*: The viruses and their replication. In: Fields BN, Knipe DM (eds) Fields virology vol 1. Lippincott-Raven, Philadelphia, pp 825–841
23. Shirako Y, Strauss JH (1994) Regulation of Sindbis virus RNA replication: Uncleaved P123 and nsP4 function in minus strand RNA synthesis whereas cleaved products from P123 are required for efficient plus strand RNA synthesis. J Virol 185: 1874–1885

24. Sjöberg EM, Suomalainen M, Garoff H (1994) A significantly improved Semliki Forest virus expression system based on translation enhancer segments from the viral capsid gene. BioTechology 12: 1127–1131
25. Sjöberg EM, Garoff H (1996) The translation-enhancing region of the Semliki Forest virus subgenome is only functional in the virus-infected cell. J Gen Virol 77: 1323–1327
26. Strauss EG, Rice CM, Strauss JH (1984) Complete nucleotide sequence of the genomic RNA of Sindbis virus. Virology 133: 92–110
27. Strauss JH, Strauss EG (1994) The alphaviruses: gene expression, replication, evolution. Microbiol Rev 58: 491–562
28. Weaver SC (1995) Evolution of alphaviruses. In: Gibbs AJ, Calisher CH, Garcia-Arenal F (eds) Molecular basis of virus evolution. Cambridge University Press, Cambridge, pp 501–530
29. Weaver SC (1997) Vector biology in viral pathogenesis. In: Nathanson N (ed) Viral pathogenesis. Lippincott-Raven, New York, pp 329–352
30. Weiss B, Rosenthal R, Schlesinger S (1980) Establishment and maintenance of persistent infection by Sindbis virus in BHK cells. J Virol 33: 463–474
31. Whitten TM, Quets AT, Schloemer RH (1991) Identification of the hepatitis B virus factor that inhibits expression of the beta interferon gene. J Virol 65: 4699–4704

Author's address: Dr. Ilya Frolov, Department of Microbiology and Immunology, University of Texas Medical Branch, 301 University Boulevard, Galveston, TX 77555-1019, U.S.A.; e-mail: ivfrolov@utmb.edu

Arch Virol (2004) [Suppl] 18: 149–159

Archives of Virology

Subversive neuroinvasive strategy of rabies virus

M. Lafon

Unité de Neuroimmunologie Virale, Institut Pasteur, Paris, France

Summary. Rabies virus (RABV) is a pathogen well-adapted to the nervous system, where it infects the neurons. RABV is transmitted by the bite of an infected animal. It enters the nervous system via a motor neuron through the neuromuscular junction, or via a sensory nerve through nerve spindles. It then travels from one neuron to the next, along the spinal cord to the brain and the salivary glands. The virions are then excreted in the saliva of the animal and can be transmitted to another host by bite. Thus, preservation of the neuronal network integrity is crucial for the virus to be transmitted. Successful invasion of the nervous system by RABV seems to be the result of a subversive strategy based on the survival of infected neurons including protection against virus-mediated apoptosis and destruction of T cells that invade the CNS in response to infection.

Introduction

Rabies virus (RABV), (family *Rhabdoviridae*, genus *Lyssavirus*) is a well-adapted pathogen of the mammalian nervous system, where it infects mostly neurons. This virus is transmitted by the bite of a RABV-infected vertebrate. It enters the nervous system via a motor neuron through the neuromuscular junction, or via a sensory nerve through nerve spindles. It then travels along the spinal cord to the brain before spreading to the salivary glands. Virions are excreted in the saliva of the vertebrate host and are transmitted to another host by bite. It is intriguing to note that RABV progression is interrupted neither by destruction of the infected neuron nor by the immune response. Thus, it is likely that RABV has developed a subversive strategy to avoid functional neuron impairment, which compromises the infectious cycle.

Viruses are notorious for their ability to subvert normal cell systems and thereby promote virus replication. Preservation of cell integrity can be obtained by manipulating the pathways that activate apoptosis in order to increase survival of infected cells [1]. Viruses also utilize diverse mechanisms in order to escape the consequences of the host defences. In particular, viruses can facilitate their own dissemination by developing stealth strategies evading cytotoxic C lymphocyte (CTL) and natural killer cell (NK) attack. Induction of apoptosis in the T cells or NK cells, supposedly to eliminate the infected cells, is one of these strategies. Killing of T cells can be obtained by the Fas/FasL apoptotic pathway. Fas is a molecule of the tumor necrosis factor-alpha (TNF-α) family. It contains a cytoplasmic death domain. Its interaction with FasL drives death signalling in cells bearing Fas, such as activated lymphocytes [3, 12]. In this article, we describe how a RABV strain that is highly adapted to the CNS has adopted invasive strategies that preserve the neuron integrity and evade the immune response.

Infection of the CNS by PV and CVS rabies virus strains

The question of how RABV has adopted a subversive strategy to evade the immune response and the protective cellular response, such as apoptosis, can be answered, at least in part, with the illustration of experiments that use the mouse as the animal model for infection with RABV. Several strains of rabies viruses with different levels of pathogenicity in mice have been selected. Both pathogenic virus strains and those with attenuated pathogenicities are neurotropic and both replicate in neurons. However, the pathogenic virus strain CVS (Challenge Virus Standard) causes fatal acute encephalitis, whereas an abortive infection strain PV (Pasteur virus) results in a non-fatal disease involving transient infection, followed by irreversible paralysis of the inoculated limbs.

After injection of 10^7 p.f.u of the CVS or PV strain of RABV in each hind leg of BALB/c mice, the development of the symptoms were analyzed as a function of viral genomic RNA in spinal cord and brain (Fig. 1). The appearance of the first signs of hind limb weakness as early as day 4 p.i. and of the paralysis of one

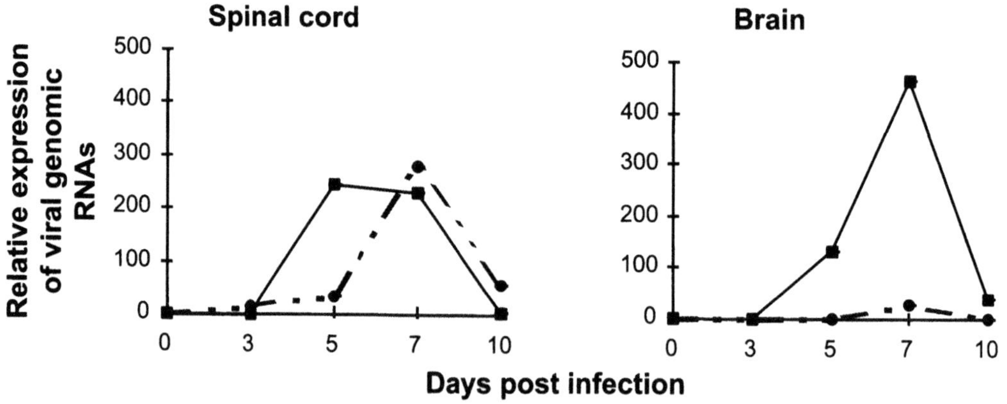

Fig. 1. Comparative infection of mouse nervous system with CVS, a highly neuroinvasive rabies virus strain and PV, an abortive rabies virus strain. BALB/c mice received an intramuscular injection of 10^7 infectious particles of rabies virus strain CVS or PV into both hindlegs. Invasion of BALB/c spinal cord (left) and brain (right) by CVS (solid line, black squares) and PV (dotted line, black circles) was measured by real-time PCR 3, 5, 7 and 10 days post-infection in the spinal cord and brain of CVS- and PV-infected mice and of non-infected mice (day 0). Each point represents the arithmetic mean of results from three or four mice. The S.D. was less than 7%

hind leg as early as day 5 p.i. occurred simultaneously with the invasion of the spinal cord by CVS. Aggravation of paralysis (intensity of the loss of mobility and number of paralysed limbs) was correlated with the progress of viral spread in the spinal cord. In PV-infected mice, the kinetics of hind limb weakness and paralysis development were similar to those in CVS-infected mice, but sequence of events began one day later. By day 9, considerable aggravation of clinical symptoms was observed in CVS-infected mice, with the onset of hunchback and prostration, followed by generalised paralysis. In contrast, the clinical symptoms improved or remained at a constant level in PV-infected mice. The PV-infected mice suffered irreversible paralysis but survived, whereas CVS-infected mice died. The disease issues correlate with differences in the invasion of the brain by the two viruses. The PV virus displayed poor brain invasion, limited to infection of the Purkinje cells in the cerebellum, whereas CVS achieved high levels of generalized brain infection by day 7 p.i. This indicates that, despite the capacity of PV to invade the spinal cord, the neuroinvasiveness of this virus into the brain is impaired, compared to that of CVS. The restricted brain invasion by PV can be a result either of the intrinsic property of the virus strain, such as induction of an antiviral state, or control of PV infection by the immune response.

Apoptosis in the PV-infected CNS is a T-cell-mediated event

To understand why PV does not invade the entire brain, we firstly addressed the possibility that PV, and not CVS, triggers apoptosis of infected neurons. The number and the nature of apoptotic cells in cerebellar sections of PV and CVS infected mouse brains were compared by immunohistochemistry. To identify whether infected cells died of apoptosis, cerebellum sections were double-stained by the TUNEL technique or with monoclonal antibody directed against activated caspase 3, and with a polyclonal antibody specific for RABV nucleocapsid. As shown in Fig. 2A and B, sections of PV-infected cerebellum contain many apoptotic cells (red staining in middle panel for TUNEL (A) and activated caspase (B)). Most of apoptotic cells were infected cells (yellow staining in right panels). CVS-infected cerebellum contains 7 time fewer apoptotic cells than PV-infected cells (data not shown). Thus, in contrast to CVS, PV infection is characterized by an important induction of apoptosis in the CNS. Surprisingly, this is inconsistent with previous observations indicating that PV is a poor apoptosis inducer in neuroblastoma cells. Indeed, we established that in contrast to ERA and SN10, two attenuated strains of rabies viruses used for oral live vaccines, the neurotropic virus strains CVS and PV induce only limited *in vitro* apoptosis in human neuroblastoma SK-N-SH cells [2, 10, 13]. The discrepancy between the *in vivo* and *in vitro* observations suggests that additional pro-apoptotic parameters, absent *in vitro*, occur in the PV-infected CNS. Since previous studies indicate that RABV-infected CNS was heavily infiltrated by T-cells [4, 5], we investigated the role of T-cells as apoptosis inducers in PV-mediated neuron apoptosis by using nude (*nu/nu*) mice. As shown in Fig. 2C, despite widespread infection of the CNS, PV-infected nude mice, which lack T-cells, exhibited very few TUNEL positive cells. This indicates that apoptosis in

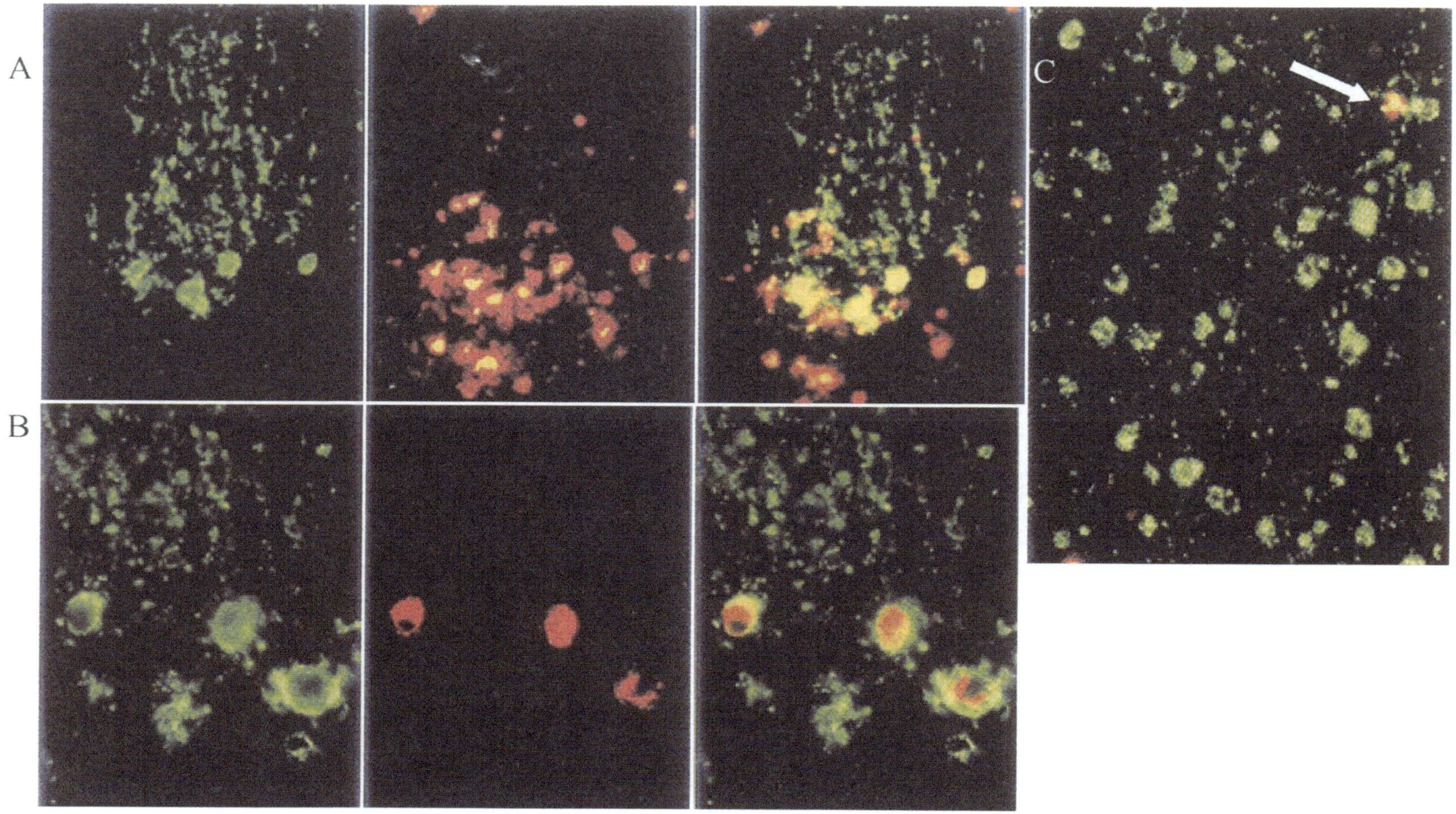

Fig. 2. Apoptosis of PV-infected neurons is a T cell-mediated mechanism. Mice were killed 7 days after injection of the PV virus into the hind legs. Cerebellum and spinal cord sections were double-stained with rabies virus nucleocapsid-specific Ab (green) and (**A** and **C**) with TUNEL (red) or **B** with an antibody specific for activated caspase (red). In the cerebellum (**A** and **B**) the rabies virus infects Purkinje cells only and not granulosa cells. In CVS-infected cerebellum apoptosis occurred in infected Purkinje cells (and also in uninfected cells). **C** Spinal cord sections of PV-infected nude mice. All the infected cells with the exception of one (arrow) escaped apoptosis (magnification ×25)

the PV-infected CNS is a T-cell-mediated event and confirms that neurotropic RABV strains do not trigger virus-mediated apoptosis [5, 2, 10, 13]. It is interesting to note that the T-cell-mediated apoptosis, the mechanisms of which have not yet been elucidated, occurred not only in infected cells but also in uninfected cells (see middle panels of Fig. 2). In addition, PV-infected nude mice developed a severe, lethal encephalitis similar to those triggered by CVS infection, suggesting that T-cell-mediated apoptosis is an important factor in controlling RABV neuroinvasion. This is consistent with other observations indicating that abortive RABV infection is replaced by a lethal rabies encephalitis in immunosuppressed mice following cyclophosphamide treatment or in β2 microglobulin knock out mice, which failed to express MHC class I molecules and as a result are deficient in CD8 T cell responses [7].

T-cells are triggered to death in the CVS-infected CNS

To understand why T-cells cannot efficiently control neuroinvasion in CVS infection, the nature of the cells that underwent apoptosis in CVS infected CNS was investigated further. CVS- and PV-infected spinal cord sections, examined 5 and 7 days p.i., were double stained to detect the fragmented nuclei (TUNEL) in lymphocytes (CD3), macrophages (CD11b), activated microglia/dendritic cells (CD11c) and infected neurons. Mean analysis of at least 20 different fields of spinal cord and cerebellum sections indicated that in the CVS infection, most TUNEL-positive cells were CD3 T-cells (Fig. 3A). This is in striking contrast to what was observed in PV-infected CNS since on day 7, 75% of CD3 T-cells in CVS-infected sections were apoptotic, compared to only 5% in PV-infected sections. Figure 3B shows a T-cell expressing surface CD3 (green) with fragmented nuclei (red) at a high magnification (X63). Kinetic analysis of T-cell mononuclear cell infiltration indicates that a decrease in the number of migratory cells coincides with the peak of apoptosis (day 7), suggesting strongly that T-cell apoptosis may account for the disappearance of migratory cells in CVS infection.

CD3 T-cell apoptosis is a Fas/FasL mediated mechanism

Interaction of Fas with its ligand FasL drives death signalling in cells that express Fas at the surface. Since T-cells are known to express Fas, we analyzed whether migratory T-cells underwent apoptosis by the Fas/FasL pathway in CVS infection. According to such a hypothesis, it was expected that FasL is up-regulated in early stages of the infection in CVS-infected but not in PV-infected CNS. The expression of FasL mRNAs was analysed by real time PCR in extracts of CVS- and PV-infected spinal cords and brains. As shown in Fig. 4A, CVS infection induced the early production of FasL mRNA (day 5 p.i) in the spinal cord, whereas PV infection did not. In the CVS-infected spinal cord, FasL and virus mRNA (see Fig. 1) reached a maximum on day 5 p.i., suggesting that the up-regulation of Fas-L is linked to spinal cord invasion. FasL mRNA expression is followed by Fas L protein expression, as detected by immunohistochemistry on CVS-infected CNS

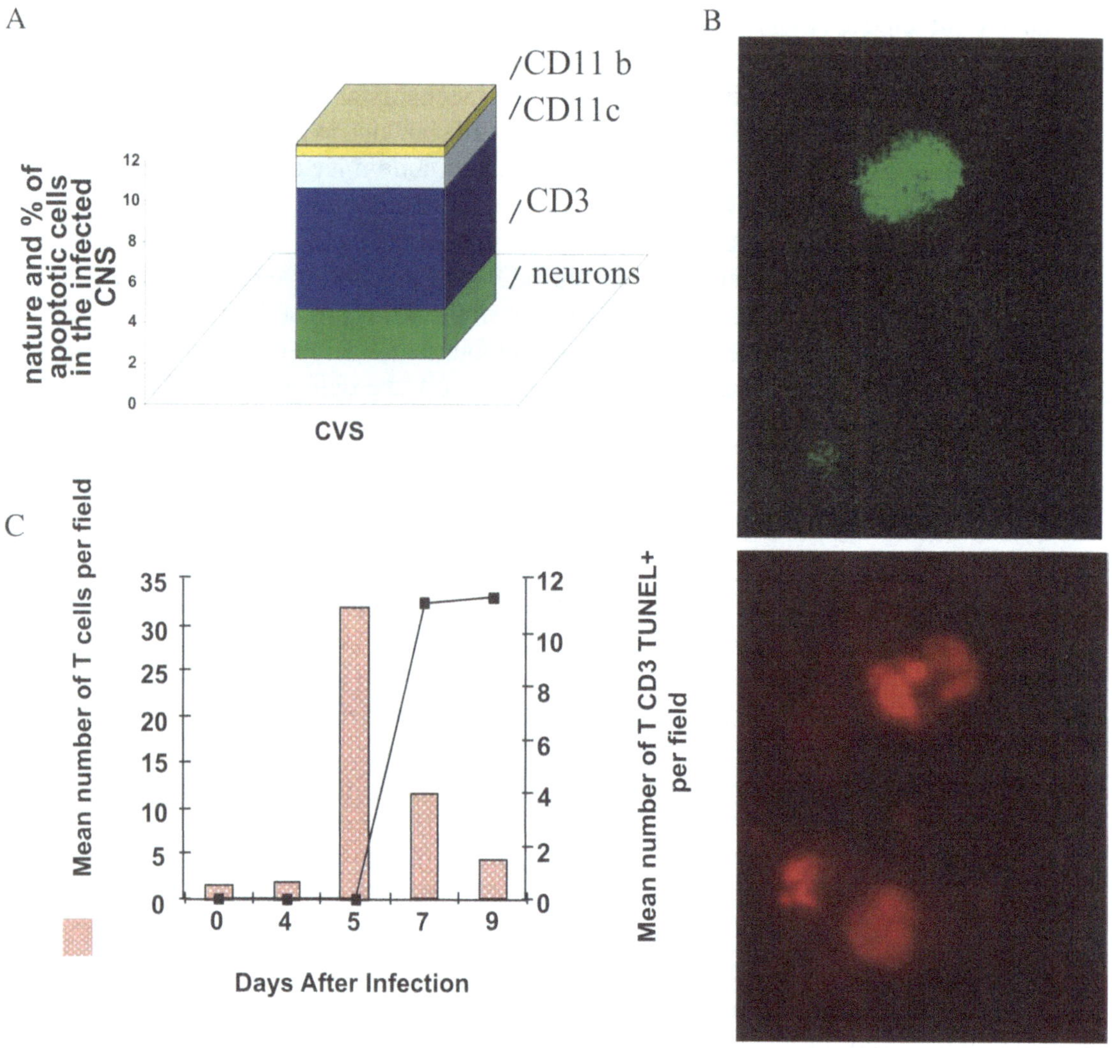

Fig. 3. Migratory CD3 T cells form the major group of TUNEL-positive cells in the CVS-infected CNS. **A** Mice were killed 7 days after injection with CVS into both hind legs. Spinal cord sections were double-stained with the TUNEL reagents and CD3 or CD11b and CD11c monoclonal antibody or antibody specific for RABV nucleocapsid. The number of TUNEL-positive cells and the percentages of CD11b, nucleocapsid or CD3 among the TUNEL-positive cells were determined in twenty microscope fields (magnification ×40) for each condition. **B** CD3 positive cells encounter apoptosis as shown by immunohistochemistry. One CD3 positive cells (green, top panel) is TUNEL positive (red low panel). This picture is representative of several observations (40× objective). **C** Kinetics of CD3 T-cells (histogram) and number of apoptotic cells (TUNEL positive cells, line) per field of spinal cord sections in CVS-infected mice. Positive cells were detected by UV microscopic analysis of at least 20 different fields

sections (Fig. 4B). In the CVS-infected CNS, infected neurons produce FasL. However, some uninfected cells also produced FasL (red cells in right panel of Fig. 4B).

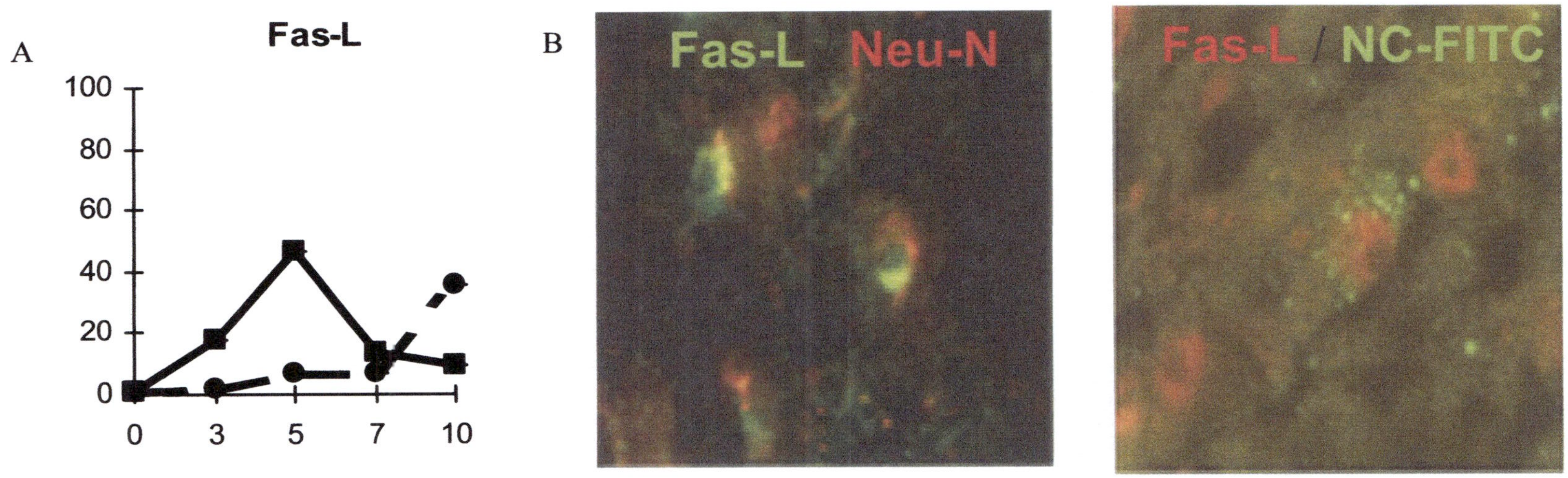

Fig. 4. Fas-L is expressed by neurons in the CNS of CVS infected mice. **A** Kinetic of Fas-L in the CNS of RABV-infected mice. Relative Fas-L mRNAs expression was evaluated at day 0, 3, 5, 7 and 10 post-infection in spinal cord (left panel) of either CVS-infected (solid line, black square) or PV-infected mice (dotted line, black circle). Each point represents the arithmetic mean of results from 3–4 mice. **B** neurons (NeuN positive cells, red) express Fas-L (green) (left panel). RABV positive cells (green) express FasL (red) (right panel). (63× objective)

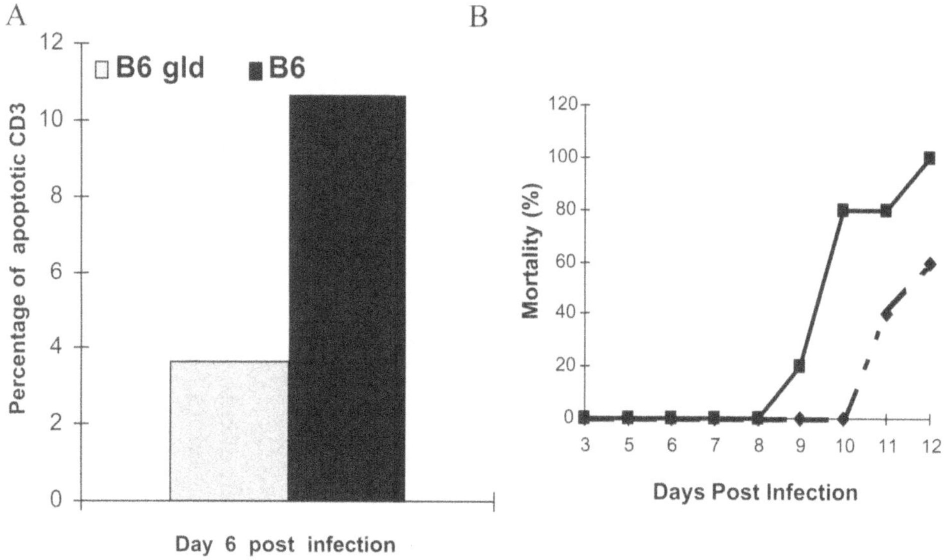

Fig. 5. Absence of FasL reduces severity of rabies disease in mice. B6 and B6 gld mice were injected i.m. with 10^7 p.f.u of CVS into each hind leg. **A** percentage of TUNEL positive CD3 cells were numbered in brain sections of gld mice (grey bar) and B6 mice (black bar) day 6 post-infection and (**B**) CVS mortality was compared in the two groups of mice. Solid line and dashed lines represent mortality of B6 and gld mice respectively. Experiments were performed twice with n = 5 and n = 8 respectively

These data indicate that migratory $CD3^+T$ cells underwent apoptosis and that FasL protein production was up-regulated during CVS virus infection. We then investigated whether $CD3^+T$ cells died by apoptosis by determining whether $CD3^+$ apoptosis occurred in the CNS of mice lacking a functional FasL (gld mice). Groups of gld mice and of normal B6 mice were infected with CVS. The number of infiltrating $CD3^+T$ cells that underwent apoptosis (TUNEL) was compared in brain sections of CVS-infected B6 and gld mice on day 5 p.i (Fig. 5A). Despite similar levels of brain infection (data not shown) we found that there were three times as many CD3/TUNEL double-positive cells in B6 mice infected with CVS than in infected gld mice. The lower levels of T cell death in mice lacking functional FasL strongly supports the notion that most of the T cells that undergo apoptosis in the CVS-infected CNS follows the Fas/FasL apoptotic pathway.

Characterisation of CVS rabies disease in C57Bl/6 (B6) and gld mice

The involvement of FasL in CVS pathogenesis was assayed by comparing the progression of CVS infection in B6 and gld mice from day 3 until day 12. The first signs of hind limb weakness appeared as early as day 5 p.i. and paralysis of one hind leg appeared as early as day 6 p.i. in B6 mice. In contrast, in CVS-infected gld mice, there was a two day-delay in the development of symptoms. Moreover, the paralysis that developed was less severe in infected gld mice than in their normal counterparts, as mean clinical scores on day 9 were 2.8 for gld mice and 4.5 for B6. On day 12 p.i. all infected B6 mice were dead, whereas only 60% of the infected gld mice had died (Fig. 5B). These data indicate that CVS infection is less severe in FasL-deficient mice than in normal mice and that the deletion of FasL favours the survival of mice infected with CVS. These data strongly support the hypothesis that FasL up-regulation is a key factor in RABV pathogenesis.

Discussion

The aim of this study was to elucidate the mechanisms that the neuroinvasive CVS virus uses to preserve the neuronal network, the integrity of which is a prerequisite for propagation within the CNS. We found that CVS is a non-lytic virus that killed the migratory T-cells. The Fas/FasL apoptotic pathway is thought to be involved in the disappearance of T-cells because fewer CD3 T-cells undergo apoptosis in gld mice (which have a *Fas-L* gene containing a point mutation resulting in a non-functional form of FasL). The upregulation of FasL is hypothesized to play a role in RABV pathogenesis because CVS-infected gld mice display lower rates of mortality, with death and symptoms occurring later, than do CVS-infected normal

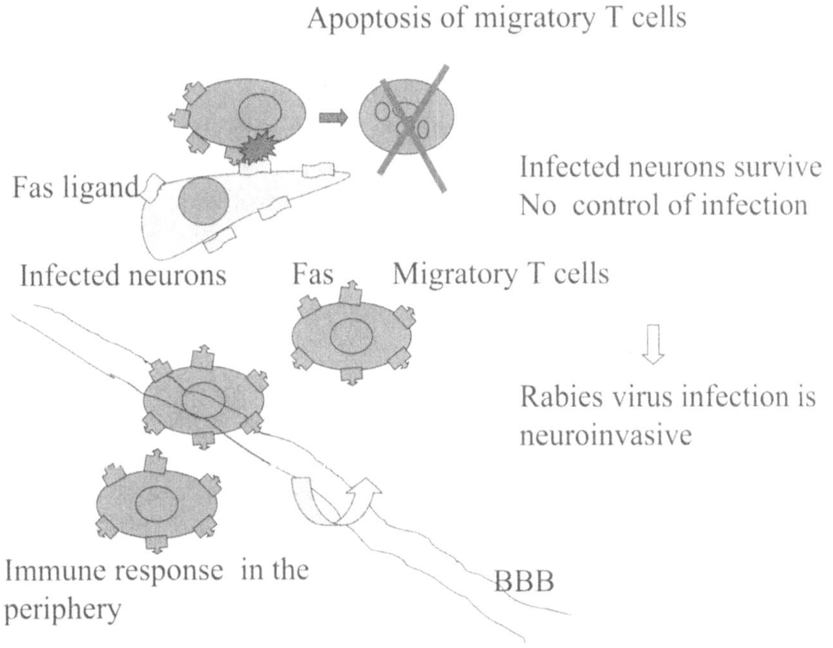

Fig. 6. Scheme of the mechanisms adopted by the neuroinvasive rabies virus strain (CVS) to subvert the host defences. Activated T-cells expressing Fas cross the blood brain barrier (BBB) attracted by cytokines, such as TNF-α, secreted locally in reaction to infection. Fas-L is up-regulated by infected neurons. T-cells undergo apoptosis after interaction of Fas with Fas-L. As a consequence, the virus subverts the host immune response and favors its propagation through the nervous system

mice. This suggests that the immunosubversive strategy of RABV involves the up-regulation of FasL production to reduce the protective immune response. Based on these results, we propose a model to account for the success of CVS infection (Fig. 6). Following injection of CVS into the hind limbs, progressive infection of the spinal cord and the brain is accompanied by production of inflammatory cytokines and chemokines which attract lymphocytes, resulting in their migration across the blood brain barrier. However, CVS upregulates FasL. The protective T-cells specifically triggered in the periphery are inefficient at controlling acute RABV encephalitis because they are destroyed by apoptosis shortly after they enter the CNS. By preserving neuron integrity, neurotropic RABV is non-cytolytic for the neurons it infects, and protection against the immune response facilitates propagation of neuroinvasive virus (CVS) through the CNS to the brain, resulting in the death of the mice.

In contrast, non-neuroinvasive PV infection did not upregulate FasL and kill migratory T-cells. As a consequence, PV-infected neurons were destroyed and propagation of the infection within the brain was interrupted. The mechanisms involved in T-cell mediated apoptosis, which interrupts the infection, are unknown. MHC class I molecules were found to be expressed in the nerve parenchyma during RABV infection [8, 11]. Thus it cannot be excluded that neuronal death is the result of direct killing by CD8+ T-cells. However, the T-cell mediated mechanisms involved not only infected neurons but non-infected cells as well, suggesting that bystander apoptosis occurred. The mechanism by which T-cells mediate injury could result of the pro-inflammatory cytokines and chemokines they secrete ([5]). IFN-γ is a potent activator of microglia, inducing up-regulation of MHC class II expression, phagocytosis and production of cytokines [9]. Cytokines may directly damage neural cells or act indirectly via activation of macrophages and microglial cells. These cells secrete factors toxic for oligodendrocytes, such as nitric oxide (NO) and TNF-α. NO has been found to be produced in the brain during RABV infection [6]. Destruction of oligodendrocytes, the cells that produce the protective sheath of myelin around neuronal axons, may participate in neuronal death by depriving them of protection.

Taken together, these observations suggest that RABV uses a subversive strategy based on the preservation of the neuronal network and the destruction of T-cells that invade the CNS in response to infection. This is achieved in part because RABV cannot induce apoptosis of the neuron it infects, and in part because it can trigger cell death in the migratory infiltrating "protective" T-cells. Moreover, RABV triggers the secretion of neuroprotective factors such as IL-6 and limits inflammation of the infected nervous system [2, 4]. It also upregulates the expression of Bcl-2 which is not only an antiapoptotic agent but may also favour virus progression through Bcl-2 action on neuron microtubule stabilization (Baloul and Lafon, in preparation).

Acknowledgments

This work was supported by institutional grants from the Institut Pasteur, Paris, France.

References

1. Alcami A, Koszinowski UH (2000) Viral mechanisms of immune evasion. Trends Microbiol 8: 410–418
2. Baloul L, Lafon M (2003) Apoptosis and rabies virus neuroinvasion. Biochimie 85: 777–788
3. Bellgrau D, Gold D, Selawry H, Moore J, Franzusoff A, Duke RC (1995) A role for CD95 ligand in preventing graft rejection. Nature 377: 630–632
4. Camelo S, Lafage M, Lafon M (2000) Absence of the p55 Kd TNF-alpha receptor promotes survival in rabies virus acute encephalitis. J Neurovirol 6: 507–518
5. Galelli A, Baloul L, Lafon M (2000) Abortive rabies virus central nervous infection is controlled by T lymphocyte local recruitment and induction of apoptosis. J Neurovirol 6: 359–372
6. Hooper DC, Ohnishi ST, Kean R, Numagami Y, Dietzschold B, Koprowski H (1995) Local nitric oxide production in viral and autoimmune diseases of the central nervous system. Proc Natl Acad Sci USA 92: 5312–5316
7. Hooper DC, Morimoto K, Bette M, Weihe E, Koprowski H, Dietzschold B (1998) Collaboration of antibody and inflammation in clearance of rabies virus from the central nervous system. J Virol 72: 3711–3719
8. Irwin DJ, Wunner WH, Ertl HC, Jackson AC (1999) Basis of rabies virus neurovirulence in mice: expression of major histocompatibility complex class I and class II mRNAs. J Neurovirol 5: 485–494
9. Merrill JE, Benveniste EN (1996) Cytokines in inflammatory brain lesions: helpful and harmful. Trends Neurosci 19: 331–338
10. Prehaud C, Lay S, Dietzschold B, Lafon M (2003) Glycoprotein of non-pathogneic viruses is a key determinant of human cell apoptosis. J Virol 77: 10537–10547
11. Prosniak M, Hooper DC, Dietzschold B, Koprowski H (2001) Effect of rabies virus infection on gene expression in mouse brain. Proc Natl Acad Sci USA 98: 2758–2763
12. Suda T, Nagata S (1994) Purification and characterization of the Fas-ligand that induces apoptosis. J Exp Med 179: 873–879
13. Thoulouze M-I, Lafage M, Yuste VJ, Baloul L, Edelman L, Kroemer G, Israel N, Lafon M (2003) High level of Bcl-2 counteracts apoptosis mediated by a live rabies virus vaccine strain and induces long-term infection. Virology 314: 549–561

Author's address: Dr. Monique Lafon, Unité de Neuroimmunologie Virale, Institut Pasteur, 25, rue du Dr. Roux, 75724 Paris Cedex 15, France; e-mail: mlafon@pasteur.fr

Arch Virol (2004) [Suppl] 18: 161–170

Archives of
Virology

Neurovirulence and host factors in flavivirus encephalitis – evidence from clinical epidemiology

T. Solomon[1,2] and **P. M. Winter**[1]

[1]Departments of Medical Microbiology and Neurology,
University of Liverpool, Liverpool, U.K.
[2]University of Texas Medical Branch, Galveston, Texas, U.S.A.

Summary. Japanese encephalitis virus (JEV) and West Nile virus (WNV) provide some of the most important examples of emerging zoonotic viral encephalitides. For these flaviviruses, only a small proportion of those infected develop clinical features, and these may range from a non-specific flu-like illness to a severe fatal meningoencephalitis, often with Parkinsonian features, or a poliomyelitis-like flaccid paralysis. The factors governing the clinical presentations, and outcome of flavivirus infections are poorly understood, but studies have looked at viral virulence determinants and the host immune response. Previous studies on JEV have suggested that the distribution of the four genotypes across Asia may relate to the differing clinical epidemiology (epidemic disease in the north, endemic disease in the south). However, new data based on the complete nucleotide sequence of a virus representing one of the oldest lineages, and phylogenetic analyses of all JEV strains for which genetic data are available, suggest that the distribution is best explained in terms of the virus' origin in the Indonesia–Malaysia region (where all genotypes have been found), and the spread of the more recent genotypes to new geographical areas. Clinical studies have shown that innate immunity, as manifested by interferon alpha levels, is important in JEV and other flaviviruses, but treatment with interferon alpha did not improve the outcome. A failure of the humoral immune response, is associated with death from encephalitis caused by JEV and WNV. Cellular immunity has been less well characterized, but CD8+ and CD4+ T cells are thought to be important.

Introduction

Mosquito-borne members of the genus *Flavivirus*, family *Flaviviridae*, provide some of the most important examples of emerging zoonotic viral encephalitides [8]. Much of the recent clinical and epidemiological data has focused on Japanese encephalitis virus (JEV), which is globally the most important of these viruses, and causes large outbreaks in southern and eastern Asia [43], and West Nile virus (WNV), which has caused recent large outbreaks in southern Europe, Russia, the Middle East and North America [36, 47]. Other members of this, the JEV complex, include St Louis encephalitis virus (SLEV), which previously was the most important flavivirus in America, and Murray Valley encephalitis virus (MVEV), which causes occasional outbreaks in Australia [8]. Flaviviruses consist of a single strand of positive sense RNA, wrapped in a nucleocapsid, surrounded by a lipid membrane containing an envelope glycoprotein, thought to be critical for entry into host cells. Members of the JEV complex are transmitted between vertebrate hosts, especially birds, by mosquitoes, principally of the *Culex* species. Humans become infected coincidentally when they encroach upon this enzootic cycle. For JEV, swine, which in Asia are often kept close to the home, are important bridging hosts bringing the virus close to humans. Most human infections with JEV are asymptomatic, or cause a mild non-specific febrile illness. When infection of the central nervous system (CNS) occurs, it may result in inflammation of the meninges (meningitis), the brain substance (encephalitis), the spinal cord (myelitis), or all three components – meningoencephalomyelitis. Beginning with

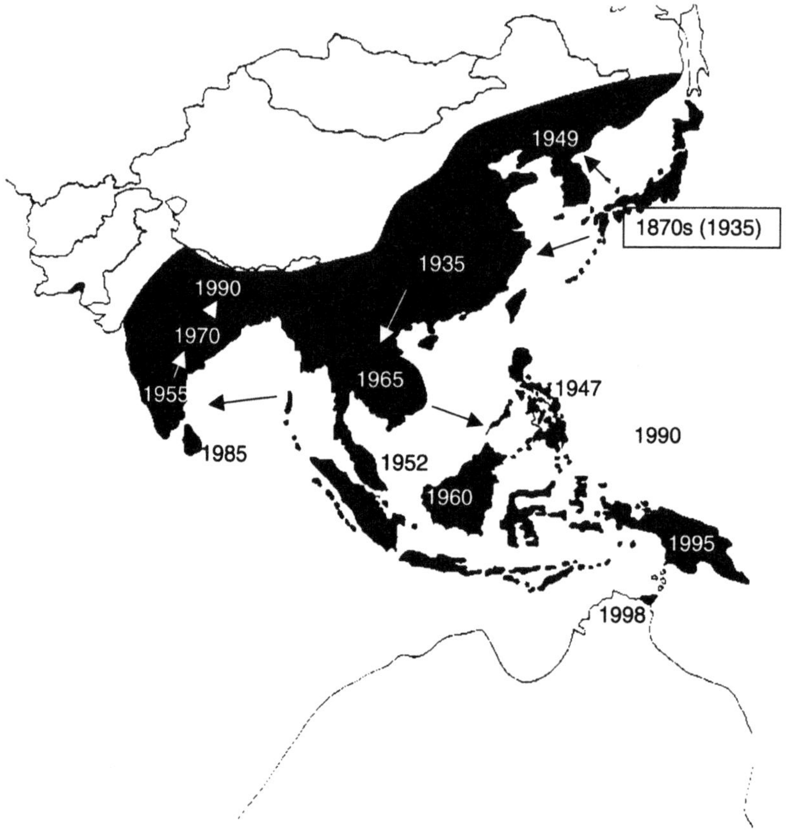

Fig. 1. Current distribution of Japanese encephalitis. The approximate dates of the first major outbreaks, or first virus isolations, since epidemics of encephalitis were first described in Japan in the 1870s, are shown [43]

recent studies on JEV, this review examines the clinical epidemiology of the mosquito-borne neurotropic flaviviruses, and then considers evidence for the relative contributions of strain virulence determinants and host factors.

Clinical epidemiology of JE

Epidemics of encephalitis were described in Japan from the 1870s onwards, and JEV was first isolated in the 1930s, and the virus has apparently spread across much of eastern and southern Asia since then (Fig. 1) [43]. Whereas in northern temperate climates JEV causes large summer epidemics, in southern tropical parts of Asia it causes year round disease (Fig. 2). In both regions, the majority of human JEV infections are asymptomatic [48]. Estimates of the ratio of symptomatic to asymptomatic infection range from about 1 in 25 (for American service personnel in Korea) [16] to about 1 in 1000 (for children in parts of China) [21]. In endemic areas most disease occurs in children, because by the time they reach adulthood nearly all the population has been exposed or been vaccinated. However, when the

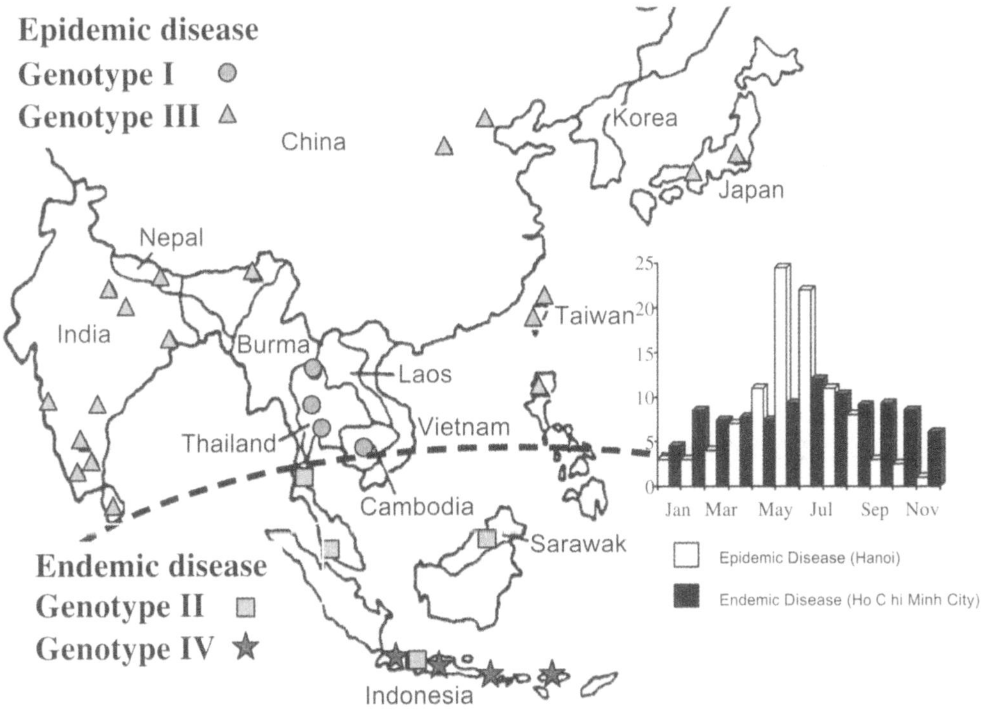

Fig. 2. Map showing the clinical epidemiology of JE in Asia, and the early postulated relationship with virus genotypes. In northern parts of Asia (above the dotted line) where JE occurs in large summer epidemics (typified by the monthly distribution of cases in Hanoi, northern Vietnam shown as a percentage of the total), genotypes I and III were found; in southern parts of Asia, where the disease occurs year round (as shown for Ho Chi Minh city in southern Viet Nam), genotypes II and IV are found [12, 13, 39]

virus spreads to new areas, adults there also develop disease, as was demonstrated when JEV reached Nepal and Australia [18, 49].

Clinicopathological correlates

Clinical features of JEV infection may range from a non-specific febrile illness to a severe meningoencephalitis. In addition, the virus may cause an aseptic meningitis, or a recently described poliomyelitis-like acute flaccid paralysis [38]. Neurophysiological studies suggest that JEV causes flaccid paralysis by invading the lower motor neurons in the anterior horns of the spinal cord; pathological observations of fatal cases support this [19, 31]. A Parkinson's disease-like syndrome, in which patients have mask-like facies, tremors and cogwheel rigidity is common in Japanese encephalitis [30, 33]. Other movement disorders include choreoathetosis (an uncontrollable writhing of the limbs), oro-facial dyskinesias (e.g. lip-smacking and chewing movements), and opisthotonus (hyperextension of the back and neck due to increased tone). In one recent study, one third of patients had movement disorders acutely or during convalescence [40]. Radiological and pathological studies suggest the movement disorders are the clinical correlate of damage of the basal ganglia and thalamus [29], the areas also affected in Parkinson's disease.

The case-fatality rate for Japanese encephalitis is about 20–30%, but approximately half of the survivors have severe neuropsychiatric sequelae, imposing a large socioeconomic burden in poor rural Asian communities that can ill afford it. There is no antiviral treatment for Japanese encephalitis, but pathophysiological studies have suggested that seizures, particularly uncontrolled seizures, and raised intracranial pressure may contribute to the poor outcome, and may be treatable [32, 40, 42].

Clinical epidemiology of other mosquito-borne zoonotic flaviviruses

The epidemiology of other mosquito-borne JEV complex flaviviruses is broadly similar to that of JEV. The majority of infections are asymptomatic, but severe CNS disease, most commonly encephalitis, can occur. Flaccid weakness, suggesting damage to the anterior horn cells of the spinal cord, is seen in comatose patients with encephalitis caused by SLEV, MVEV and WNV [1, 4, 9]. In addition a

Fig. 3. Current beliefs on the origin and distribution of JE virus. Top, Neighbor joining phylogenetic tree of envelope protein genes, showing the genetic relationship between the different genotypes of virus – genotypes IV and V are thought to represent the oldest lineages. Indonesian strains are boxed, and genotypes are shown on the right,; (see ref [46] for more details). Bottom, map showing the postulated origin of JE virus from a JE-serogroup ancestor in the Indonesia–Malaysia region. The ancestral JE strain is then believed to have evolved here into the different genotypes of virus, all of which have been found in this region. Only the more recent genotypes have subsequently spread to new geographical areas

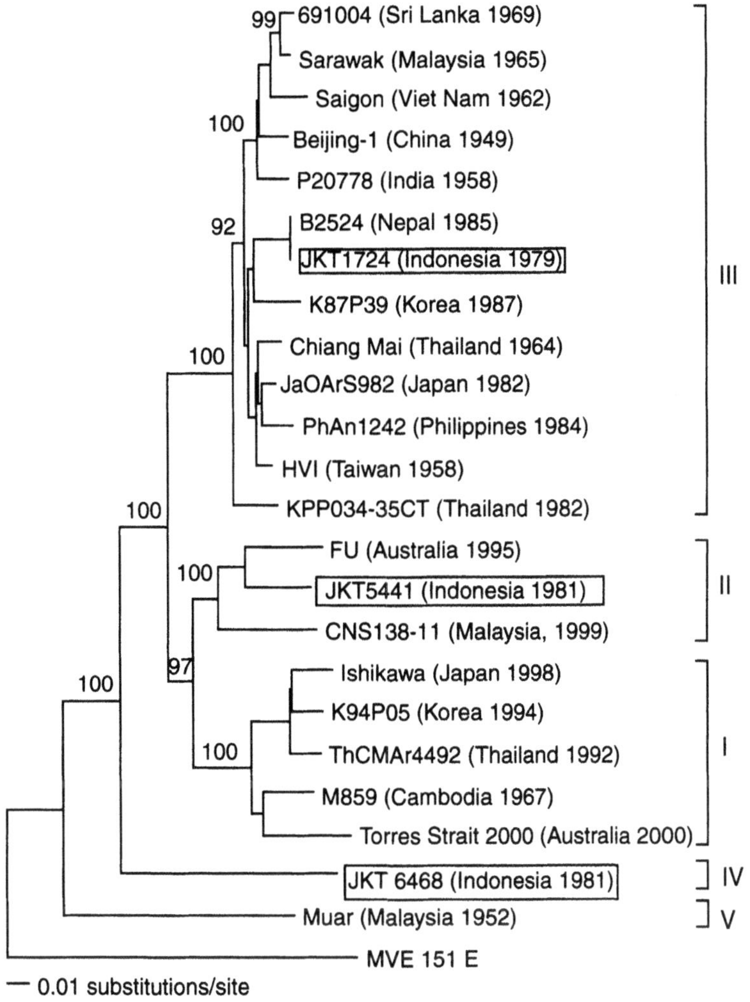
99
691004 (Sri Lanka 1969)
Sarawak (Malaysia 1965)
Saigon (Viet Nam 1962)
100
Beijing-1 (China 1949)
P20778 (India 1958)
92
B2524 (Nepal 1985)
JKT1724 (Indonesia 1979)
K87P39 (Korea 1987)
100
Chiang Mai (Thailand 1964)
JaOArS982 (Japan 1982)
PhAn1242 (Philippines 1984)
HVI (Taiwan 1958)
100
KPP034-35CT (Thailand 1982)
FU (Australia 1995)
100
JKT5441 (Indonesia 1981)
CNS138-11 (Malaysia, 1999)
97
100
Ishikawa (Japan 1998)
K94P05 (Korea 1994)
100
ThCMAr4492 (Thailand 1992)
M859 (Cambodia 1967)
Torres Strait 2000 (Australia 2000)
JKT 6468 (Indonesia 1981)
Muar (Malaysia 1952)
MVE 151 E
III
II
I
IV
V
0.01 substitutions/site

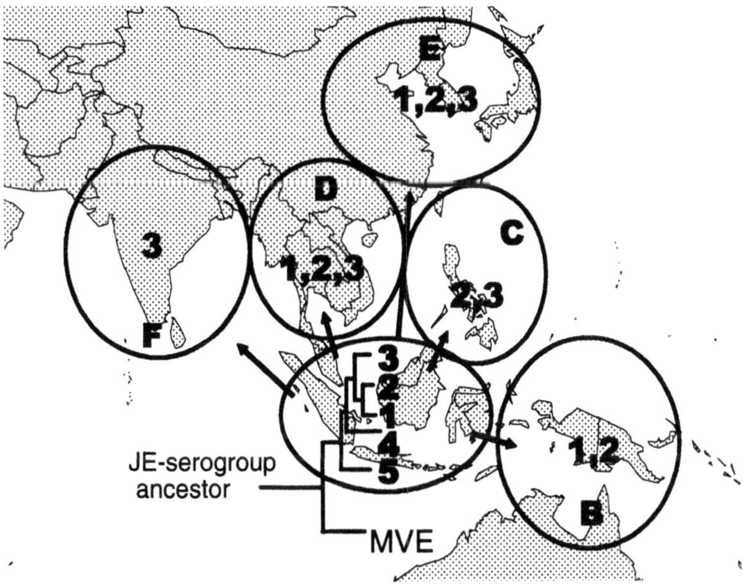
E
1,2,3
D
1,2,3
3
F
C
2,3
3
2
1
4
5
1,2
B
JE-serogroup
ancestor
MVE

poliomyelitis-like acute flaccid paralysis presentation is also being recognised increasingly in fully conscious patients infected with WNV [15, 26]. Parkinsonian features, and other movement disorders associated with basal ganglia damage are seen in infections caused by MVEV and SLEV, and are have also been described in WNEV [2, 11, 45].

Pathogenesis of flavivirus encephalitis

The factors governing who, of those infected with neurotropic flaviviruses, develops neurological disease, and the pattern of clinical features and outcome are poorly understood. In animal models of flavivirus encephalitis, strain virulence determinants and the host immune response both have a role, but how important these are in human disease is not yet understood, but being investigated.

Virulence determinants and human disease

Studies in the early 1990s indicated that there are at least four genotypes of JEV, distinguished by at least 10% divergence at the nucleotide level [12, 13]. The observation that genotypes I and III were found in northern areas of Asia, where the disease is epidemic, whereas genotypes II and IV circulate in the endemic southern areas, led to the suggestion that differences in the genotype distribution might explain the differing clinical epidemiology and thus support the hypothesis of a role for different genotypic virulence. However recent observations suggest that this paradigm may not be correct: for example in Australia a "northern" genotype I isolate was isolated in 2000 [37], and in Viet Nam the same genotype (genotype III) has been found to circulate in northern epidemic and the southern endemic areas [22, 39]. This question was addressed recently with the determination of the complete nucleotide sequence of a genotype IV strain (the only genotype that had not been fully sequenced), and examination of the distribution of all known genotypes [41]. The data suggested that the distribution of JEV is best explained in terms of the virus' origin in the Indonesia–Malaysia region (where all genotypes have been found), and the spread of the more recent genotypes to new geographical areas (Fig. 3). Why the older genotypes have not spread is unclear, but a critical change in the envelope protein that might affect vector or host susceptibility was suggested as one possibility. This analysis of genotypes does not preclude the possibility that more subtle strain differences may be important in human virulence, and more work is needed to examine this question.

Similar questions have been raised regarding the clinical epidemiology of WNV. This virus tends to cause a febrile illness in Africa, but its spread to Europe and America has been associated with large outbreaks of CNS disease. Two lineages of WNV (broadly equivalent to the genotypes of JEV) have been identified by phylogenetic studies. And the fact that only lineage I strains have been isolated during the large CNS outbreaks led to the suggestion that this lineage is more virulent than lineage II [25]. However any explanation linking lineage to virulence must take account of the fact that lineage I strains also circulate in Africa, where large outbreaks of CNS disease have not been observed [10].

The host immune response in flavivirus encephalitis

A variety of studies indicate that the host immune response may be critical in determining the outcome of human flavivirus infection. Following the bite of an infected mosquito, virus is thought to replicate peripherally, causing a transient viraemia before crossing the blood brain barrier. The means of crossing into the CNS is uncertain, though pathological studies indicating wide dissemination through the brain parenchyma suggest passage across the cerebrovascular endothelium is more likely than via the olfactory nerves (where the blood brain barrier is scant) [23, 34]. The human immune response may control this process at various stages. A role for innate immunity in JEV infection is suggested by the elevated interferon alpha levels found in plasma and CSF [7]. However, a recent clinical trial showed treatment with interferon alfa-2a did not improve the outcome [44]. The humoral immune response to JEV, and to WNV infection is characterized by early production of IgM antibodies in both serum and CSF, followed by production of IgG [6, 28]. A failure of antibody production is associated with an increased likelihood of virus isolation, and of a fatal outcome in Japanese encephalitis [5], and probably also in encephalitis caused by WNV [20, 45]. Patients on immunosuppressive drugs were especially vulnerable to severe disease caused by WNV in recent outbreaks in North America. In areas where more than one flavivirus co-circulates, prior infection with a different flavivirus may afford some protection, presumably through cross reacting neutralizing antibodies. Thus, in Thailand Japanese encephalitis patients whose antibody pattern suggests prior dengue infection, are less likely to die [27]. Similarly in an outbreak of SLEV infection in Florida, prior exposure to dengue viruses was thought to explain why older patients were relatively spared [3]. The cellular immune response to flaviviruses has been less well characterized. However, histological studies of fatal cases show perivascular cuffing with infiltration of macrophages and T cells, some of which were CD8+ cells [23]. In non-fatal cases, JEV-specific T cell proliferation of peripheral T lymphocytes (including CD4+ and CD8+ responses) have been demonstrated during convalescence [24]. Non-immunological host factors that disrupt the blood brain barrier, may also be important in determining outcome. For example, co-infection with cysticercosis (caused by the parasite *Taenia solium*) appears to be disproportionately common in fatal Japanese encephalitis cases [14]. Similarly, the increased severity of WNV infection in the elderly may be due to subclinical cerebrovascular disease [17, 35].

Concluding comments

Encephalitides caused by zoonotic flaviviruses are spreading globally, and becoming increasingly important. Although vaccines are available for some of these viruses, the majority of the population in endemic areas remain at risk. A better understanding of the viral and host factors may be critical to developing appropriate therapy for these conditions.

Acknowledgements

Some of the work described was funded by the Wellcome Trust of Great Britain. TS is wellcome Trust Career Development Fellow.

References

1. Asnis DS, Conetta R, Teixeira AA, Waldman G, Sampson BA (2000) The West Nile virus outbreak of 1999 in New York: the Flushing Hospital experience. Clin Infect Dis 30: 413–418
2. Bennet NM (1976) Murray Valley encephalitis, 1974. Clinical features. Med J Aust 2: 446–454
3. Bond JO, Hammon WM (1970) Epidemiological studies of possible cross protection between dengue and St Louis encephalitis arboviruses in Florida. Am J Epidemiol 92: 321–329
4. Brinker KR, Monath TP (1980) The Acute Disease. In: Monath TP (ed) St Louis encephalitis. American Public Health Associates, Washington DC, pp 503–534
5. Burke DS, Lorsomrudee W, Leake CJ, Hoke CH, Nisalak A, Chongswasdi V, Laorakpongse T (1985) Fatal outcome in Japanese encephalitis. Am J Trop Med Hyg 34: 1203–1210
6. Burke DS, Nisalak A, Ussery MA, Laorakpongse T, Chantavibul S (1985) Kinetics of IgM and IgG responses to Japanese encephalitis virus in human serum and cerebrospinal fluid. J Infect Dis 151: 1093–1099
7. Burke DS, Morill JC (1987) Levels of interferon in the plasma and cerebrospinal fluid of patients with acute Japanese encephalitis. J Infect Dis 155: 797–799
8. Burke DS, Monath TP (2001) Flaviviruses. In: Knipe DM, Howley PM (eds) Fields Virology, vol 1. Lippincott Williams & Wilkins, Philadelphia, USA, pp 1043–1126
9. Burrow JN, Whelan PI, Kilburn CJ, Fisher DA, Currie BJ, Smith DW (1998) Australian encephalitis in the Northern Territory: clinical and epidemiological features, 1987–1996. Aust N Z J Med 28: 590–596
10. Burt FJ, Grobbelaar AA, Leman PA, Anthony FS, Gibson GV, Swanepoel R (2002) Phylogenetic relationships of southern African West Nile virus isolates. Emerg Infect Dis 8: 820–826
11. Cerna F, Mehrad B, Luby JP, Burns D, Fleckenstein JL (1999) St. Louis encephalitis and the substantia nigra: MR imaging evaluation. AJNR Am J Neuroradiol 20: 1281–1283
12. Chen WR, Tesh RB, Rico-Hesse R (1990) Genetic variation of Japanese encephalitis virus in nature. J Gen Virol 71: 2915–2922
13. Chen WR, Rico-Hesse R, Tesh RB (1992) A new genotype of Japanese virus from Indonesia. Am J Trop Med Hyg 47: 61–69
14. Desai A, Shankar SK, Jayakumar PN, Chandramuki A, Gourie-Devi M, Ravikumar BV, Ravi V (1997) Co-existence of cerebral cysticercosis with Japanese encephalitis: a prognostic modulator. Epidemiol Infect 118: 165–171
15. Gadoth N, Weitzman S, Lehmann EE (1979) Acute anterior myelitis complicating West Nile fever. Arch Neurol 36: 172–173
16. Halstead SB, Grosz CR (1962) Subclinical Japanese encephalitis. I. Infection of Americans with limited residence in Korea. Am J Hyg 75: 190–201

17. Han LL, Popovici F, Alexander Jr J, Laurentia V, Tengelsen LA, Cernescu C, Gary Jr HE, Ion-Nedelcu N, Campbell GL, Tsai TF (1999) Risk factors for West Nile virus infection and meningoencephalitis, Romania, 1996. J Infect Dis 179: 230–233
18. Hanna J, Ritchie S, Phillips DA, Shield J, Bailey MC, Mackenzie JS, Poidinger M, McCall BJ, Mills PJ (1996) An outbreak of Japanese encephalitis in the Torres Trait, Australia, 1995. Med J Aust 165: 256–260
19. Haymaker W, Sabin AB (1947) Topographic distribution of lesions in central nervous system in Japanese B encephalitis. Nature of the lesions with report of a case on Okinawa. Arch Neurol Psychiatry 57: 673–692
20. Huang C, Slater B, Rudd R, Parchuri N, Hull R, Dupuis M, Hindenburg A (2002) First isolation of *West Nile virus* from a patient with encephalitis in the United States. Emerg Infect Dis [serial online] 8: 11/21/02
21. Huang CH (1982) Studies of Japanese encephalitis in China. Adv Virus Res 27: 71–101
22. Huong VTQ, Ha DQ, Deubel V (1993) Genetic study of Japanese encephalitis viruses from Vietnam. Am J Trop Med Hyg 49: 538–544
23. Johnson RT, Burke DS, Elwell M, Leake CJ, Nisalak A, Hoke CH, Lorsomrudee W (1985) Japanese encephalitis: immunocytochemical studies of viral antigen and inflammatory cells in fatal cases. Ann Neurol 18: 567–573
24. Konishi E, Mason PW, Innis BI, Ennis FA (1995) Japanese encephalitis virus-specific proliferative responses of human peripheral blood T lymphocytes. Am J Trop Med Hyg 53: 278–283
25. Lanciotti RS, Ebel GD, Deubel V, Kerst AJ, Murri S, Meyer R, Bowen M, McKinney N, Morrill WE, Crabtree MB, Kramer LD, Roehrig JT (2002) Complete genome sequences and phylogenetic analysis of West Nile virus strains isolated from the United States, Europe, and the Middle East. Virology 298: 96–105
26. Leis AA, Stokic DS, Polk JL, Dostrow V, Winkelmann M (2002) A poliomyelitis-like syndrome from West Nile virus infection. N Engl J Med 347: 1279–1280
27. Libraty DH, Nisalak A, Endy TP, Suntayakown S, Vaughn DW, Innis BL (2000) Risk factors for severe disease in Japanese encephalitis. In: 49th Annual Meeting of the American Society of Tropical Medicine and Hygiene, Houston, Texas, pp 228–229
28. Martin DA, Biggerstaff BJ, Allen B, Johnson AJ, Lanciotti RS, Roehrig JT (2002) Use of immunoglobulin m cross-reactions in differential diagnosis of human flaviviral encephalitis infections in the United States. Clin Diagn Lab Immunol 9: 544–549
29. Misra UK, Kalita J, Jain SK, Mathur A (1994) Radiological and neurophysiological changes in Japanese encephalitis. J Neurol Neurosurg Psych 57: 1484–1487
30. Misra UK, Kalita J (1997) Movement disorders in Japanese encephalitis. J Neurol 244: 299–303
31. Misra UK, Kalita J (1997) Anterior horn cells are also involved in Japanese encephalitis. Acta Neurol Scand 96: 114–117
32. Misra UK, Kalita J (2001) Seizures in Japanese encephalitis. J Neurol Sci 190: 57–60
33. Misra UK, Kalita J (2002) Prognosis of Japanese encephalitis patients with dystonia compared to those with parkinsonian features only. Postgrad Med J 78: 238–241
34. Monath TP, Cropp CP, Harrison AK (1983) Mode of entry of a neurotropic virus into the central nervous system. Reinvestigation of an old controversy. Lab Invest 48: 399–410
35. Nash D, Mostashari F, Fine A, Miller J, O'Leary D, Murray K, Huang A, Rosenberg A, Greenberg A, Sherman M, Wong S, Layton M (2001) The outbreak of West Nile virus infection in the New York City area in 1999. N Engl J Med 344: 1807–1814
36. Petersen LR, Roehrig JT, Hughes JM (2002) West Nile virus encephalitis. N Engl J Med 347: 1225–1226

37. Pyke AT, Williams DT, Nisbet DJ, van den Hurk AF, Taylor CT, Johansen CA, Macdonald J, Hall RA, Simmons RJ, Mason RJ, Lee JM, Ritchie SA, Smith GA, Mackenzie JS (2001) The appearance of a second genotype of Japanese encephalitis virus in the Australasian region. Am J Trop Med Hyg 65: 747–753
38. Solomon T, Kneen R, Dung NM, Khanh VC, Thuy TTN, Ha DQ, Day NPJ, Nisalak A, Vaughn DW, White NJ (1998) Poliomyelitis-like illness due to Japanese encephalitis virus. Lancet 351: 1094–1097
39. Solomon T, Dung NM, Kneen R, Gainsborough M, Vaughn DW, Khanh VT (2000) Japanese encephalitis. J Neurol Neurosurg Psych 68: 405–415
40. Solomon T, Dung NM, Kneen R, Thao LT, Gainsborough M, Nisalak A, Day NP, Kirkham FJ, Vaughn DW, Smith S, White NJ (2002) Seizures and raised intracranial pressure in Vietnamese patients with Japanese encephalitis. Brain 125: 1084–1093
41. Solomon T, Ni H, Beasley DW, Ekkelenkamp M, Cardosa MJ, Barrett AD (2002) The origin and evolution of Japanese encephalitis virus: new evidence from the fourth genotype. ICID 2002
42. Solomon T, Vaughn DW (2002) Pathogenesis and clinical features of Japanese encephalitis and West Nile virus infections. In: Mackenzie JS, Barrett AD, Deubel V (eds) Japanese encephalitis and West Nile virus infections. Current topics in microbiology and immunology. vol 267. Springer-Verlag, pp 171–194
43. Solomon T (2003) Recent advances in Japanese encephalitis. J Neurovirol 9: 1–10
44. Solomon T, Dung NM, Wills B, Kneen R, Gainsborough M, Diet TV, Thuy TTN, Loan HT, Khanh VC, Vaughn DW, White NJ, Farrar JJ (2003) Interferon alfa-2a in Japanese encephalitis: a randomised double-blind placebo-controlled trial. Lancet 361: 821–826
45. Solomon T, Fisher AF, Beasley DWC, Mandava P, Granwehr BP, Langsjoen H, Travassos Da Rosa AP, Barrett ADT, Tesh RB (2003) Natural and nosocomial infection in a patient with West Nile encephalitis and extrapyramidal movement disorders. Clin Infect Dis 36: E140–145
46. Solomon T, Ni H, Beasley DW, Ekkelenkamp M, Cardosa MJ, Barrett AD (2003) Origin and Evolution of Japanese Encephalitis Virus in Southeast Asia. J Virol 77: 3091–3098
47. Tsai TF, Popovici F, Carnescu C, Campbell GL, Nedelcu NI (1998) West Nile encephalitis epidemic in Southeastern Romania. Lancet 352: 767–771
48. Vaughn DW, Hoke CH (1992) The epidemiology of Japanese encephalitis: prospects for prevention. Epidemiol Rev 14: 197–221
49. Zimmerman MD, Scott RM, Vaughn DW, Rajbhandari S, Nisalak A, Shrestha MP (1997) Short report: an outbreak of Japanese encephalitis in Kathmandu, Nepal. Am J Trop Med Hyg 57: 283–284

Author's address: Dr. Tom Solomon, Department of Medical Microbiology, University of Liverpool, Duncan Building, Daulby Street, Liverpool, L69 3GA, U.K.; e-mail: tsolomon@liv.ac.uk

Arch Virol (2004) [Suppl] 18: 171–178

Archives of Virology

Regulation of apoptosis by viruses that infect insects

P. M. Irusta[1], **E. Lamos**[1], **H. L. Galonek**[2], **M. A. Vander Maten**[3], **M. C. H. Boersma**[1], **Y.-B. Chen**[4], and **J. M. Hardwick**[1,2,4]

[1]W. Harry Feinstone Department of Molecular Microbiology and Immunology, Johns Hopkins School of Public Health, Baltimore, Maryland, U.S.A.
[2]Department of Neurology, Johns Hopkins School of Medicine, Baltimore, Maryland, U.S.A.
[3]Northern Virginia Community College, Annandale, Virginia, U.S.A.
[4]Department of Pharmacology and Molecular Sciences, Johns Hopkins School of Medicine, Baltimore, Maryland, U.S.A.

Summary. Orthobunyviruses and alphaviruses cause encephalitis, neuronal apoptosis and mortality in mammals, but fail to kill the mosquitoes that transmit these viruses. Therefore, host cell factors, as well as viral factors, regulate the outcome of infection. *Drosophila* Reaper is a pro-death factor in the *Drosophila* nervous system during development but homologues were not previously known to exist outside flies. Recent discovery of a Reaper protein encoded by orthobunyaviruses provides interesting insight into the mechanisms by which viruses modulate the death pathway in both vertebrate and invertebrate hosts.

Introduction

Members of diverse families of viruses infect and replicate in insects. Whereas some of these viruses cause disease by inducing apoptotic cell death, others must inhibit apoptosis to be pathogenic for their host organism. For example, the ability of large DNA baculoviruses to be pathogenic for caterpillars requires that the virus inhibit programmed cell death that is triggered in the insect host upon infection. In contrast, the plus-strand RNA Sindbis virus (family *Togaviridae*, genus *Alphavirus*) is highly pathogenic in newborn laboratory mice because it induces neuronal apoptosis. However, these same pro-apoptotic mechanisms apparently are suppressed in the mosquitoes that transmit the virus, as Sindbis virus is largely nonpathogenic for mosquitoes. Similarly, the mosquito-borne negative-strand RNA orthobunyaviruses (family *Bunyaviridae*, genus *Orthobunyavirus*) cause neuronal apoptosis and can be highly pathogenic for humans. Exploring the molecular mechanisms by which viruses control host cell death draws attention to the similarities between programmed cell death pathways in mammals and insects.

Mitochondria are central players in regulating cell death

Mitochondria are essential organelles for cell survival, although many of their functions and even their genomes are superfluous under some circumstances [18]. Thus, mitochondria are conceivably the logical target to facilitate cell death. Furthermore mitochondria contribute to the cell death pathway by releasing specific factors into the cytosol of cells destined to die [19]. Although all multi-cellular, and perhaps unicellular, organisms are capable of undergoing programmed cell death, convincing evidence that mitochondria actively participate in programmed cell death was limited to mammals, but recently extended the nematode *C. elegans* [36]. The most widely studied mitochondrial factor known to be involved in cell death is, ironically, cytochrome *c*, previously known only for its role in the electron transport chain that ultimately generates ATP via oxidative phosphorylation. Cytochrome *c* is encoded by a nuclear, not mitochondrial, gene and is imported into mitochondria where it is assembled into its heme-containing form, holocytochrome *c*. It remains associated with the inner membrane in healthy cells but, following a death stimulus, is translocated from the intermembrane space into the cytosol by a yet unknown mechanism involving the Bcl-2 family proteins Bax and Bak. Cytochrome *c* takes on a new function when released into the cytosol, where it promotes cell death by serving as a cofactor for apoptosome assembly and caspase activation [19]. Thus, cytochrome *c*, which normally carries electrons between complex III and complex IV in the respiratory chain on the inner mitochondrial membrane, becomes a cofactor for caspase-9 activation in the cytosol [22]. This cytosolic function of cytochrome *c* may turn out to be unique to mammals, but there are many other death-inducing factors that are released from mitochondria during cell death. AIF (apoptosis inducing factor) and the endonuclease endoG are released from mitochondria and contribute to cell death in both mammals and *C. elegans* [36]. However, the mechanisms by which these factors move through the outer mitochondrial membrane into the cytosol are unknown. In addition to changes in mitochondrial permeability, changes in mitochondrial morphology and decreased mitochondrial membrane potential are also associated with programmed cell death, although the order of

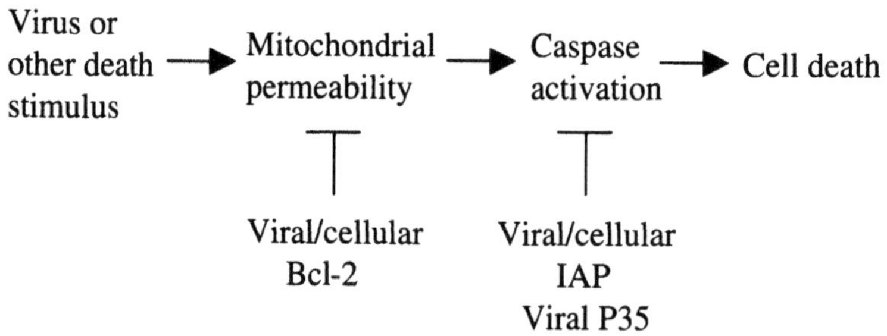

Fig. 1. Viruses interfere with mitochondria-dependent and caspase-dependent programmed cell death. Both cells and viruses encode Bcl-2 (B cell lymphoma/leukemia number 2) proteins and IAP (inhibitor of apoptosis) proteins, but cellular homologues of the baculovirus caspase inhibitor P35 are not obvious

events and the significance of their individual contributions are still debated [25]. Cellular Bcl-2 family proteins are key regulators of mitochondrial permeability and probably other architectural changes associated with programmed cell death, but their biochemical mechanisms of action are not understood [1]. While many mammalian viruses encode Bcl-2 homologues that potently inhibit apoptosis [10, 15], there are a variety of other viral factors that regulate cell death and act on mitochondria or that regulate caspase activation [5] (Fig. 1).

Viral inhibitors of caspases

Caspases are intracellular cysteine proteases that are activated following a cell death stimulus and that facilitate apoptotic cell death by cleaving key cellular proteins [33]. Although caspases may be unique to metazoans (multicellular organisms), a related yeast metacaspase has also been implicated in promoting cell death, suggesting that these mechanisms may be ancient in origin [23]. While caspases undoubtedly contribute to the death of a cell by dismantling the components of a cell, arguably their most important contribution is to dispose of cell corpses. That is, caspases facilitate disposal of the dying cell by packaging the cell corpse, such that it is recognized by neighboring cells or professional phagocytes that will engulf and degrade the corpse [14]. Undoubtedly there are other ways for cells to commit suicide without caspases, but these other pathways are less well characterized.

Baculovirus AcMNPV of moth caterpillars encodes a caspase inhibitor P35 that is capable of inhibiting a number of mammalian caspases. Baculovirus P35 is required for efficient late virus gene expression and progeny virus production [6, 7]. Infected caterpillars literally melt away as the baculovirus "consumes" its host. However, P35-deficient viruses trigger rapid apoptotic cell death in virus-infected caterpillar, cells thereby severely impairing production of progeny virus. This reduced replication by mutant viruses prevents spread of the virus and the caterpillar is resistant to disease [26]. This is a clear example of a viral anti-apoptotic strategy that significantly alters/promotes viral pathogenesis. There is little doubt that equivalent roles for virus-induced apoptosis will be uncovered in human disease, but the processes are much more complex. The three-dimensional structure of P35 in a complex with mammalian caspase-8 reveals a covalent thioester linkage between P35 and the caspase active site [38]. The activity and specificity of P35 indicates that P35 more potently inhibits the downstream "effector" caspases rather than the upstream caspases that "initiate" the caspase cascade.

A second type of anti-death protein identified in baculoviruses was dubbed the inhibitor of apoptosis protein (IAP). The viral gene encoding IAP was identified in a screen for viral genes that could functionally substitute for P35. Of the three known baculovirus IAP proteins (Cp-IAP, Op-IAP, Ac-IAP) encoded by three different viruses, only two of them (Cp-IAP, Op-IAP) have anti-apoptotic activity. Interestingly, these anti-apoptotic IAPs are encoded by viruses that lack a gene for P35. Mutagenesis studies indicate that the N-terminal BIRs (baculovirus

IAP repeats) and the C-terminal RING finger domains of baculovirus IAPs are important for their anti-apoptotic functions [7].

Shortly after the identification of baculovirus IAPs, cellular IAP proteins were identified in the genomes of *Drosophila*, nematodes, yeast and humans, as defined by the presence of one-to-three BIR motifs each with a single zinc finger [32]. Humans are now known to encode eight IAP-related proteins: c-IAP1, c-IAP2, XIAP, ILP-2, ML-IAP, NAIP, Survivin and Apollon (the human homologue of Bruce = BIR repeat ubiquitin conjugating enzyme), but these do not all possess the ability to inhibit apoptosis but have other functions [11, 27]. Of these, XIAP is the strongest caspase inhibitor supported by structure determination in which XIAP was found to bind in the active site of caspases. However, it is clear that the E3 ubiquitin ligase activity of the C-terminal RING domain also plays a critical role in regulating cell death and probably other cellular functions.

Sindbis virus induces caspase-mediated apoptosis in a variety of cell types in the laboratory [30]. While XIAP inhibits Sindbis virus-induced apoptosis, Sindbis virus induces caspase cleavage of c-IAP1 resulting in release of a C-terminal fragment of c-IAP1 that is potently pro-apoptotic [8, 12]. Thus, the relative levels of XIAP and c-IAP1 may be important determinants of the outcome of a Sindbis virus infection.

Viral and cellular reaper proteins

Caspases are held in check by IAP proteins during developmental cell death in *Drosophila*, as suppression of these proteins results in cell death of many if not all cells [31]. Similar mechanisms may operate in mosquitoes, but the function of mosquito IAP homologues is only beginning to be delineated [3]. To successfully activate caspases in *Drosophila* cells during development, it is necessary to inactivate the IAPs via the concerted effort of the death-inducing proteins Reaper, Hid and Grim [24, 37]. Reaper is a 65 amino acid protein that directly binds to *Drosophila* IAP (DIAP) and inhibits its function in part by promoting autoubiquitination and degradation of DIAP [24, 31]. In an in vitro cell extract system, *Drosophila* Reaper also induces release of cytochrome *c* from mitochondria [35] and, curiously, also inhibits cellular translation [17, 29]. Mammalian homologues of Reaper have not yet been reported, but the orthobunyavirus NSs proteins share amino acid sequence similarity with fly Reaper [9]. These Reaper-like viral factors are non-structural proteins encoded by the smallest viral genome segment (NSs). The C-terminal ~50 amino acids of vReaper share sequence similarity with the C-terminal region of *Drosophila* Reaper. However, the 14 amino acid N-terminal RHG motif that is conserved among the *Drosophila* cell death regulators Reaper, Hid, Grim, Jafrac2 and Sickle, and with mammalian proteins SMAC/Diablo and Omi/HtrA2 are lacking in the vReaper proteins. This RHG motif is required for direct binding to IAP proteins and is important but not essential for the pro-death function of *Drosophila* Reaper. Even without this RHG motif, vReaper proteins apparently possess all of the functions associated with the C-terminal domain of their *Drosophila* counterpart [31].

Like *Drosophila* Reaper, the San Angelo orthobunyavirus vReaper protein (SA-vRpr) efficiently inhibits protein translation in cell extracts prepared from rabbit reticulocytes and from *Xenopus* eggs [17]. Expression of SA-vRpr in mammalian cells is difficult to achieve with conventional promoters (perhaps because SA-vRpr inhibits its own translation), but with the aid of a viral IRES element, expression of SA-vRpr results in severe global inhibition of host protein synthesis. The vReaper proteins appear to be important for efficient orthobunyavirus replication and for full virulence in mice, as an engineered mutant Bunyamwera virus lacking vReaper is impaired in these functions [4].

Again like *Drosophila* Reaper, vReaper accelerates cytochrome *c* release from mitochondria in an in vitro system employing *Xenopus* egg extracts. Release of cytochrome *c* in this assay results in caspase activation and nuclear fragmentation [13]. A role for endogenous cytochrome *c* in *Drosophila* cell death was recently reported [2]. The mechanism by which *Drosophila* Reaper promotes release of cytochrome *c* requires its ability to bind Scythe. Scythe normally suppresses the protein folding function of Hsc/Hsp70, and inhibition of this folding function by *Drosophila* Reaper is thought to contribute to cell death [34, 35]. The orthobunyavirus vReaper also binds efficiently to Scythe and is as effective as *Drosophila* Reaper at reversing the inhibition of protein folding by Scythe [9]. Thus, vReaper and *Drosophila* Reaper appear to induce mitochondrial permeability and subsequent caspase activation by the same Scythe-regulated mechanism. In addition, both viral and *Drosophila* Reaper proteins conceivably have the potential to promote death by inhibiting the synthesis of short-lived anti-apoptotic factors like IAPs.

Orthobunyaviruses are transmitted by mosquitoes in nature. One of these viruses, La Crosse virus, is the major cause of viral pediatric encephalitis in North America. In mammals, orthobunyaviruses can infect neurons in the brain. To assess the function of vReaper in neurons in an animal without the contribution of other orthobunyavirus factors, vReaper from San Angelo orthobunyavirus was expressed via the neuronotropic Sindbis alphavirus vector [16].

This Sindbis virus vector (dsTE12Q) can induce neuronal apoptosis and mortality in newborn mice but fails to kill mice by the time they reach one week of age [16, 21]. In 1-week-old and older mice, Sindbis virus usually establishes a persistent infection that can be mimicked in cultures of mature primary neurons [20] (Fig. 2). However, Sindbis virus encoding SA-vRpr causes 100% mortality in 2-week-old mice, while control viruses caused 0% mortality [9]. Therefore, the orthobunyavirus vReaper significantly enhanced the pathogenesis of Sindbis virus in mice, consistent with the potent virulence of orthobunyavirus infections of humans. Analysis of mice infected with recombinant Sindbis virus encoding SA-vRpr revealed an abundance of apoptotic neurons detected by TUNEL staining. In contrast, the *Drosophila* Reaper had little effect on Sindbis virus-infected mice (Irusta and Hardwick, unpublished data). Unlike mammalian cell lines, Sindbis virus fails to induce significant apoptosis in mosquito cell lines [28]. Interestingly, infection of mosquito C7/10 cells with Sindbis virus encoding SA-vRpr also failed to induce programmed cell death (Irusta and Hardwick, unpublished data).

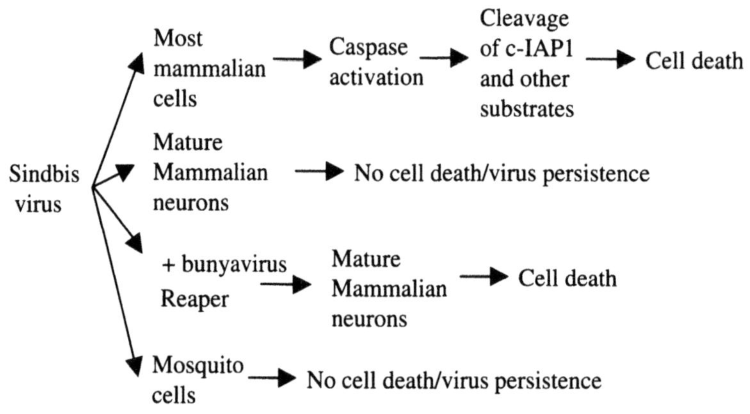

Fig. 2. Orthobunyavirus Reaper relieves block to Sindbis virus-induced apoptosis of mature neurons. Sindbis virus induces apoptosis in immature neurons but causes a persistent infection in mature neurons and mosquito cells. However, infection with a recombinant Sindbis virus encoding the orthobunyavirus Reaper (vSA-Rpr) results in apoptosis of infected neurons

Perhaps this is not surprising, given that orthobunyaviruses must successfully replicate without killing the mosquito that vectors this virus between vertebrate hosts. Therefore, unlike the large DNA baculoviruses that must inhibit apoptosis to cause disease, orthobunyaviruses and the alphavirus Sindbis virus likely cause disease in vertebrate hosts by inducing apoptosis. Orthobunyaviruses may have captured the mosquito Reaper gene to aid its existence in mosquitoes, with the unfortunate consequence of becoming pathogenic in mammalian neurons. Alternatively, orthobunyaviruses may have captured a yet unreported vertebrate homologue of Reaper to regulate mammalian cell apoptosis.

References

1. Adams JM, Cory S (1998) The Bcl-2 protein family: arbiters of cell survival. Science 281: 1322–1326
2. Arama E, Agapite J, Steller H (2003) Caspase activity and a specific cytochrome C are required for sperm differentiation in Drosophila. Dev Cell 4: 687–697
3. Blitvich BJ, Blair CD, Kempf BJ, Hughes MT, Black WC, Mackie RS, Meredith CT, Beaty BJ, Rayms-Keller A (2002) Developmental- and tissue-specific expression of an inhibitor of apoptosis protein 1 homologue from Aedes triseriatus mosquitoes. Insect Mol Biol 11: 431–442
4. Bridgen A, Weber F, Fazakerley JK, Elliott RM (2001) Bunyamwera bunyavirus nonstructural protein NSs is a nonessential gene product that contributes to viral pathogenesis. Proc Natl Acad Sci USA 98: 664–669
5. Chen Y, Fannjiang Y, Hardwick JM (2003) Cell death in viral infections. In: Lockshin R, Zakeri Z (eds), When cells die II. Wiley & Sons, New York, NY
6. Clem RJ, Fechheimer M, Miller LK (1991) Prevention of apoptosis by a baculovirus gene during infection of insect cells. Science 254: 1388–1390
7. Clem RJ, Miller LK (1994) Induction and inhibition of apoptosis by insect viruses. In: Cope FO, Tomei LD (eds), apoptosis II: The molecular basis of cell death. Cold Spring Harbor Laboratory Press, Plainview, NY, pp 89–110

8. Clem RJ, Sheu T-T, Richter BWM, He W-W, Thornberry NA, Duckett CS, Hardwick JM (2001) c-IAP1 is cleaved by caspases to produce a proapoptotic C-terminal fragment. J Biol Chem 276: 7602–7608
9. Colón-Ramos DA, Irusta P, Gan E, Olson M, Song S, Morimoto RI, Elliott R, Lombard M, Hollingsworth R, Hardwick JM, Smith G, Kornbluth S (2003) Translation inhibition and apoptotic induction by bunyaviral small non-structural proteins bearing sequence similarity to reaper. Mol Biol Cell 14: 4162–4172
10. Cuconati A, White E (2002) Viral homologs of BCL-2: role of apoptosis in the regulation of virus infection. Genes Devel 16: 2465–2478
11. Deveraux QL, Takahashi R, Salvesen GS, Reed JC (1997) X-linked IAP is a direct inhibitor of cell-death proteases. Nature (London) 388: 300–304
12. Duckett CS, Nava VE, Gedrich RW, Clem RJ, Van Dongen JL, Gilfillan MC, Shiels H, Hardwick JM, Thompson CB (1996) A conserved family of apoptosis inhibitors related to the baculovirus *iap* gene. EMBO J 15: 2685–2694
13. Evans EK, Kuwana T, Strum SL, Smith JJ, Newmeyer DD, Kornbluth S (1997) Reaper-induced apoptosis in a vertebrate system. EMBO J 16: 7372–7381
14. Fadok VA, Chimini G (2001) The phagocytosis of apoptotic cells. Semin Immunol 13: 365–372
15. Hardwick JM, Bellows DS (2003) Viral versus cellular BCL-2 proteins. Cell Death Differ 10: S68–S76
16. Hardwick JM, Levine B (2000) Sindbis virus vector system for functional analysis of apoptosis regulators. Methods Enzymol 322: 492–508
17. Holley CL, Olson MR, Colon-Ramos DA, Kornbluth S (2002) Reaper eliminates IAP proteins through stimulated IAP degradation and generalized translational inhibition. Nat Cell Biol 4: 439–444
18. Jacobson MD, Burne JF, King MP, Miyashita T, Reed JC, Raff MC (1993) Bcl-2 blocks apoptosis in cells lacking mitochondrial DNA. Nature (London) 361: 365–369
19. Kroemer G, Reed JC (2000) Mitochondrial control of cell death. Nat Med 6: 513–519
20. Levine B, Huang Q, Isaacs JT, Reed JC, Griffin DE, Hardwick JM (1993) Conversion of lytic to persistent alphavirus infection by the bcl-2 cellular oncogene. Nature (London) 361: 739–742
21. Lewis J, Wesselingh SL, Griffin DE, Hardwick JM (1996) Sindbis virus-induced apoptosis in mouse brains correlates with neurovirulence. J Virol 70: 1828–1835
22. Li P, Nijhawan D, Budihardjo I, Srinivasula SM, Ahmad M, Alnemri ES, Wang X (1997) Cytochrome c and dATP-dependent formation of Apaf-1/caspase-9 complex initiates an apoptotic protease cascade. Cell 91: 479–489
23. Madeo F, Herker E, Maldener C, Wissing S, Lachelt S, Herlan M, Fehr M, Lauber K, Sigrist SJ, Wesselborg S, Frohlich KU (2002) A caspase-related protease regulates apoptosis in yeast. Mol Cell 9: 911–917
24. Martin SJ (2002) Destabilizing influences in apoptosis: sowing the seeds of IAP destruction. Cell 109: 793–796
25. Martinou JC, Green DR (2001) Breaking the mitochondrial barrier. Nat Rev Mol Cell Biol 2: 63–67
26. Means JC, Muro I, Clem RJ (2003) Silencing of the baculovirus Op-iap3 gene by RNA interference reveals that it is required for prevention of apoptosis during Orgyia pseudotsugata M nucleopolyhedrovirus infection of Ld652Y cells. J Virol 77: 4481–4488
27. Miller LK (1999) An exegesis of IAPs: salvation and surprises from BIR motifs. Trends Cell Biol 9: 323–328

28. Miller ML, Brown DT (1992) Morphogenesis of Sindbis virus in three subclones of Aedes albopictus (mosquito) cells. J Virol 66: 4180–4190
29. Muro I, Hay BA, Clem RJ (2002) The Drosophila DIAP1 protein is required to prevent accumulation of a continuously generated, processed form of the apical caspase DRONC. J Biol Chem 277: 49644–49650
30. Nava VE, Rosen A, Veliuona MA, Clem RJ, Levine B, Hardwick JM (1998) Sindbis virus induces apoptosis through a caspase-dependent, CrmA-sensitive pathway. J Virol 72: 452–459
31. Palaga T, Osborne B (2002) The 3D's of apoptosis: death, degradation and DIAPs. Nat Cell Biol 4: E149–E151
32. Salvesen GS, Duckett CS (2002) IAP proteins: blocking the road to death's door. Nat Rev Mol Cell Biol 3: 401–410
33. Thornberry NA, Lazebnik Y (1998) Caspases: enemies within. Science 281: 1312–1316
34. Thress K, Henzel W, Shillinglaw W, Kornbluth S (1998) Scythe: a novel reaper-binding apoptotic regulator. EMBO J 17: 6135–6143
35. Thress K, Song J, Morimoto RI, Kornbluth S (2001) Reversible inhibition of Hsp70 chaperone function by Scythe and Reaper. EMBO J 20: 1033–1041
36. Wang X, Yang C, Chai J, Shi Y, Xue D (2002) Mechanisms of AIF-mediated apoptotic DNA degradation in Caenorhabditis elegans. Science 298: 1587–1592
37. White K, Grether ME, Abrams JM, Young L, Farrell K, Steller H (1994) Genetic control of programmed cell death in *Drosophila*. Science 264: 677–683
38. Xu G, Cirilli M, Huang Y, Rich RL, Myszka DG, Wu H (2001) Covalent inhibition revealed by the crystal structure of the caspase-8/p35 complex. Nature 410: 494–497

Author's address: J. Marie Hardwick, Department of Molecular Microbiology and Immunology, Johns Hopkins School of Public Health, 615 N Wolfe St., Baltimore, MD 21205, U.S.A.; e-mail: hardwick@jhu.edu

Arch Virol (2004) [Suppl] 18: 179–190

Archives of Virology

Semliki Forest virus infection of laboratory mice: a model to study the pathogenesis of viral encephalitis

J. K. Fazakerley

Center for Infectious Diseases, College of Medicine and Veterinary Medicine, University of Edinburgh, Edinburgh, U.K.

Summary. Semliki Forest virus (SFV) infection of the laboratory mouse provides an experimental system to study the pathogenesis of viral encephalitis. Following extraneural inoculation the virus is efficiently neuroinvasive and crosses the blood-brain barrier to initiate perivascular foci of infection in neurons and oligodendrocytes. The outcome of infection ranges from clinically unapparent mild encephalitis to fatal panencephalitis. SFV infections of the developing nervous system are always highly destructive and are generally fatal. In contrast, SFV infections of the mature nervous system can result in persistent infection with no apparent cell loss. This dramatic difference is attributable to developmental changes in the interactions between virus and CNS cells. Antibody responses clear the systemic infection and control the CNS infection. $CD8^+$ T-cells are required to generate the lesions of inflammatory demyelination which can be a feature of the neuropathology. This article reviews the pathogenesis of SFV encephalitis, describing the neuropathology and the mechanisms which underlie it and which may be fundamental to many viral encephalitides.

Introduction

Given the inaccessibility of the CNS and the problems of invasive investigations in humans, our understanding of pathogenic processes in viral encephalitis has been derived primarily from studies of animal model systems. Of the viruses which cause encephalitis in humans or in economically important animal species, many of these or closely related viruses, can be studied as infections of laboratory rodents. The major questions include, how do viruses enter the CNS, is the blood-brain barrier compromised during infection, what is the tropism of the virus, what is the neuropathology, what are the mechanisms of this neuropathology and how can these be prevented, ameliorated or repaired? Semliki Forest virus (SFV) infection of the laboratory mouse provides one experimental model system to address these questions. SFV is highly neuroinvasive and depending upon the combination of virus and host genetics, host age and immune responses the outcome of infection ranges from clinically unapparent mild encephalitis to fulminant fatal encephalitis. The virus is closely related to the equine encephalitides, Eastern, Western and Venezuelan equine encephalitis viruses. Understanding SFV pathogenesis provides not only an understanding of the equine encephalitides but also of interactions between neurotropic viruses, CNS cells and immune responses which are likely to be fundamental to the pathogenesis of viral encephalitis.

SFV strains

SFV is a member of the *Alphavirus* genus of the family *Togaviridae*. It occurs naturally in Africa where it infects horses and humans. In a survey in Central Africa, virus was isolated from 22 patients with fever, severe persistent headaches myalgia and arthralgia [40]. Virus could also be isolated from locally caught *Aedes aegypti*. Experimentally the virus has been studied in the laboratory mouse as a model system to understand viral and immunologic events in viral encephalitis and demyelinating disease [7, 21, 22]. A number of SFV strains vary in their virulence, including the virulent L10 and related prototype strains and the avirulent A7 and A7(74) strains [13]. The prototype, A7 and A7(74) strains have been sequenced [27, 28, 60, 61, 73]. Transcription of the complete virus cDNA gives rise to RNA [27, 38]. When generated within cells or transfected into cells this can be translated and replicated to generate infectious virus particles. The cDNA of the prototype strain is known as SFV4 and infectious virus generated from this has the same designation. The ease with which this system works has allowed the rapid generation of virus mutants for reverse genetics studies and is the basis for the SFV vector system, which is used for recombinant protein expression in eukaryotic cells and which has much promise for use as transient gene therapy and vaccine systems [10, 24, 45].

When working with commonly used strains of SFV, seroconversion of laboratory workers is not uncommon [75]. SFV is relatively stable in aerosols of tissue homogenates [9], although aerosols of purified virus are rapidly inactivated [15]. SFV can be inactivated by acetylethyleneimine and this inactivation preserves antigenicity; the preparation can stimulate a sufficiently good immune response to protect mice from infection with virulent SFV strains [3]. There is one reported case of a fatal neurological disease in a scientist working with the Osterrieth strain of the virus [75]. At postmortem, a typical viral meningoencephalomyelitis was observed and SFV was isolated from both cerebrospinal fluid and brain. It is known that the scientist was working with virus supernatant from SFV-infected BHK-21 cells. This strain is no longer used and the genotype and route and dose of this infection remain unknown.

Virus replication

The SFV genome consists of a single approximately 11 Kb strand of positive sense RNA which acts directly as the message. The isolated RNA is infectious.

Replication is via a negative sense intermediate that gives rise to full-length genomic RNA and a subgenomic message that represents the 3′ third of the genomic RNA and contains the genes for the viral structural proteins. Both the genomic and subgenomic RNAs are capped and polyadenylated. From 5′ to 3′ along the genomic RNA there are 4 non-structural proteins designated nsP1 to nsP4, followed by the capsid protein C and the envelope glycoproteins E2, E3 and E1. The virion is enveloped and on budding has a spike composed of three E1/E2 heterodimers. Within the lipid envelope the RNA is associated with the capsid protein. This protein also interacts with the C-terminal domain of the E2 protein to drive virion budding.

The receptor has been suggested to be MHC-I molecules [31], though this has not been confirmed and remains controversial. Virus entry and fusion is via the endosomal system and has been much studied [39]. Fusion of the viral and endosomal membranes is a function of the E1 glycoprotein and is dependent upon the presence of cholesterol and reorganization of the envelope spike at acid pH [35, 49, 74]. Upon fusion the capsid is liberated into the cytoplasm. RNA replication is associated with smooth membrane structures termed cytopathic vacuoles, type-I and type-II [29, 30]. These structures can be separated from other subcellular structures on sucrose gradients [26, 41]. Replication in cultured cells has been studied in most detail. NsP1-4 mediated RNA replication occurs on specialised structures termed spherules, surface invaginations of the type-I cytopathic vacuoles [37]. In neurons, replication also occurs on the type-I vacuoles and virions bud at the plasma membrane and into the lumen of the type-II vacuoles, which then fuse with the plasma membrane to release mature virions [55]. Early in neuronal infection clumps of electron dense material are observed throughout the cytoplasm. These contain capsid protein and are probably RNA capsid complexes. They migrate to underlie membranes including the plasma membrane and are also sites of virus budding [55–58].

Entry into host

SFV was originally isolated from mosquitoes and, by analogy with other alphaviruses, it can be assumed that mosquitoes are the vectors for natural transmission. Experimentally the virus can infect mice by many routes of inoculation [13, 14]. Following intraperitoneal (ip) inoculation into adult mice SFV replicates in muscles, including skeletal, smooth and cardiac muscle [6, 59]. A high titer plasma viremia is detectable by 24 hours, peaks at 48 hours and rapidly declines thereafter to become undetectable by infectivity assay at 4 days [20, 59]. All strains that have been studied (L10, SFV4, A7, A7(74)) are neuroinvasive. The much studied A7(74) and L10 strains have been shown to replicate in and probably enter the brain across cerebral endothelial cells and to initiate perivascular foci of CNS infection [20, 54, 68]. It is likely that other SFV strains also enter the CNS predominantly by this route. Infection of the brain via the olfactory nerves can also occur [33, 48]. Once in the CNS all strains of the virus appear to infect neurons and oligodendrocytes but not astrocytes and only rarely meningeal, ependymal or choroid plexus cells [8, 57].

Productive and restricted replication

From the original perivascular foci the virulent L10 strain spreads rapidly throughout the brain, producing a panencephalitis and death within 3 or 4 days [20]. This occurs irrespective of age. In contrast, the dynamics of A7(74) infection of the CNS varies according to age. In neonatal mice the virus spreads rapidly throughout the brain and infection is fatal often within 48 h. As the mice age the virus is less able to spread in the CNS and there is a sharp age-related virulence [17, 25, 48]. In weaned mice the A7(74) strain has a restricted replication in the CNS and remains confined to perivascular foci [17, 20, 56]. These strain and age differences are illustrated in Fig. 1. The age-related virulence is not a function of the maturity of specific immune responses, as this strain of SFV remains confined to small, predominantly perivascular foci in athymic *nu/nu* mice or mice with severe combined immunodeficiency [6, 17, 20]. Comparison of the extent of spread in the

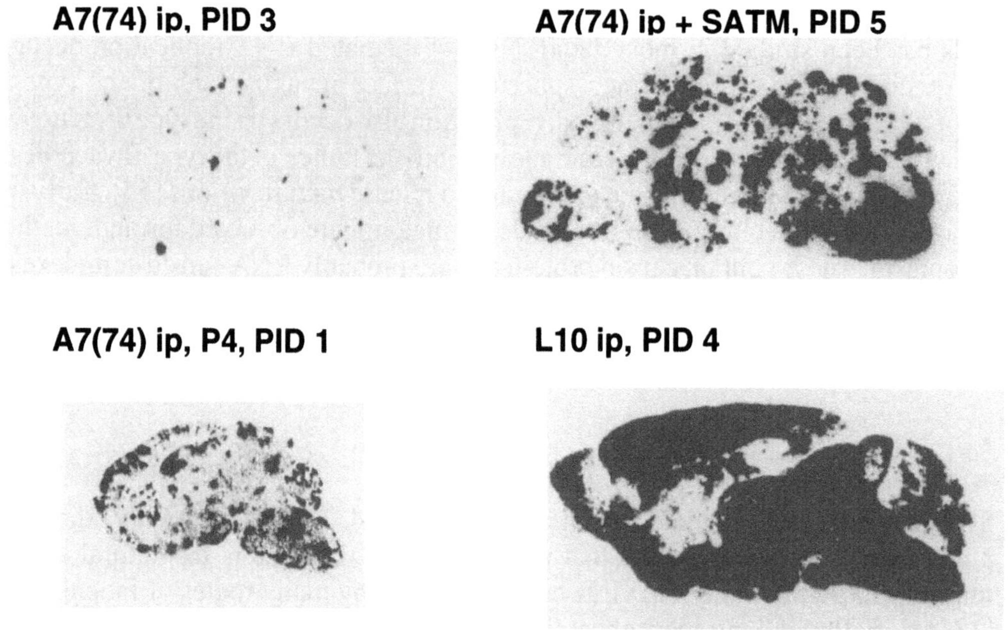

Fig. 1. Distribution of SFV RNA in the mouse brain as determined by in situ hybridization followed by autoradiography. Each panel shows a sagital half brain section cut close to the midline. P4 indicates the post natal age at inoculation, all other sections are taken from the brains of mice infected >3 weeks of age. In all cases inoculation was intraperitoneal. The time of sampling post inoculation (days) (PID) is given. GSTM indicates that this mouse received 10 mg of gold sodium aurothiomalate 4 h prior to inoculation with virus. Note the contrast between focal nature of the A7(74) infection in weaned mice, and the widespread infection of this virus in suckling (neonatal) mice, and in weaned mice treated with gold compounds. The latter have a widespread distribution similar to the L10 strain in weaned mice. For more details of these studies see Refs. [36, 44, 49]

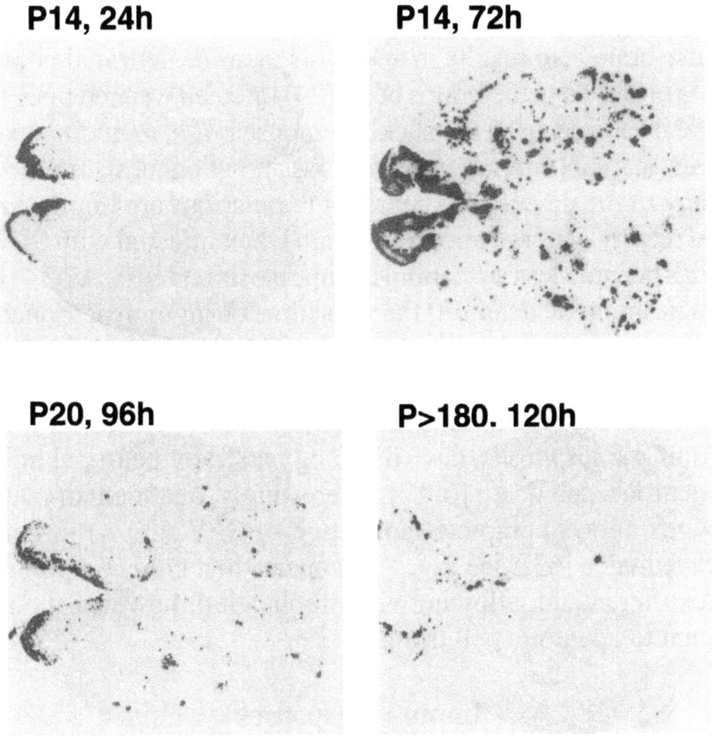

Fig. 2. Age-related change in the distribution of SFV A7(74) with time post-infection and with age (*P*) at inoculation following intranasal inoculation. Black areas indicate the presence of viral RNA as determined by in situ hybridization followed by autoradiography. Each panel shows two sagital half brain sections following bisection of the brain down the midline. Note the infection of the olfactory bulb at 24 h which then spreads to deeper but connected brain areas by 72 h. The extent of infection is greater at 72 h in the animal inoculated at P14, than at 96 h in the animal inoculated at P20 and this is greater than at 120 h in the animal inoculated at P > 180. For more details of these studies see Ref. [40]

CNS following intraperitoneal and intranasal inoculations in mice of different ages from birth to weaning indicates that as neuronal circuits mature in the postnatal brain so the infection becomes restricted [47, 48]. This is illustrated in Fig. 2.

Electronmicroscopic studies show that L10 infection of CNS cells from mice of all ages is productive, as is A7(74) infection of cells in neonatal or suckling mouse brain. In contrast, no infectious virions, budding virus or cytopathic vacuoles have ever been observed in the brains of weaned mice infected with A7(74) virus [55–57]. However, large aggregates of dense fibrous material that can be immunolabelled with antibody to viral capsid protein have been observed in neurons and oligodendrocytes of 3–4 week old mice infected with A7(74). These are probably RNA capsid complexes [58]. Replication of this virus appears to be restricted in the neurons of 3–4 week old mice [20]. This restriction is age-related since A7(74) replicates well in neurons of neonatal and suckling mice [20, 55, 56].

SFV infection of cells in continuous culture and primary neuronal cultures results in apoptotic cell death [1, 62]. Apoptosis can be delayed by over expression of bcl-2 or by treatment with caspase inhibitors. SFV A7(74) infection of the neonatal mouse brain also results in apoptosis of some neuronal populations, but no apoptosis is observed in the brains of A7(74) infected weaned mice [1, 2]. Many developmental events occur in the suckling mouse brain, including axonogenesis, synaptogenesis, axonal pruning, neuronal loss and gliogenesis [reviewed in reference 23]. Many of the developing neuronal populations are highly susceptible to apoptosis and readily undergo apoptotic death when infected with SFV. With maturation neurons become less susceptible to apoptosis and SFV A7(74) infection no longer results in apoptotic death. At the same time changes in neuronal physiology dictate that the virus can no-longer replicate productively and neuronal infection becomes restricted. This change is most readily observed in athymic *nu/nu* and mice with severe combined immunodeficiency (SCID) which, in the absence of appropriate immune responses, have life-long persistent neuronal infections with no apparent neuronal cell death [6, 17]. Interestingly, treatment of adult mice with gold compounds allows complete replication of SFV A7(74) in neurons in the brains of weaned mice [63]; see Fig. 1. It appears that gold compounds cause the neurons to dedifferentiate, allowing virus replication; however the neurons still remain resistant to apoptotic cell death.

Immune responses

In the absence of specific immune responses SFV A7(74) can persist in CNS cells, apparently without damaging them, and essentially for the life of the mouse [6, 17]. In immunocompetent mice the virus is eliminated from the CNS; infectious virus cannot be detected by plaque assay after day 8, by virus isolation in neonatal mice after day 11 or by in situ hybridization after day 14 [20, 32, 71]. RT-PCR indicates that viral RNA can be detected for several weeks after infection [16]. After the clearance of infectious virus by day 14, immunocompetent mice develop lesions of inflammatory demyelination. These lesions are apparent throughout the CNS white matter [34, 71]. No demyelinating lesions are observed in SCID mice [6] despite virus persistence in the CNS. Specific immune responses therefore have a role both in clearing this infection and in generating the lesions of demyelination. The roles of the different effector functions of the immune response have been dissected by studying naturally arising mutant mice, genetically engineered mice, mice treated with general or selective immunosuppressants and by adoptive transfer of immune effector functions.

a) Interferons

Following ip inoculation of virus, plasma levels of type-I interferons parallel the viremia [14]. In mice with a genetic deletion of functional type-I interferon receptors [46], infection with SFV A7(74) is fatal within 48 h. Study of the infection in these mice [Fazakerley et al., unpublished] indicates that this is due to a widespread infection of several organs and tissues, most notably the exocrine pancreas. By the time of death infection of the CNS is minimal. Intracerebral inoculation of virus results in a widespread infection of meningeal, oligodendrocyte and ependymal cells. Presumably, these cell types normally become refractory to virus replication following interferon activation, since in the presence of interferon infection is rarely observed in meningeal and ependymal cells. The limited infection of neurons remained unchanged, indicating that restricted replication in mature neurons is independent of type-I interferon responses.

b) Antibodies

SFV A7(74) infection of BALB/c mice results in a rapid neutralizing serum IgM antibody response followed by an IgG2a response and a late IgG1 response [20]. During the normal course of infection in immunocompetent mice, the blood-brain barrier is leaky from day 4 to 10, allowing passage of immunoglobulins [50, 68]. Antibody-producing plasma cells can be detected in the brain and intrathecal antibodies are detectable for months [52, 53]. In contrast, SCID mice have no antibodies, a persistent viremia and a persistent CNS infection [6]. Athymic *nu/nu* mice control the viremia but have a persistent CNS infection. The *nu/nu* mice produce serum anti-viral IgM but not IgG [72]. These data suggest IgM antibodies control the viremia but are unable to eliminate the CNS infection. This was confirmed by resolution of persistent viremia but not CNS infection by transfer of day 7 *nu/nu* mouse serum to infected SCID mice [6]. In contrast, transfer of high titer IgG sera from immunocompetent BALB/c mice to SCID mice abolished both the viremia and titers of infectious virus from the CNS [6]. Whether this was complete eradication of all viral sequences was not determined by more sensitive techniques such as RT-PCR. Antibodies are also highly effective in protecting mice from challenge with a lethal dose of SFV. A number of B-cell epitopes have been mapped on the viral glycoproteins [12, 65] and even a single monoclonal antibody directed to the E2 envelope glycoprotein has been shown to protect mice against a virulent infection [11]. Combinations of linear B-cell epitopes and B and T cell epitopes have been tested as potential vaccine candidates [65, 66].

c) Cytokines

Quantitative measurements of the levels of cytokine transcripts, including interleukins (IL), 1α, 1β, 2, 4, 5, 6, 7, 10, 12, interferon-γ, TNF-α, TNF-β, TGF-β, and GMCSF indicate that in the CNS of BALB/c mice an early Th1 response is rapidly followed by a Th2 response [Fazakerley et al., unpublished data]. This is consistent with the rapid early IgG2a response followed by a slow IgG1 response [20]. Studies on the expression of cytokines by immunohistology are difficult but IL-1α, IL-1β, IL-3, IL-6, IL-10 and TNFα are readily detected in the CNS during SFV infection (44).

d) Cellular immune responses

Beginning 3 days after ip infection of BALB/c mice with SFV A7(74), perivascular and infiltrating mononuclear cells can be observed in the CNS [51]. Histopathologic studies indicate that the areas of inflammatory infiltrates and the areas of infection coincide [44, 70]. Perivascular cuffs are a prominent feature of the neuropathology and are maximal between 7 and 10 days. Cerebral endothelial cells upregulate adhesion molecule expression after infection [44, 69]. CD49d (VLA-4) is expressed in lesions and antibodies to this can reduce both infiltration of inflammatory cells and the extent of demyelination [64]. Most infiltrating mononuclear cells, some ependymal cells, some meningeal cells, and some parenchymal cells with a dendritic morphology are MHC-I$^+$ [44]. The MHC-I$^+$ cells surround the foci of infection and levels of MHC-I staining decrease with distance from each focus. MHC-II$^+$ cells, probably pericytes can be observed adjacent to blood vessels in areas of infection and MHC-II$^+$ astrocytes and microglial cells are present in areas of infection and inflammation [44]. A few infiltrating inflammatory cells are MHC-II$^+$, the majority are not. CD45/B220$^+$ B-cells are also present in the inflammatory lesions, including lesions with demyelination.

Mice infected with SFV have strong delayed-type hypersensitivity responses to the virus [36]. T-cell epitopes for some haplotypes have been mapped [65–67]. Adoptive transfer studies in SCID and *nu/nu* mice and studies in mice immunosuppressed with total body irradiation, cyclophosphamide, cyclosporin or cycloleucine all demonstrate that T-cell responses are pathogenic and are required to generate the lesions of demyelination [4, 6, 17–20]. Depletion of CD8$^+$ cells with a monoclonal antibody greatly reduces demyelination whereas depletion of CD4$^+$ cells reduces the extent of the inflammation but not the demyelination [70]. Given the requirement for CD8$^+$ cells for demyelination and given that the virus infects oligodendrocytes, CD8$^+$ T-cell destruction of virally infected oligodendrocytes would seem to be the most likely mechanism of demyelination, though this remains to be proven. Mice infected with SFV A7(74) generate CNS specific autoantibodies, are more sensitive to the induction of experimental allergic encephalomyelitis and have T-cells that cross react with both viral and CNS epitopes [5, 42, 43]. The role if any, of autoimmune responses in the generation of SFV A7(74)-induced demyelinating lesions remains unclear.

In summary, SFV infection of the laboratory mouse provides an excellent model for studying neuroinvasion, blood-brain barrier events and immune responses in viral encephalitis. Two other features of this model system are of particular note, firstly the age-related virulence the analysis of which is providing fundamental information on the relationships between neurons and viruses and secondly, the lesions of demyelination, where an understanding of the pathogenesis may be highly relevant to diseases such as multiple sclerosis.

References

1. Allsopp TE, Scallan MF, Williams A, Fazakerley JK (1998) Virus infection induces neuronal apoptosis: A comparison with trophic factor withdrawal. Cell Death Differ 5: 50–59
2. Allsopp TE, Fazakerley JK (2000) Altruistic cell suicide and the specialized case of the virus-infected nervous system. Trends Neurosci 23: 284–290
3. Amor S, Webb HE (1986) Use of normal-acetylethyleneimine [AEI] for the inactivation of Semliki Forest virus in vitro. J Med Virol 19: 367–376

4. Amor S, Webb HE (1987) The effect of cycloleucine on SFV A7(74) infection in mice. Br J Exp Pathol 68: 225–235
5. Amor S, Webb HE (1988) CNS pathogenesis following a dual viral infection with Semliki Forest (alphavirus) and Langat (flavivirus). Br J Exp Pathol 69: 197–208
6. Amor S, Scallan MF, Morris MM, Dyson H, Fazakerley JK (1996) Role of immune responses in protection and pathogenesis during Semliki Forest virus encephalitis. J Gen Virol 77: 281–291
7. Atkins GJ, Sheahan BJ, Liljestrom P (1999) The molecular pathogenesis of Semliki Forest virus: a model virus made useful. J Gen Virol 80: 2287–2297
8. Balluz IM, Glasgow GM, Killen HM, Mabruk MJ, Sheahan BJ, Atkins GJ (1993) Virulent and avirulent strains of Semliki Forest virus show similar cell tropism for the murine central nervous system but differ in the severity and rate of induction of cytolytic damage. Neuropathol Appl Neurobiol 19: 233–239
9. Benbough JE (1969) The effect of relative humidity on the survival of airborne Semliki Forest virus. J Gen Virol 4: 473–477
10. Berglund P, Tubulekas I, Liljestrom P (1996) Alphaviruses as vectors for gene delivery. Trends Biotechnol 14: 130–134
11. Boere WA, Benaissa-Trouw BJ, Harmsen M, Kraaijeveld CA, Snippe H (1983) Neutralizing and non-neutralizing monoclonal antibodies to the E2 glycoprotein of Semliki Forest virus can protect mice from lethal encephalitis. J Gen Virol 64: 1405–1408
12. Boere WA, Harmsen T, Vinje J, Benaissa-Trouw BJ, Kraaijeveld CA, Snippe H (1984) Identification of distinct antigenic determinants on Semliki Forest virus by using monoclonal antibodies with different antiviral activities. J Virol 52: 575–582
13. Bradish CJ, Allner K, Maber HB (1971) The virulence of original and derived strains of Semliki Forest virus for mice, guinea-pigs and rabbits. J Gen Virol 12: 141–160
14. Bradish CJ, Allner K, Fitzgeorge R (1975) Immunomodification and the expression of virulence in mice by defined strains of Semliki Forest virus: the effects of cyclophosphamide. J Gen Virol 28: 225–237
15. de Jong JC, Harmsen M, Plantinga AD, Trouwbrost T (1976) Inactivation of Semliki Forest virus in aerosols. Appl Environ Microbiol 32: 315–319
16. Donnelly SM, Sheahan BJ, Atkins GJ (1997) Long-term effects of Semliki Forest virus infection in the mouse central nervous system. Neuropathol Appl Neurobiol 23: 235–241
17. Fazakerley JK, Webb HE (1987a) Semliki Forest virus-induced, immune-mediated demyelination: adoptive transfer studies and viral persistence in nude mice. J Gen Virol 68: 377–385
18. Fazakerley JK, Webb HE (1987b) Cyclosporine enhances virally induced T-cell-mediated demyelination. The effect of cyclosporine on a demyelinating virus infection. J Neurol Sci 78: 35–50
19. Fazakerley JK, Webb HE (1987c) Semliki Forest virus induced, immune mediated demyelination: the effect of irradiation. Br J Exp Pathol 68: 101–113
20. Fazakerley JK, Pathak S, Scallan M, Amor S, Dyson H (1993) Replication of the A7(74) strain of Semliki Forest virus is restricted in neurons. Virology 195: 627–637
21. Fazakerley JK, Buchmeier MJ (1993) Pathogenesis of virus-induced demyelination. Adv Virus Res 42: 249–324
22. Fazakerley JK, Amor S, Nash AA (1997) Animal model systems of MS. In: Russell WC (ed), Molecular biology of multiple sclerosis. John Wiley & Sons, Clinchster, pp 255–273
23. Fazakerley JK (2001) Neurovirology and developmental neurobiology. Adv Virus Res 56: 73–124
24. Fleeton MN, Chen M, Berglund P, Rhodes G, Parker SE, Murphy M, Atkins GJ, Liljestrom P (2001) Self-replicative RNA vaccines elicit protection against influenza

A virus, respiratory syncytial virus, and a tickborne encephalitis virus. J Infect Dis 183: 1395–1398
25. Fleming P (1977) Age-dependent and strain-related differences of virulence of Semliki Forest virus in mice. J Gen Virol 37: 93–105
26. Friedman RM, Levin JG, Grimley PM, Berezesky IK (1972) Membrane-associated replication complex in arbovirus infection. J Virol 10: 504–515
27. Garoff H, Frischauf A, Simons K, Lehrach H, Delius H (1980) Nucleotide sequence of cDNA coding for Semliki Forest virus membrane glycoproteins. Nature 288: 236–241
28. Glasgow GM, Killen HM, Liljestrom P, Sheahan BJ, Atkins GJ (1994) A single amino acid change in the E2 spike protein of a virulent strain of Semliki Forest virus attenuates pathogenicity. J Gen Virol 75: 663–668
29. Grimley PM, Berezesky IK, Friedman RM (1968) Cytoplasmic structures associated with an arbovirus infection: loci of viral ribonucleic acid synthesis. J Virol 2: 1326–1338
30. Grimley PM, Friedman RM (1970) Development of Semliki Forest virus in mouse brain – An electron microscopic study. Exp Mol Pathol 12: 1–13
31. Helenius A, Morein B, Fries E, Simons K, Robinson P, Schirmaer V (1978) Human (HLA-A and HLA-B) and murine (H-2k) and (H-2d) histocompatibility antigens are cell surface receptors for Semliki Forest virus. Proc Natl Acad Sci USA 75: 3846–3850
32. Jagelman S, Suckling AJ, Webb HE, Bowen ETW (1978) The pathogenesis of avirulent Semliki Forest virus infections in athymic nude mice. J Gen Virol 41: 599–607
33. Kaluza G, Lell G, Reinacher M, Stitz L, Willems WR (1987) Neurogenic spread of Semliki Forest virus in mice. Arch Virol 93: 97–110
34. Kelly WR, Blakemore WF, Jagelman S, Webb HE (1982) Demyelination induced in mice by avirulent Semliki Forest virus. II. An ultrastructural study of focal demyelination in the brain. Neuropathol Appl Neurobiol 8: 43–53
35. Kielian M, Helenius A (1984) pH-induced changes in the fusogenic spike protein of Semliki Forest virus. J Cell Biol 101: 2284–2291
36. Kraaijeveld CA, Harmsen M, Khader Boutahar-Trouw B (1979) Delayed-type hypersensitivity against Semliki Forest virus. Infect Immun 23: 217–221
37. Kujala P, Ikaheimonen A, Ehsani N, Vihinen H, Auvinen P, Kaariainen L (2001) Biogenesis of the Semliki Forest virus RNA replication complex. J Virol 75: 3873–3884
38. Liljestrom P, Garoff H (1991) A new generation of animal cell expression vectors based on the Semliki Forest virus replicon. Biotechnology 9: 1356–1361
39. Marsh M, Kielian MC, Helenius A (1984) Semliki Forest virus entry and the endocytic pathway. Biochem Soc Trans 12: 981–983
40. Mathiot CC, Grimaud G, Garry P, Bouquety JC, Mada A, Daguisy AM, Georges AJ (1990) An outbreak of human Semliki Forest virus infections in Central African Republic. Am J Trop Med Hyg 42: 386–393
41. Mehta S, Pathak S, Webb HE (1990) Induction of membrane proliferation in mouse CNS by gold sodium thiomalate with reference to increased virulence of the avirulent Semliki Forest virus. Biosci Rep 10: 271–279
42. Mokhtarian F, Swoveland P (1987) Predisposition to EAE induction in resistant mice by prior infection with Semliki Forest virus. J Immunol 138: 3264–3268
43. Mokhtarian F, Shi Y, Grob D (1996) Epitopes of Semliki Forest virus which evoke cross-reactive immune-response with peptides of the central-nervous-system. FASEB J 10: 2092–2092
44. Morris MM, Dyson H, Baker D, Harbige LS, Fazakerley JK, Amor S (1997) Characterization of the cellular and cytokine response in the central nervous system following Semliki Forest virus infection. J Neuroimmunol 74: 185–197

45. Morris-Downes MM, Phenix KV, Smyth J, Sheahan BJ, Lileqvist S, Mooney DA, Liljestrom P, Todd D, Atkins GJ (2001) Semliki Forest virus-based vaccines: persistence, distribution and pathological analysis in two animal systems. Vaccine 19: 1978–1988
46. Muller U, Steinhoff U, Reis L, Hemmi S, Pavlovic J, Zinkernagel RM, Aguet M (1994) Functional role of type I and type II interferons in antiviral defense. Science 264: 1918–1921
47. Oliver KR, Scallan MF, Dyson H, Fazakerley JK (1997) Susceptibility to a neurotropic virus and its changing distribution in the developing brain is a function of CNS maturity. J Neurovirol 3: 38–48
48. Oliver KR, Fazakerley JK (1998) Transneuronal spread of Semliki Forest virus in the developing mouse olfactory system is determined by neuronal maturity. Neurosci 82: 867–877
49. Omar A, Koblet H (1988) Semliki Forest virus-particles containing only the E1 envelope glycoprotein are infectious and can induce cell cell-fusion. Virology 166: 17–23
50. Parsons LM, Webb HE (1982a) Blood brain barrier disturbance and immunoglobulin G levels in the cerebrospinal fluid of the mouse following peripheral infection with the demyelinating strain of Semliki Forest virus. J Neurol Sci 57: 307–318
51. Parsons LM, Webb HE (1982b) Virus titers and persistently raised white cell counts in cerebrospinal-fluid in mice after peripheral infection with demyelinating Semliki Forest virus. Neuropathol Appl Neurobiol 8: 395–401
52. Parsons LM, Webb HE (1984) Specific immunoglobulin G in serum and cerebrospinal fluid of mice infected with the demyelinating strain of Semliki Forest virus. Microbios Letters 25: 135–140
53. Parsons LM, Webb HE (1989) Identification of immunoglobulin-containing cells in the central nervous system of the mouse following infection with the demyelinating strain of Semliki Forest virus. Br J Exp Pathol 70: 247–255
54. Pathak S, Webb HE (1974) Possible mechanisms for the transport of Semliki Forest virus into and within mouse brain: An electron microscopic study. J Neurol Sci 23: 175–184
55. Pathak S, Webb HE, Oaten SW, Bateman S (1976) An electron-microscopic study of the development of virulent and avirulent strains of Semliki Forest virus in mouse brain. J Neurol Sci 28: 289–300
56. Pathak S, Webb HE (1978) An electron-microscopic study of avirulent and virulent Semliki Forest virus in the brains of different ages of mice. J Neurol Sci 39: 199–211
57. Pathak S, Webb HE (1983a) Semliki Forest virus multiplication in oligodendrocytes in mouse-brain with reference to demyelination. J Physiol (London) 339: 17
58. Pathak S, Webb HE (1988a) Cytoplasmic viral core aggregates and budding of mature virus and spherules in mouse brain following Semliki Forest virus infections. Inst Phys Conf Ser 3: 213–214
59. Pusztai R, Gould E, Smith H (1971) Infection pattern in mice of an avirulent and virulent strain of Semliki Forest virus. Br J Exp Pathol 52: 669–677
60. Santagati MG, Maatta JA, Itaranta PV, Salmi AA, Hinkkanen AE (1995) The Semliki Forest virus E2 gene as a virulence determinant. J Gen Virol 75: 47–52
61. Santagati MG, Maatta JA, Roytta M, Salmi AA, Hinkkanen AE (1998) The significance of the 3′-nontranslated region and E2 amino acid mutations in the virulence of Semliki Forest virus in mice. Virology 243: 66–77
62. Scallan MF, Allsopp TE, Fazakerley JK (1997) bcl-2 acts early to restrict Semliki Forest virus replication and delays virus-induced programmed cell death. J Virol 71: 1583–1590
63. Scallan MF, Fazakerley JK (1999) Aurothiolates enhance the replication of Semliki Forest virus in the CNS and the exocrine pancreas. J Neurovirol 5: 392–400

64. Smith JP, Morris-Downes M, Brennan FR, Wallace GJ, Amor S (2000) A role for alpha4-integrin in the pathology following Semliki Forest virus infection. J Neuroimmunol 106: 60–68
65. Snijders A, Benaissa-Trouw BJ, Oosterlaken TA, Puijk WC, Posthumus WP, Meloen RH, Boere WA, Oosting JD, Kraaijeveld CA, Snippe H (1991) Identification of linear epitopes on Semliki Forest virus E2 membrane protein and their effectiveness as a synthetic peptide vaccine. J Gen Virol 72: 557–565
66. Snijders A, Benaissatrouw BJ, Snippe H, Kraaijeveld CA (1992a) Immunogenicity and vaccine efficacy of synthetic peptides containing Semliki Forest virus B-cell and T-cell epitopes. J Gen Virol 73: 2267–2272
67. Snijders A, Benaissatrouw BJ, Visservernooy HJ, Fernandez I, Snippe H, Kraaijeveld CA (1992b) A delayed-type hypersensitivity-inducing T-cell epitope of Semliki Forest virus mediates effective T-helper activity for antibody-production. Immunol 77: 322–329
68. SoiluHanninen M, Eralinna JP, Hukkanen V, Roytta M, Salmi AA, Salonen R (1994) Semliki-Forest virus infects mouse-brain endothelial-cells and causes blood-brain-barrier damage. J Virol 68: 6291–6298
69. SoiluHanninen M, Roytta M, Salmi AA, Salonen R (1997) Semliki Forest virus infection leads to increased expression of adhesion molecules on splenic T-cells and on brain vascular endothelium. J Neurovirol 3: 350–360
70. Subak-Sharpe I, Dyson H, Fazakerley JK (1993) In vivo depletion of $CD8^+$ T cells prevents lesions of demyelination in Semliki Forest virus infection. J Virol 67: 7629–7633
71. Suckling AJ, Pathak S, Jagelman S, Webb HE (1978) Virus associated demyelination: a model using avirulent Semliki Forest virus infection of mice. J Neurol Sci 36: 147–154
72. Suckling AJ, Jagelman S, Webb HE (1982) Immunoglobulin synthesis in nude (nu/nu), nu/+ and reconstituted nu/nu mice infected with a demyelinating strain of Semliki Forest virus. Clin Exp Immunol 47: 283–288
73. Takkinen K (1986) Complete nucleotide-sequence of the nonstructural protein genes of Semliki Forest virus. Nucleic Acids Res 14: 5667–5682
74. Wahlberg JM, Garoff H (1992) Membrane-fusion process of Semliki Forest virus.1. Low pH-induced rearrangement in spike protein quaternary structure precedes virus penetration into cells. J Cell Biol 116: 339–348
75. Willems WR, Kaluza G, Boschek GB, Bauer H (1979) Semliki Forest virus: Cause of a fatal case of human encephalitis. Science 203: 1128–1129

Author's address: Dr. John K. Fazakerley, Center for Infectious Diseases, University of Edinburgh, Summerhall, Edinburgh, EH9 1QH, UK; e-mail: John.Fazakerley@ed.ac.uk

Arch Virol (2004) [Suppl] 18: 191–200

Archives of Virology

A novel principle of attenuation for the development of new generation live flavivirus vaccines

R. M. Kofler, F. X. Heinz, and **C. W. Mandl**

Institute of Virology, University of Vienna, Vienna, Austria

Summary. The genus *Flavivirus* includes a number of important human pathogens that impose major health problems in large regions of the world. The emergence of flaviviruses in new geographic regions (e.g., West Nile virus in North America) and rapid socioeconomic changed in many developing countries where flaviviruses such as dengue virus and Japanese encephalitis virus and endemic demand the development of new vaccines against these diseases. Using tick-borne encephalitis virus as a model we have established a new method to generate attenuated flavivirus strains that may be useful for generating cost-effective and safe live vaccines. This method relies on the specific introduction of deletions into one of the structural proteins, the capsid protein C. These deletions remove parts or all of an internal stretch of hydrophobic amino acid residues that probably is involved in virion assembly. We observed that remarkably long deletions were tolerated, yielding viable viral mutants that were highly attenuated in the mouse model but efficiently induced protective immunity. Biochemical analyses suggested that attenuation was caused by an assembly defect of infectious virions but the mutants produced ample amounts of non-infections subviral particles. The generation of viable mutants with deletions longer that 16 amino acid residues depended on additional, spontaneously emerging mutations within protein C that increased the hydrophobicity of the mutant protein. Although the second-site mutations increased infectivity, they did not restore neuroinvasiveness. Mouse experiments demonstrated excellent safety and immunogenicity profiles for these mutants.

Introduction

Members of the genus *Flavivirus*, family *Flaviviridae*, such as tick-borne encephalitis virus (TBEV), yellow fever virus (YFV), Japanese encephalitis virus (JEV), West Nile virus (WNV), and Dengue virus (DENV), cause major human health problems in large areas of the world [4]. In spite of the availability of safe and effective vaccines for some of these pathogens, there is a continuing and growing demand for new flavivirus vaccines. Particular challenges for flavivirus vaccine development are imposed by the complex pathogenesis of some flavivirus diseases (e.g., dengue hemorrhagic fever), the socioeconomic situation of developing countries where flaviviral diseases are most common, or the introduction of flaviviruses into new populations, such as the recent emergence of WNV in North America.

Flaviviruses are small, round, enveloped particles that contain only three structural proteins, i.e., the membrane-anchored surface proteins M and E and the capsid protein C [11]. The genome consists of a single approximately 11-kb-long RNA molecule that encodes all of the viral proteins in a single long open reading frame flanked by a rather short 5′ noncoding region (NCR) and a somewhat longer 3′NCR.

For a number of flaviviruses, including all of the above mentioned important pathogens, infectious cDNA clones have been assembled during the past years making it possible to specifically mutate the genomes of these viruses [21]. Generation of viral mutants by site-specific mutagenesis techniques and biological characterization of these mutants have revealed a number of strategies to generate attenuated flavivirus strains that might serve as vaccine candidates. In particular, point mutations in the large envelope protein E and deletions introduced into the 3′-NCR have been demonstrated to be useful for the purpose of creating attenuated strains [8, 12, 16, 17, 20]. In addition, chimeric viruses containing the surface proteins of a flavivirus inserted into the genetic backbone of another flavivirus have been constructed and introduced as vaccine candidates [2].

Here we summarize and discuss work that has identified and characterized a novel target for generating attenuated flavivirus strains: the capsid protein C [9, 10]. Using TBEV as a model, the effects of introducing various deletions into this protein were studied. The availability of an excellent animal model (adult mice) for this virus makes TBEV a superior system for studying both attenuation of flaviviruses and induction of protective immunity. Viable virus progeny were obtained in spite of significant deletions introduced to the central region of this protein that removed parts or all of a distinct hydrophobic domain that may be an important determinant of the assembly process of virus particles. Additional spontaneously emerging mutations were observed in the cases of the larger deletions and their significance was assessed. Viable mutants remained genetically stable in subsequent passages. Most importantly, the deletion mutants were shown to be highly attenuated in the mouse model and to efficiently elicit a protective immune response. Second-site mutations in protein C apparently re-established infectivity but did not restore neuroinvasiveness of the deletion mutants. The biochemical and biological properties of these mutants indicate that the introduction of deletions into the capsid protein C represents a more favourable technique to generate attenuated flaviviruses than did previously tested procedures.

Deletion mutagenesis of the TBEV protein C

Protein C of TBE virus is a 116-residue-long highly basic protein containing a total of 11 arginine and 12 lysine residues [14]. Secondary structure prediction revealed

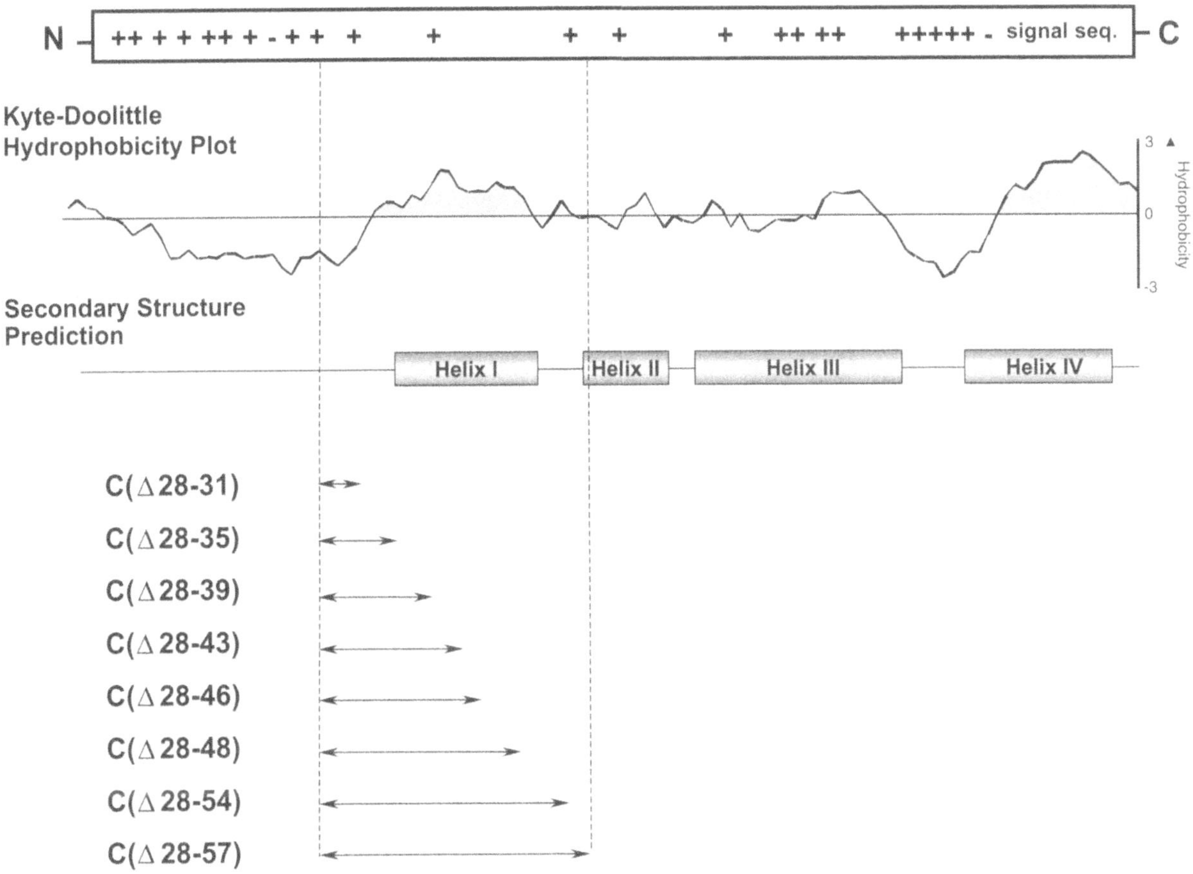

Fig. 1. Characteristics of the TBEV capsid protein C and deletion mutants. The distribution of positively charged amino acid residues (+) is shown on top. A hydrophobicity plot calculated by the Kyte-Doolittle algorithm reveals the presence of two hydrophobic regions (shaded areas). Secondary-structure prediction shows four alpha-helical domains, labelled as helix I to IV. The locations of the introduced deletion mutations are shown as two-headed arrows together with the designations of the corresponding mutants. The numbers in the names of the mutants indicate which residues were deleted

two particularly noteworthy features of this sequence, which are illustrated in Fig. 1. First, in spite of the high overall proportion of charged amino acid residues, there are two distinct hydrophobic domains. Second, the secondary structure conformation of protein C is predicted to be largely alpha-helical. The employed algorithm strongly predicted the formation of four helices that would take up most of the carboxy-terminal two thirds of the protein. Helix IV corresponds to the hydrophobic carboxy-terminal end of protein C, which is known to act as an internal signal sequence for the translocation of the subsequent protein, prM, into the lumen of the endoplasmic reticulum and which is removed from mature protein C by the action of the viral protease NS2B/3 [11]. Helix I is part of the central hydrophobic domain, which extends approximately from residue 34 to 51 (exact boundaries vary depending on the algorithm and parameter settings used).

Part of this region was previously proposed to anchor mature protein C into the viral membrane [18].

A set of deletion mutations of increasing sizes (removing 4, 8, 12, 16, 19, 21, 27, or 30 amino acid residues) was engineered into the infectious cDNA clone of TBEV strain Neudoerfl [13–15]. As shown in Fig. 1, these deletions begin shortly before the central hydrophobic domain (at position 28) and, with increasing size, remove incremental parts of or this entire domain. The designations of the resultant mutants as listed in Fig. 1 refer to the locations of the deletions within protein C.

Isolation of viable mutant virus progeny

Full-length RNAs of the deletion mutants were transcribed in vitro and introduced into BHK-21 cells by electroporation. All the mutants were found to express protein E and to export this protein from the cells. This indicated that the deletions did not significantly impair replication or translation of the viral genome. The amount of protein E present in the cell culture supernatants was quantified by ELISA [7] and equal quantities were used to inoculate fresh BHK-21 cells to determine whether infectious virus mutants were present in these samples. Mutant C(Δ28–43) and all mutants carrying smaller deletions efficiently infected these cells and were carried readily through subsequent passages. In contrast, only a few cells were infected in the first passage of mutant C(Δ28–46), but in the course of subsequently cycles of transferring supernatants to fresh cells, this mutant was successfully propagated with apparently increasing efficiency [10]. Mutant C(Δ28–48), however, and the larger deletion mutants, were found to be non-infectious in this cell culture system.

Intracranial inoculation of suckling mice is known to be approximately 100-fold more sensitive growth system for TBE virus than the common cell culture systems [16]. Therefore, supernatants from all of the mutants were also inoculated into litters of suckling mice to assess the viability of capsid deletion mutants in this system. In contrast to the corresponding experiment in BHK-21 cells, all of the mutants successfully infected these mice, as revealed by the appearance of typical encephalitic symptoms [10]. Virus harvested from the brains of these animals was subjected to a second round of infection of suckling mice and then used to infect BHK-21 cells. It was determined that all the mutants also were viable in cell culture and were efficiently and readily passaged. Thus it was possible to obtain viable virus progeny from mutants carrying deletions in the capsid proteins ranging in size from 4 to 30 amino acid residues [9, 10].

The capsid protein coding regions of these derived and viable virus progeny were sequenced to confirm the presence of the engineered deletions and to determine whether additional mutations might have arisen during the isolation process. The presence of the desired deletions was confirmed in all cases. No additional mutations were identified in the mutants carrying deletions as large as 16 amino acid residues. However, all mutants carrying larger deletions were found to have additional mutations in protein C that apparently had arisen during the passages [10]. Independent additional passaging experiments were performed, and the derived virus isolates were also sequenced. In summary, these experiments yielded

Table 1. Spontaneous mutations emerging during the isolation of viable mutants with large deletions in protein C

Mutants	Spontaneous mutations[a]	
	Nucleotides	Amino acids
C(Δ28–46)	C302U	P57L
C(Δ28–48)	Du364–387	Du78–85
	A341U	Q70L
	C374U	T81M
	Du302–363	Du59-K77
	C347U	T72I
	A299U, G328U, C347U	N56I, V66F, T72I
	G328U, C374U	V66F, T81M
C(Δ28–54)	A368U	K79I
	C374U	T81M
	A401U	Q90L
C(Δ28–57)	Du202–213 + 304–366	Du24–27 + 58–78
	A368U	K79I

[a]The position numbers correspond to the RNA genomic sequence (nucleotides) of Tick-borne encephalitis virus strain Neudoerfl (GenBank accession number U27495) or the amino acid position in protein C. Each line represents the result of an individual passaging experiment. *Du*, duplication

a multitude of different spontaneous mutations, all of which are summarized in Table 1. In most of the cases, single point mutations had evolved, and there were also combinations of two or three point mutations or in-frame sequence duplications.

To investigate whether these additional mutations were responsible for restoring the viability of the deletion mutants, mutants were constructed that contained a deletion and a second-site mutation in protein C in a defined wild-type genetic background [10, 13]. Transfection of RNAs transcribed in vitro from a clone containing only the deletion, or containing the deletion in combination with a second site-mutation, and subsequent passaging experiments yielded clear evidence that the second-site mutations alone were capable of restoring viability. Mutants carrying both a deletion and a second-site mutation remained genetically stable during subsequent passages, indicating that there was no selection pressure towards the generation of additional compensatory mutations in protein C.

Biochemical and biological characterizations

The production and export of protein E was investigated on a quantitative level for some of the deletion mutants [9]. This revealed that mutants produced approximately equal quantities of protein E per cell at various time-points postelectroporation, indicating that the deletions in protein C did not measurably diminish protein expression. The export of protein E was found to be somewhat reduced and delayed compared to the wild-type control, suggesting influence of the mutations on particle formation and release.

Flavivirus structural proteins can, in principle, assemble into two different forms of particles: (i) nucleocapsid-containing virions with a diameter of 50 nm and a buoyant density of 1.19 g/cm^3 and (ii) capsid-less, non-infectious subviral particles with a diameter of about 30 nm and a buoyant density of only 1.14 g/cm^3 [5, 22]. Rate zonal centrifugation using a discontinuous sucrose gradient can be used to distinguish between these two particle forms [9]. The analysis of particles released from BHK-21 cells transfected with the mutant RNAs indicated that mutants carrying only small deletions exhibited peaks corresponding to both viral and subviral particles. With increasing deletion sizes, however, larger proportions of subviral particles and decreasing amounts of viral particles were detected in this type of analysis [9]. To better define the physical forms of these particles, similar separations were performed on continuous sucrose gradients and the particle-containing fractions were further analyzed by equilibrium sucrose density centrifugation. This analysis showed that the subviral particle fractions had, corresponding to the established density of non-infectious recombinant subviral particles, a buoyant density of 1.14 g/cm^3 [1, 22].

To characterize the in vitro growth properties of the mutant viruses C(Δ28–35), C(Δ28–39), and C(Δ28–43), plaque assays were performed using PS cells and single-step growth curves were prepared on CE-cells [9, 10]. The mutants were found to produce clear but significantly smaller plaques compared to the wild-type virus. Mutants with larger deletions formed smaller plaques than did mutants with smaller deletions. Growth curves were obtained by infecting CE cells with an m.o.i. of approximately 1 and monitoring the release of newly formed viral infectivity during 1-h periods. The results demonstrated that the release of infectious particles was reduced and delayed compared to that of the wild-type virus control. Nevertheless, the mutants ultimately reached plateau values that were only 10- to 100-fold lower than the wild-type level.

The pathogenicity and immunogenicity of mutant C(Δ28–43) and the two "double-mutants" C(Δ28–48/Q70L) and C(Δ28–48/Du78–85) were assessed in the adult mouse model [12, 16, 17]. Upon peripheral inoculation, virulent, neuroinvasive TBEV strains caused lethal encephalitis in almost all infected animals, whereas attenuated strains caused asymptomatic infections that induced a specific antibody response. Inoculation of groups of mice with different doses of virus allows the determination of the 50-percent-lethal dose (LD_{50}) as a parameter of neuroinvasiveness. Also, from the determination of seroconversion in surviving mice the 50-percent-infectious dose (ID_{50}) can be calculated. With regard to vaccine development, strains with a high LD_{50} and a low ID_{50}, conveniently expressed as the LD_{50}/ID_{50} ratio (called the attenuation index) are most desirable. All three of the mutants were found to be highly attenuated [9, 10]. Mutant C(Δ28–48/Du78–85) did not cause disease in any of the mice up to the highest inoculation dose of 10^6 p.f.u. In the case of mutant C(Δ28–48/Q70L), one mouse out of 10 was killed at the maximum inoculation dose, and none at any lower dose.

With respect to peripheral infectivity (subcutaneous route), and thus the induction of seroconversion, mutants C(Δ28–48/Q70L) and C(Δ28–48/Du78–85) were found to be superior to mutant C(Δ28–43), yielding attenuation index values of more than 10^5. A challenge experiment using a lethal dose of virulent TBEV strain Hypr [24] showed that all of the mice that had seroconverted after being inoculated with the mutants were fully protected against disease, indicating that the observed antibody response correspond with the development of a protective immunity.

Discussion and conclusions

TBEV is capable of forming infectious particles with a capsid protein that lacks major portions including up to almost one-third of its amino acid sequence. This finding reveals a remarkable and previously unexpected structural and functional flexibility of this protein and of the assembly process. Because flaviviruses share a high degree of structural homology, it is likely that viable capsid deletion mutants can be generated for other members of the genus *Flavivirus*, perhaps for viruses of the other two genera of the family *Flaviviridae* [23]. This prediction, however, as yet awaits experimental verification.

The viability of deletion mutants with 19 amino acid residues or more depended on the emergence of additional mutations [10]. These were generally located downstream of the introduced deletion. A striking common feature of all the observed spontaneously arising point mutations was that they replaced polar with nonpolar residues or a hydrophobic residue that has a small side chain with a larger residue such as Phe, which is known to particularly favor membrane integration [3]. It had been proposed previously that the central hydrophobic domain of protein C mediates membrane association of the nucleocapsid [18], and this may be an important determinant for the assembly of infectious particles. Our results now rule out the possibility that individual residues of this domain are essential for viability, e.g., by providing specific protein interactions with the membrane-spanning domains of the viral surface proteins. The results do, however, corroborate the notion that hydrophobic interactions play an important role in the assembly process.

A central aspect of these investigations is the identification of protein C as a target for the development of live flavivirus vaccines. Although attenuation of flaviviruses has been achieved by numerous approaches, including the specific introduction of mutations in structural and non-structural proteins and the terminal NCRs [8, 12, 16, 17, 20], the potential to generate viable mutants carrying deletions in the capsid protein and to achieve significant attenuation by this method had not been recognized previously. Moreover, our data indicated that this attenuation principle has a number of important advantages over other approaches. Unlike most of the mutations that have been introduced into non-structural proteins or NCRs, this attenuation apparently does not rely on diminished translation or replication, but maintains a high level of protein expression. This should be beneficial for achieving an efficient and protective immune response. Moreover, all of the known major antigenic targets required for the development of a protective immune response, in particular glycoprotein E, which contains important neutralizing epitopes, are fully intact in these constructs. This should mean a major advantage over approaches that achieve attenuation through various protein E mutations. Our data suggest that the attenuation achieved by deletions within the central hydrophobic domain of the capsid protein is mainly based on the disturbance of the assembly of infectious virus particles, while still allowing subviral particles to be formed. These subviral particles, however, are non-infectious and known to be excellent immunogens [6, 19]. A general advantage of deletion mutations compared to point mutations resides in the practical impossibility of reversion to a wild-type sequence. Although second-site mutations were observed in the case of mutants carrying particularly large deletions, it is important to note that these were capable of increasing infectivity but they did not restore the neuroinvasiveness of TBEV and remained genetically stable during subsequent passages [10]. The data obtained in the mouse experiments demonstrated the excellent safety and immunogenicity of both mutants with or without second-site mutations. The attenuation index values determined for the two double mutants were the highest obtained so far for any mutant of TBEV. It is very likely that the principle of introducing deletions into the capsid protein to achieve stable attenuation is also applicable to other flaviviruses. Combining such deletions with second-site mutations in protein C or mutations elsewhere in the genome that have been shown to attenuate virulence by a different molecular mechanism may help to further improve the safety and efficacy of such vaccine candidates. Capsid deletion mutants represent a promising starting point for the development of live flavivirus vaccines.

References

1. Allison SL, Stadler K, Mandl CW, Kunz C, Heinz FX (1995) Synthesis and secretion of recombinant tick-borne encephalitis virus protein E in soluble and particulate form. J Virol 69: 5816–5820
2. Arroyo J, Guirakhoo F, Fenner S, Zhang ZX, Monath TP, Chambers TJ (2001) Molecular basis for attenuation of neurovirulence of a yellow fever virus/Japanese encephalitis virus chimera vaccine (ChimeriVax-JE). J Virol 75: 934–942
3. Braun P, von Heijne G (1999) The aromatic residues Trp and Phe have different effects on the positioning of a transmembrane helix in the microsomal membrane. Biochem 38: 9778–9782
4. Burke DS, Monath TP (2001) Flaviviruses, pp 1043–1125. In: Knipe DM, Howley PM et al. (eds), Fields virology, 4th edn. Lippincot Williams & Wilkins, Philadelphia, Pa
5. Ferlenghi I, Clarke M, Ruttan T, Allison SL, Schalich J, Heinz FX, Harrison SC, Rey FA, Fuller SD (2001) Molecular organization of a recombinant subviral particle from tick-borne encephalitis virus. Mol Cell 7: 593–602
6. Heinz FX, Allison SL, Stiasny K, Schalich J, Holzmann H, Mandl CW, Kunz C (1995) Recombinant and virion-derived soluble and particulate immunogens for vaccination against tick-borne encephalitis. Vaccine 13: 1636–1642
7. Heinz FX, Tuma W, Guirakhoo F, Kunz C (1986) A model study of the use of monoclonal antibodies in capture enzyme immunoassays for antigen quantification

exploiting the epitope map of tick-borne encephalitis virus. J Biol Stand 14: 133–141
8. Holzmann H, Stiasny K, Ecker M, Kunz C, Heinz FX (1997) Characterization of monoclonal antibody-escape mutants of tick-borne encephalitis virus with reduced neuroinvasiveness in mice. J Gen Virol 78: 31–37
9. Kofler RM, Heinz FX, Mandl CW (2002) Capsid protein C of tick-borne encephalitis virus tolerates large internal deletions and is a favorable target for attenuation of virulence. J Virol 76: 3534–3543
10. Kofler RM, Leitner A, O'Riordain G, Heinz FX, Mandl CW (2003) Spontaneous mutations restore the viability of tick-borne encephalitis virus mutants with large deletions in protein C. J Virol 77: 443–451
11. Lindenbach BD, Rice CM (2001) *Flaviviridae*: The viruses and their replication. In: Knipe DM, Howley PM et al. (eds), Fields Virology. 4th edn. Lippincott Williams & Wilkins, Philadelphia, pp 991–1041
12. Mandl CW, Allison SL, Holzmann H, Meixner T, Heinz FX (2000) Attenuation of tick-borne encephalitis virus by structure-based site-specific mutagenesis of a putative flavivirus receptor binding site. J Virol 74: 9601–9609
13. Mandl CW, Ecker M, Holzmann H, Kunz C, Heinz FX (1997) Infectious cDNA clones of tick-borne encephalitis virus European subtype prototypic strain Neudoerfl and high virulence strain Hypr. J Gen Virol 78: 1049–1057
14. Mandl CW, Heinz FX, Kunz C (1988) Sequence of the structural proteins of tick-borne encephalitis virus (Western subtype) and comparative analysis with other flaviviruses. Virology 166: 197–205
15. Mandl CW, Heinz FX, Stockl E, Kunz C (1989) Genome sequence of tick-borne encephalitis virus (Western subtype) and comparative analysis of nonstructural proteins with other flaviviruses. Virology 173: 291–301
16. Mandl CW, Holzmann H, Meixner T, Rauscher S, Stadler PF, Allison SL, Heinz FX (1998) Spontaneous and engineered deletions in the 3′ noncoding region of tick-borne encephalitis virus: construction of highly attenuated mutants of a flavivirus. J Virol 72: 2132–2140
17. Mandl CW, Kroschewski H, Allison SL, Kofler R, Holzmann H, Meixner T, Heinz FX (2001) Adaptation of tick-borne encephalitis virus to BHK-21 cells results in the formation of multiple heparan sulfate binding sites in the envelope protein and attenuation in vivo. J Virol 75: 5627–5637
18. Markoff L, Falgout B, Chang A (1997) A conserved internal hydrophobic domain mediates the stable membrane integration of the dengue virus capsid protein. Virology 233: 105–117
19. Mason PW, Pincus S, Fournier MJ, Mason TL, Shope RE, Paoletti E (1991) Japanese encephalitis virus-vaccinia recombinants produce particulate forms of the structural membrane proteins and induce high levels of protection against lethal JEV infection. Virology 180: 294–305
20. McMinn PC (1997) The molecular basis of virulence of the encephalitogenic flaviviruses. J Gen Virol 78: 2711–2722
21. Ruggli N, Rice CM (1999) Functional cDNA clones of the flaviviridae: strategies and applications. Adv Virus Res 53: 183–207
22. Schalich J, Allison SL, Stiasny K, Mandl CW, Kunz C, Heinz FX (1996) Recombinant subviral particles from tick-borne encephalitis virus are fusogenic and provide a model system for studying flavivirus envelope glycoprotein functions. J Virol 70: 4549–4557
23. van Regenmortel MHV, Faquet CM, Bishop DHL, Carstens EB, Estes MK, Lemon SM, Maniloff J, Mayo MA, McGeoch DJ, Pringle CR, Wickner RB (eds) (2000)

Virus taxonomy: classification and nomenclature of viruses. Academic Press, London, pp 859–878

24. Wallner G, Mandl CW, Ecker M, Holzmann H, Stiasny K, Kunz C, Heinz FX (1996) Characterization and complete genome sequences of high- and low-virulence variants of tick-borne encephalitis virus. J Gen Virol 77: 1035–1042

Author's address: Dr. Christian W. Mandl, Institute of Virology, Kinderspitalgasse 15, A-1095 Vienna, Austria; e-mail: christian.mandl@univie.ac.at

Arch Virol (2004) [Suppl] 18: 201–205

Archives of Virology

Tick-borne encephalitis and the impact of vaccination

F. X. Heinz and **C. Kunz**

Institute of Virology, University of Vienna, Vienna, Austria

Summary. TBE virus is endemic in many parts of Europe and Northern Asia, and in these regions it causes more than 10,000 severe cases of central nervous system disease in humans each year. The virus is primarily transmitted to humans when infected ticks take a blood meal, but infections due to the consumption of unpasteurized milk, primarily from goats, occur in certain regions. Based on genetic analyses, three closely related subtypes can be distinguished and are designated European, Siberian, and Far Eastern subtype according to their primary geographic distribution. Consistent with their close antigenic relationships, immunization studies in animals have revealed a high degree of cross-protection between virus strains belonging to different subtypes. The commercially available vaccines in Europe consist of highly purified inactivated whole TBE virus. Austria is the country with the highest coverage of TBE vaccination (86% of the total population) and this has led to a dramatic reduction in the annual number of clinical cases and proves under field conditions that vaccination is an effective means for the prophylaxis of TBE.

Introduction

Tick-borne encephalitis virus (TBEV), a member of the genus *Flavivirus* in the family *Flaviviridae*, can cause severe central nervous system diseases in humans. Based on its molecular characteristics, TBEV is closely related to several other important arthropod-transmitted human pathogens, such as yellow fever, dengue, Japanese encephalitis, and West Nile viruses [15]. Virions contain the genomic plus-stranded RNA with a length of approximately 11,000 nucleotides, which is enclosed in an isometric capsid composed of the single capsid protein C. This is surrounded by a viral envelope containing two membrane-associated proteins, E and M. The E protein is the major constituent of the virion surface. It is the viral hemagglutinin that mediates both receptor-binding and low pH-triggered fusion after virus uptake by receptor-mediated endocytosis, induces hemagglutination-inhibiting and neutralizing antibodies that are protective, and is the most important diagnostic antigen [9]. The atomic structure of the E protein from TBEV, determined by X-ray crystallography [16], revealed a new class of viral envelope proteins, now designated class II viral fusion proteins (see also article by Heinz, Stiasny and Allison, this volume).

The disease and its diagnosis

In Europe, TBE has been described as a disease that takes a biphasic course in most patients presenting with symptoms of the central nervous system (CNS) [12, 7]. The first phase starts after an incubation period of about one week, with relatively undifferentiated symptoms, including fever, headache, muscle pain, and malaise, which last for less than a week. For 70 to 80% of those infected the disease terminates at this stage, but 20 to 30% of the patients develop an overt disease of the CNS that starts a median of 8 days after the end of first phase [7]. This stage of the disease presents as meningitis, encephalitis, meningoencephalomyelitis, or meningoencephaloradiculitis. Children usually have a milder course of the disease than do older people, and meningitis in children usually resolves completely. A significant proportion of the patients, however, suffer from long-lasting sequelae, including headache, depression, cognitive dysfunction, hearing defects, and flaccid paralysis, primarily of the shoulder girdle and arms [12, 7]. The case-fatality rate in Europe has been reported to be in the range of 0.5 to 1.5%.

In Russia, significantly higher case-fatality rates have been reported, ranging from 6 to 8% in Siberia to 20 to 40% in the Far East [6]. This could be due to differences in the virulence of virus strains circulating in different geographic regions, but it is not clear at present to which extent these figures are biased by differences in the diagnosis and hospitalization rates, especially of milder cases. Chronic and progressive forms of TBE have been described in Eastern Russia but not in Europe, and even seronegative patients have been reported to suffer from this syndrome. Recently, a virus isolated from such a patient was characterized in detail, and attempts have been made to link specific genetic differences to the proposed biological properties [6].

Due to its relatively ill-defined clinical picture, the diagnosis of TBE depends on the determination of specific laboratory findings [10]. The method of choice is the detection of TBEV-specific antibodies in patient's sera by enzyme immunoassays. Usually, both IgM and IgG antibodies are already present when a patient is admitted to the hospital, i.e. at the beginning of the CNS disease, and in most cases the laboratory diagnosis can be based on the analysis of a single serum sample. Diagnostic problems can be caused by cross-reactive antibodies due to other flavivirus infections or to TBE-specific antibodies induced by vaccination.

Distribution and molecular epidemiology

TBE is endemic in certain parts of most European countries (except Great Britain, the Benelux countries, Portugal, and Spain), and large parts of northern Asia including Russia, northern China, and northern Japan. Annually more the 10,000 hospitalized cases are officially recorded in TBEV-endemic countries, but it is likely that a significant proportion of TBE cases remains undetected due to underdiagnosis. The main tick vectors are *Ixodes ricinus* (primarily Europe) and *Ixodes persulcatus* (primarily Asia) and most natural foci of TBE virus are found within the 7 °C isotherm [12, 19]. Insectivores such as shrews, moles, and hedgehogs act as the main virus reservoirs, but also larger animals such as goats, sheep, and cattle also can be infected. These animals secrete the virus into their milk and infections due to the consumption of unpasteurized milk (primarily of goats) or milk products are quite common in certain endemic regions [19]. Humans do not play any role in the maintenance of the virus in nature.

TBE virus is a member of the tick-borne serocomplex of flaviviruses that includes louping ill, Omsk hemorrhagic fever, Langat, Kyasanur Forest disease, and Powassan viruses. The E proteins of these viruses have at least 37% amino acid identity with mosquito-borne flaviviruses, such as yellow fever or dengue viruses. Conserved amino acid residues constitute the structural basis for the serological cross-reactivity often observed in hemagglutination-inhibition and enzyme immunoassays, whereas virus neutralization is much more specific and may allow individual flaviviruses to be distinguished [3]. By sequence analysis of TBEV isolates from the different endemic areas, three genetic lineages were identified; the corresponding strains are now designated as members of either the European, Siberian, or Far Eastern subtype [4, 6, 7, 14, 16, 19]. TBE virus is remarkably stable in nature. Based on the amino acid sequence of the E protein, the maximum degree of variation found within subtypes is about 2%, and even the difference between subtypes does not exceed 5 or 6% [9]. The latter is within the range of variation found for natural isolates of individual mosquito-borne flaviviruses, such as yellow fever and Japanese encephalitis virus, and it is therefore justified to define TBEV as a single virus species [15].

TBE vaccines and vaccination

The history of TBE vaccination has witnessed the use of different preparations of formalin-inactivated and also experimental live vaccines [2]. In Europe, two manufacturers (Baxter and Chiron-Behring) currently produce a highly purified formalin-inactivated whole virus vaccine that is in widespread use. With respect to TBE vaccination, Austria is in an exceptional situation, because a mass vaccination campaign was started at the beginning of the eighties, and since then 86% of the total population of Austria has been vaccinated [13]. In some highly endemic areas, vaccination coverage is as high as 95%, resulting in a dramatic decrease of annual numbers of hospitalized cases of TBE from as many as 700 to about 60 in recent years. This impressively proves the effectiveness of TBE vaccination in the field, since the disease incidence in neighbouring countries (e.g. the Czech Republic) with similar ecological and climatic conditions, but a low vaccination coverage, has remained high (Fig. 1). Based on the Austrian data, the protection rate of the TBE vaccine has been estimated to be about 98% [13]. Since the difference between the 3 TBE virus subtypes does not exceed that observed between isolates

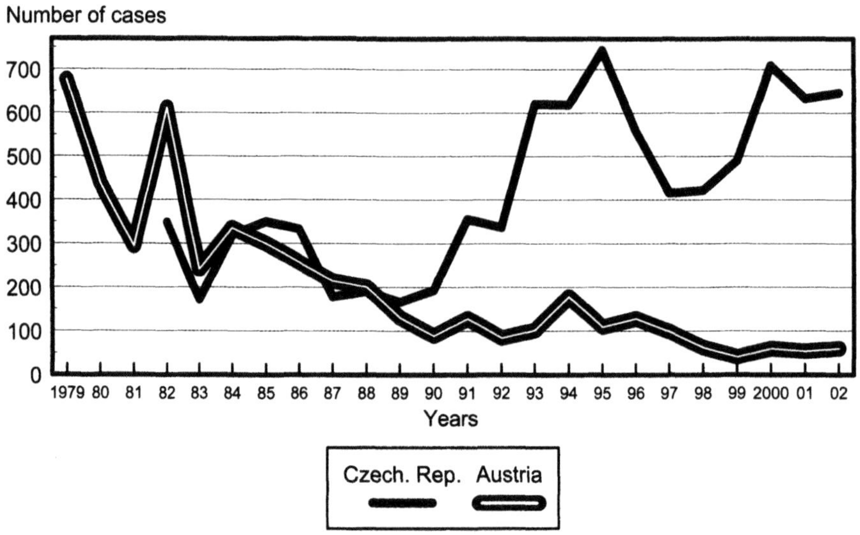

Fig. 1. Annual numbers of TBE cases in the years 1979 to 2002 in the Czech Republic and in Austria

of other flaviviruses for which vaccines are available (yellow fever, Japanese encephalitis), it is justified to assume that vaccines based on the European subtype will also protect against the other subtypes and vice versa. This has indeed been proven in animal immunization studies, which have revealed a high degree of cross-protection when the immunity induced by one subtype was challenged with a different subtype [4, 11, 17, 18, 20]. The available data clearly show that TBE can be effectively prevented by the use of the inactivated vaccines currently in use.

References

1. Bakhvalova VN, Rar VA, Tkachev SE, Matveev VA, Matveev LE, Karavanov AS, Dobrotvorsky AK, Morozova OV (2000) Tick-borne encephalitis virus strains of Western Siberia. Virus Res 70: 1–12
2. Barrett PN, Schober-Bendixen S, Ehrlich HJ (2003) History of TBE vaccines. Vaccine 21[Suppl 1]: S41–S49
3. Calisher CH, Karabatsos N, Dalrymple JM, Shope RE, Porterfield JS, Westaway EG, Brandt WE (1989) Antigenic relationships between flaviviruses as determined by cross-neutralization tests with polyclonal antisera. J Gen Virol 70(Pt 1): 37–43
4. Chiba N, Osada M, Komoro K, Mizutani T, Kariwa H, Takashima I (1999) Protection against tick-borne encephalitis virus isolated in Japan by active and passive immunization. Vaccine 17: 1532–1539
5. Ecker M, Allison SL, Meixner T, Heinz FX (1999) Sequence analysis and genetic classification of tick-borne encephalitis viruses from Europe and Asia. J Gen Virol 80: 179–185
6. Gritsun TS, Lashkevich VA, Gould EA (2003) Tick-borne encephalitis. Antiviral Res 57: 129–146
7. Haglund M, Gunther G (2003) Tick-borne encephalitis-pathogenesis, clinical course and long-term follow-up. Vaccine 21[Suppl 1]: S11–S18

8. Hayasaka D, Ivanov L, Leonova GN, Goto A, Yoshii K, Mizutani T, Kariwa H, Takashima I (2001) Distribution and characterization of tick-borne encephalitis viruses from Siberia and far-eastern Asia. J Gen Virol 82: 1319–1328
9. Heinz FX (2003) Molecular aspects of TBE virus research. Vaccine 21[Suppl 1]: S3–S10
10. Holzmann H (2003) Diagnosis of tick-borne encephalitis. Vaccine 21[Suppl 1]: S36–S40
11. Holzmann H, Vorobyova MS, Ladyzhenskaya IP, Ferenczi E, Kundi M, Kunz C, Heinz FX (1992) Molecular epidemiology of tick-borne encephalitis virus: cross-protection between European and Far Eastern subtypes. Vaccine 10: 345–349
12. Kunz C (1992) Tick-borne encephalitis in Europe. Acta Leidensia 60: 1–14
13. Kunz C (2003) TBE vaccination and the Austrian experience. Vaccine 21[Suppl 1]: S50–S55
14. Lundkvist K, Vene S, Golovljova I, Mavtchoutko V, Forsgren M, Kalnina V, Plyusnin A (2001) Characterization of tick-borne encephalitis virus from Latvia: evidence for co-circulation of three distinct subtypes. J Med Virol 65: 730–735
15. Family *Flaviviridae*. In: van Regenmortel MHV, Fauquet CM, Bishop DHL et al. (eds), Virus taxonomy: Seventh Report of the International Committee on Taxonomy of Viruses. Academic Press, San Diego
16. Rey FA, Heinz FX, Mandl C, Kunz C, Harrison SC (1995) The envelope glycoprotein from tick-borne encephalitis virus at 2 Å resolution. Nature 375: 291–298
17. Schmaljohn C, Custer D, VanderZanden L, Spik K, Rossi C, Bray M (1999) Evaluation of tick-borne encephalitis DNA vaccines in monkeys. Virology 263: 166–174
18. Schmaljohn C, VanderZanden L, Bray M, Custer D, Meyer B, Li D, Rossi C, Fuller D, Fuller J, Haynes J, Huggins J (1997) Naked DNA vaccines expressing the prM and E genes of Russian spring summer encephalitis virus and Central European encephalitis virus protect mice from homologous and heterologous challenge. J Virol 71: 9563–9569
19. Suss J (2003) Epidemiology and ecology of TBE relevant to the production of effective vaccines. Vaccine 21[Suppl 1]: S19–S35
20. Vorob'eva MS, Rasshchepkina MN, Ladyzhenskaia IP, Gorbunov MA, Pavlova LI, Bektimirov TA (1996) Comparative study of inactivated cultured vaccines against tick-borne encephalitis manufactured in Russia and in Austria by the "Immuno" firm. Vopr Virusol 41: 221–224
21. Zlobin VI, Demina TV, Belikov SI, Butina TV, Gorin OZ, Adel'shin RV, Grachev MA (2001) Genetic typing of tick-borne encephalitis virus based on an analysis of the levels of homology of a membrane protein gene fragment. Vopr Virusol 46: 17–22

Author's address: Dr. Franz X. Heinz, Institute of Virology, Kinderspitalgasse 15, A-1095 Vienna, Austria; e-mail: franz.x.heinz@univie.ac.at

Arch Virol (2004) [Suppl] 18: 207–220

Archives of Virology

Future vaccines against emerging encephalitides

R. E. Johnston and **N. L. Davis**

Carolina Vaccine Institute, University of North Carolina,
Chapel Hill, North Carolina, U.S.A.

Summary. The vaccine strategies available for control of emerging encephalitides range in a continuum from traditional approaches to those utilizing new technologies. In this report, we explore the use of live attenuated vaccines where the attenuating mutations have been selected in a rational way to improve attenuation without sacrificing effectiveness. A strategy for paired lethal and resuscitating mutations is presented that will greatly reduce the possibility of reversion to virulence. Finally, we describe an example of a vaccine vector system that could be rapidly adapted for use against these virus diseases as they emerge.

Introduction

Developing vaccines against existing and emerging encephalitis viruses is a challenging task. Outbreaks of these viruses are unpredictable and usually affect relatively small populations, or occur in poorer areas of the world. The market aspects of both circumstances make these diseases relatively unattractive targets for vaccine development by commercial enterprises. However, the scientific possibilities available for vaccine development for emerging diseases include a rich variety of new technologies as well as old approaches augmented with new technology. In this chapter, we will explore three of these approaches using an alphavirus, Venezuelan equine encephalitis virus (VEEV), as both a specific example and a model for other diseases.

Live attenuated vaccines

Traditional live attenuated vaccines have provided the most successful and cost effective interventions for the prevention of human disease. Most often, these have been derived by blind passaging in cell culture followed by empirical testing for attenuation and immunogenicity. Such vaccines can suffer from insufficient attenuation leading to moderate or severe reactions in vaccine recipients, or conversely they may be overattenuated leading to poor immunization. With the advent of molecular technologies and the careful explication of viral pathogenesis in animal models, it is possible to design live attenuated vaccines to avoid these problems and retain the low cost and long term efficacy advantages associated with this vaccine approach.

Informed selection of attenuating point mutations

VEEV causes encephalitis in mice, in horses and in a small proportion of human infections. Examination of the molecular genetics of VEEV and its pathogenesis in the mouse model suggest several principles upon which to base new live virus vaccines.

Optimal live virus vaccines are not simply viruses less fit for replication

The pathogenetic effects of virus replication are manifest through a series of complex interactions within the infected animal or human. In many ways, each differentiated tissue or organ system with which the virus interacts constitutes a different host, especially at the level of individual infected cells. Thus, a given virus is genetically optimized to replicate in a series of very different host cell types within the body during the course of disease. In contrast, passage of viruses in cell culture for the derivation of attenuated viruses amounts to a selection for virus mutants optimized for replication in a single type of cultured cell. It is not too surprising then, that many such cell-adapted mutants are attenuated, being less able to accomplish the intricate series of interactions with multiple cell types required to cause disease *in vivo*.

This understanding suggests that rather than blind passage in cell culture, one should actually select for viruses that can grow more rapidly in cell culture [4, 5]. Sindbis virus, the prototype alphavirus, was passaged in baby hamster kidney (BHK) cells under selection for rapid growth, and this led to the selection of highly attenuated mutants within fewer than 10 passages. Comparison of over a hundred attenuated isolates from this passage series revealed that all the attenuated isolates shared the property of accelerated penetration of BHK cells [14, 36]. When VEEV was passaged under a specific selection for rapid penetration, 70% of isolates from as early as the fourth passage were rapidly penetrating and attenuated in mice [29]. In contrast, 83 blind passages were required for attenuation of the VEE vaccine strain, TC-83. Sequencing of the glycoprotein genes of a number of these isolates revealed a panel of candidate attenuating mutations [16].

In parallel, a full-length cDNA clone of the VEEV single-stranded RNA genome was constructed [18]. Genetic manipulations could be performed at the level of the cDNA clone, and those changes could be recovered in the genomes of infectious virus generated from the clone. This process depends on the fact that the genomic RNA of VEEV is message sense, and purified genomic RNA extracted from the virus can initiate viral infection when introduced artificially into susceptible cells. In the cDNA clone, the VEEV genomic sequences were placed downstream of a phage T7 promoter. Linearization of the clone at a unique restriction site downstream of the VEEV sequences and *in vitro* transcription with

T7 polymerase generated VEEV genomic replicas that could be electroporated into cells, resulting in the production of infectious virus. Using this genetic system, individual candidate attenuating mutations could be introduced into the clone at the DNA level and then tested as viruses for their effect on virulence and pathogenesis in the mouse model [22]. In addition, the balance between attenuation and overattenuation could be sustained by judicious choice of individual mutations, which were attenuated to different degrees, and by combining two or three attenuating mutations in the same virus to minimize the chances of reversion [29].

Selection of attenuating point mutations based on studies of viral pathogenesis

Our studies of VEEV pathogenesis have utilized molecularly cloned viruses to minimize the effects of cell culture passage on pathogenetic phenotypes and to take advantage of specific mutations introduced into the clone. This approach has defined a pathogenetic pathway in the mouse, by which VEEV infection progresses from the site of inoculation in the left rear footpad to eventual death by encephalitis [15, 22]. The virus infects Langerhans cells at the site of inoculation, and these cells move rapidly to the draining lymph node where replication can be detected within 4 hrs post-inoculation [35]. Near peak levels of viremia are established by 12 hrs, and all the lymphoid tissues are infected by 18 hrs. Replication in the brain is detectable between 24 and 36 hrs. Invasion of the central nervous system (CNS) occurs by infection of peripheral nerves and subsequent progression into the CNS. Virus appears first in olfactory tracts of the brain, suggesting that the earliest entry into the CNS is via infection of olfactory nerves in the nasal neuroepithelium [10]. If the nasal neuroepithelium is artificially ablated, then the earliest appearance of virus in the CNS is in areas enervated by the trigeminal nerve.

The kinetics with which the virus progresses through the mouse suggested a tentative pathogenetic pathway leading ultimately to encephalitis and death. By analogy to a biochemical pathway, further elucidation of the pathogenetic pathway has been accomplished through the study of avirulent mutants. By definition, these mutants must be defective or blocked in one or more of the steps required to complete the progression to lethal encephalitis. Identification of the block points for multiple mutants make possible the elucidation of the pathogenetic pathway for the wild-type virus. An illustration of this is the avirulent mutant V3010, in which the glutamic acid at E2 amino acid 76 is replaced with lysine. The defect in this virus was manifested as a delay in reaching the draining lymph node [1, 22]. Virus could be isolated sporadically from serum, from other lymphoid organs and from the brain. However, half of the isolates from the draining lymph node and all the isolates downstream of the draining lymph node in the pathogenetic pathway were revertant, in that they contained sequence changes associated with reversion or displayed a virulence phenotype distinct from the inoculated V3010 virus.

There were important differences among the different revertants depending on the tissue source of the isolation [1]. Revertants from serum and the contralateral lymph node were second-site revertants which retained the original V3010 mutation at E2 76 but had an additional substitution of glutamic acid for lysine at E2 116. To demonstrate that the E2 116 mutation altered the V3010 phenotype, the putative second-site mutation was introduced into the V3010 clone, and the cloned virus containing both mutations (V3533) was compared to the biological isolates. The cloned and biological viruses shared the same pathogenesis phenotype. In V3533, the original block in reaching the draining lymph node was abrogated, spread and replication in the periphery was analogous to that of wild-type virus, the CNS was invaded, all the animals displayed clinical disease with neurological signs, but all recovered from the infection without obvious sequelae. Therefore, selective pressure for reversion *in vivo* is actually at the point of the genetic blockade imposed by the attenuating mutation. There is no selective pressure *per se* for reversion to wild-type virulence.

Vaccine lessons from studies of viral pathogenesis

What lessons can be learned from these experiments with respect to design of live attenuated vaccines? First, one should select point mutations which block early in the pathogenetic pathway, as the selection pressure to bypass the mutant block would not necessarily lead to a fully virulent revertant. However, if the block point imposed by a given mutation is close to the end of the pathway, i.e. near the point of death, then selection for reversion to bypass the block will be tantamount to selection for reversion to wild-type virulence. Second, genetic stability will be improved if multiple point mutations are used. Third, multiple defined mutations which block at different steps in the pathway may provide a more stable attenuated phenotype than multiple mutations blocking the same step. Finally, vaccine mutations which block too early in the pathway may result in too little replication for induction of protective immune responses.

Lethal mutations in vaccine design

Alphaviruses provide an example where a lethal mutation is an important element in a live attenuated vaccine strategy. These viruses initially produce one of their structural glycoproteins (E2) as a precursor molecule, PE2, which consists of E2 with an amino terminal extension of 55–65 amino acids (E3; Fig. 1). At the carboxy terminus of E3 is a canonical furin protease cleavage site, and in infected cells PE2 is cleaved by furin in the late Golgi resulting in the quantitative incorporation of E2 into budded virions. One of the attenuating mutations identified from a rapid penetration selection with another alphavirus, South African Arbovirus 86 (S.A.Ar-86) involved the addition of an N-linked carbohydrate chain at the first amino acid residue of E2, and this sterically prevented maturation of the E2 glycoprotein through furin-mediated cleavage of PE2 [41]. The resulting virus particles contained PE2 in place of E2, they infected cultured cells more efficiently, but they were highly attenuated in mice. This mutation was reproduced in a full-length cDNA clone of the prototype alphavirus, Sindbis. Electroporation of

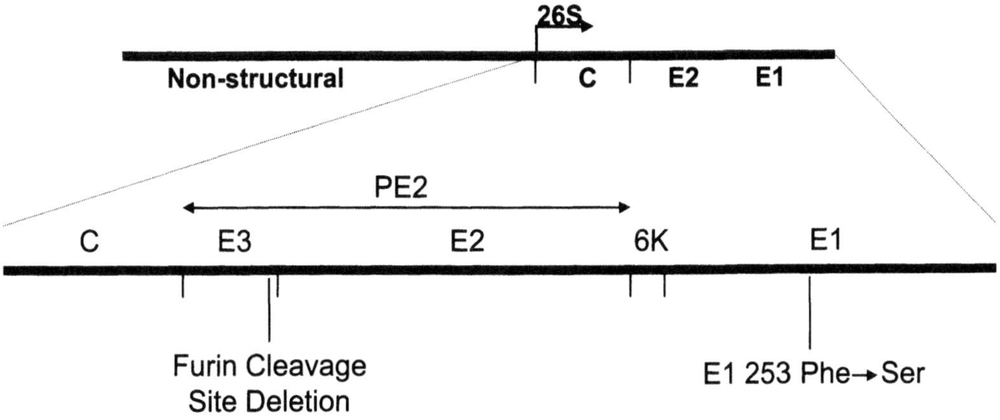

Fig. 1. VEE 3526: resuscitated cleavage site deletion vaccine. Resuscitated for cell culture growth attenuated in mice and non-human primates

the resulting *in vitro* RNA transcripts into cells programmed normal viral RNA synthesis, production of viral structural proteins and the release of viral particles containing PE2 instead of E2 [6, 24–26, 37]. Unlike the S.A.Ar-86 mutants, prototype Sindbis virions containing PE2 were not viable due to their failure to penetrate BHK cells. However, it was possible to select revertants of two types which were viable in cell culture: 1) Mutation in the N-X-S/T N-linked glycosylation signal abrogated glycosylation at position 1 of E2 and allowed PE2 cleavage. And 2) Second-site mutants retained the N-terminal glycosylation and the consequent inclusion of PE2 in the mutant virions but had a second mutation in either E2 or E3 that resuscitated the virus for the cell culture penetration phenotype. These resuscitated viruses were uniformly attenuated in animals.

In the VEEV clone, the four amino acid furin cleavage site was deleted by site directed mutagenesis, resulting in a virus genome which, when electroporated into permissive cells, could program normal RNA replication and the assembly of PE2-containing virions [12]. However, these virions were not infectious due to their inability to effect penetration of other new cells. As with Sindbis virus, VEEV second-site revertants in either E2 or E1 were isolated in cell culture. In the case of the V3526 double mutant, the lethal furin cleavage site deletion was retained, and a second, independently attenuating mutation in E1 restored cell culture infectivity (Fig. 1). V3526 was highly attenuated in mouse and primate models and it induced protection against both parenteral and aerosol challenge with virulent VEEV [38]. Compared to the existing vaccine TC-83, V3526 induced higher levels of antibody in a larger percentage of animals, was less reactogenic, and conferred stronger protection against challenge. The genetic stability of V3526 stems from the very low probability that the deletion will be repaired and from the fact that reversion of the second-site mutation to the wild-type amino acid will result in a dead virus rather than a virulent one. A cGMP lot of V3526 has been prepared and is currently undergoing requisite toxicity testing in anticipation of a phase I human trial.

The lethal deletion/resuscitating mutation strategy may be generally applicable in designing effective live attenuated vaccines with very low propensities for reversion to virulence. Recently, Kofler et al. [30, 31] created lethal deletions in the capsid protein of tick-borne encephalitis virus and found resuscitating mutations which render the double mutant viable in cell culture yet highly attenuated in animal models of disease.

Use of VEEV as a vaccine vector

Using viruses to express exogenous genes has its roots in bacteriophage transduction experiments, in which phage are used to move bacterial genes from one strain to another. In terms of vaccination, a number of different animal and human viruses have been utilized to express an immunizing gene from a given pathogen to effect protective immunization against that pathogen. Zoonotic viruses are especially well suited for this purpose. These viruses have the capacity for replication in humans, but unlike viruses that widely infect humans (e.g. adenovirus) or viruses to which large human populations are vaccinated (e.g. polio and vaccinia), only a small number of persons are already immune to most zoonotic viruses. This suggests that vectors derived from zoonotic viruses will express well in humans but that interference due to pre-existing immunity will be minimal.

The VEEV genome, when introduced into a cell, can be viewed as inducing a highly efficient machine for the production of large amounts of its own structural proteins through the transcription of a subgenomic mRNA encoding these proteins. If a gene of interest is substituted for the structural protein genes,

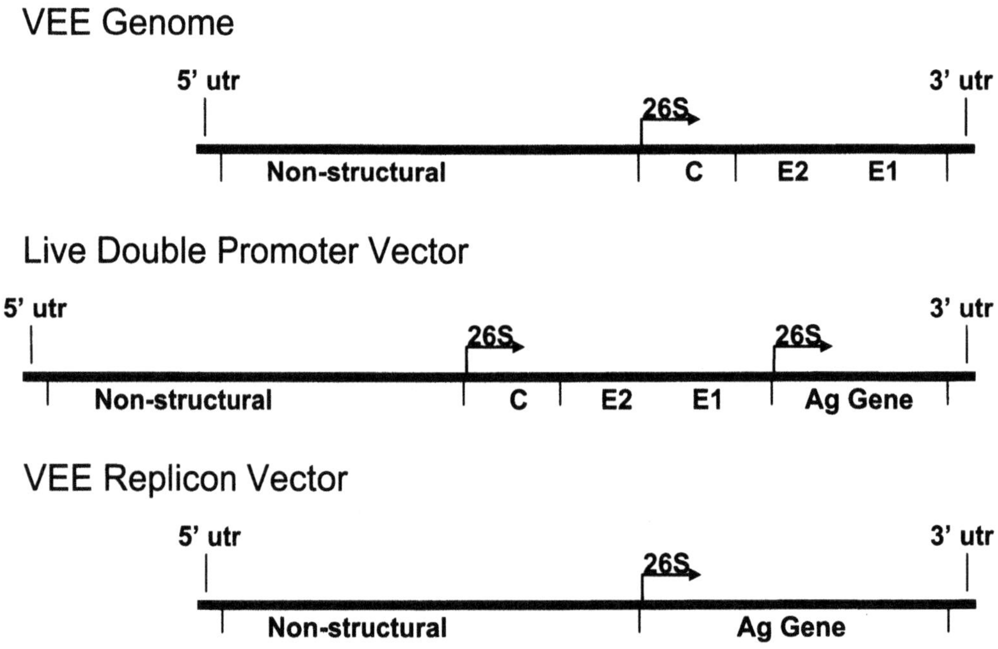

Fig. 2. VEE vaccine vectors

or if a gene of interest is inserted downstream of a second copy of the subgenomic promoter, then the VEEV machine expresses high levels of that gene product.

The vaccine vectors based on VEEV are of two types. In the first, an additional copy of the 26S subgenomic promoter was inserted between the end of the E1 gene and the start of the 3′ untranslated region (Fig. 2) [8, 9, 11]. This results in a viable virus capable of expressing an immunizing gene inserted after the second 26S promoter.

VEEV replicon vaccine vectors

The second type is a Areplicon vector in which the immunizing gene is substituted for the structural protein genes (Fig. 2) [40]. Upon electroporation of replicon RNA into cells, the VEEV replication machinery produces high levels of the subgenomic mRNA and hence high levels of the vectored gene product. However, as the structural protein genes are not present in the replicon RNA, no new virus particles are released. One can package the replicon in virions by supplying the structural proteins *in trans* from helper RNAs. The replicase proteins encoded on the replicon RNA provide the machinery for replicating the helper RNAs (Fig. 3), while the structural proteins encoded by the helpers encapsidate and envelope the replicon RNA. Only the replicon RNA is packaged into VEEV replicon particles (VRP), because the replicon RNA contains the *cis*-acting packaging signal deleted from

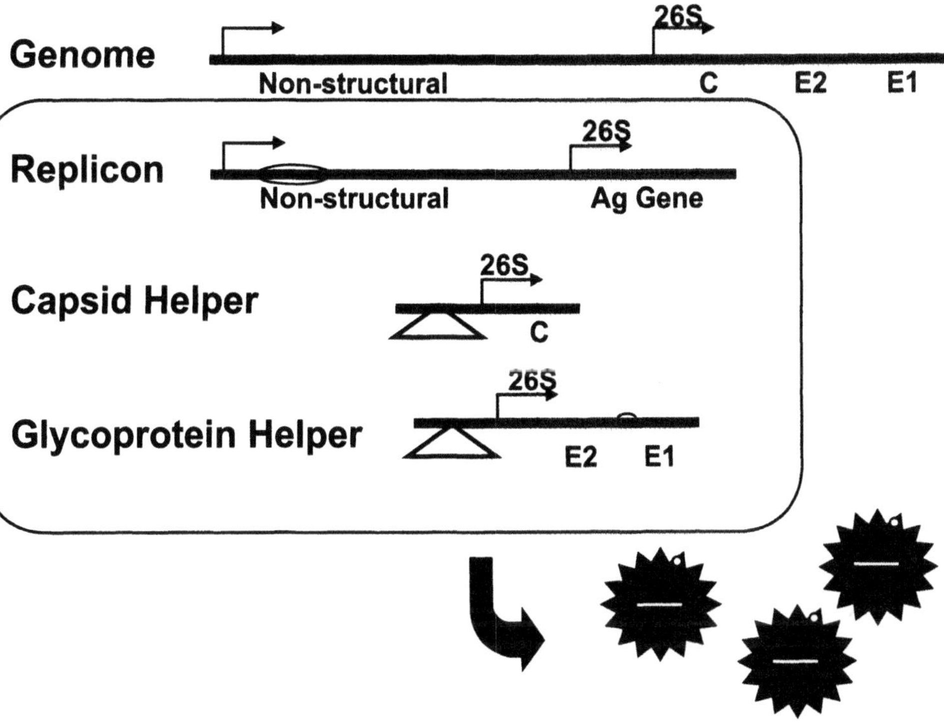

Fig. 3. Replicon expression and packaging

the helper RNAs. When the resulting VRP infect another cell, either in culture or *in vivo*, the replicon RNA is introduced, high levels of the immunizing gene product are synthesized, but no new particles are formed due the absence of the structural genes. Therefore, these single-cycle vectors are expressed only in the first cell infected.

Features of VEEV replicon vaccines

Safety in animal models: Redundant safety features are embedded into the replicon system. First, over one-third of the genome has been deleted. Second, helper functions are supplied in the form of two separate RNAs to greatly reduce the chances of re-generating an infectious virus through RNA-RNA recombination. Third, attenuating mutations can be introduced into the glycoprotein helper RNA so that even if a viable recombinant were formed, the worst case scenario would be a live attenuated VEEV vaccine strain. Finally, sensitive tests have been devised for the detection of propagation competent viruses in VRP preparations. After parenteral inoculation of various VRP in a number of laboratories, no clinical signs have been observed in experimental animals (rabbits, guinea pigs, macaques), in natural hosts of VEEV (adult mice, horses), in animals lacking a fully developed immune system (neonatal mice, one-day-old chickens, 19 day chicken embryos), and animals genetically incapable of cellular and humoral immune responses ($RAG^{-/-}$ mice).

High level gene expression: In a VEEV infected cell, the virus produces large amounts of its structural proteins to facilitate assembly of new virus particles. In a cell infected by VRP, this same machinery produces high levels of the immunizing gene product. For cytoplasmic proteins, such as the Lassa N protein [40] or the Gag protein of human (HIV) or simian (SIV) immunodeficiency viruses [13], up to 20% of the total cell protein is from the replicon expressed gene.

Induction of humoral and cellular immune responses: Both the humoral and cellular arms of the immune response are induced upon immunization with VRP [13]. This has been demonstrated in a variety of animal models including rodents and non-human primates.

Induction of mucosal immunity after parenteral inoculation: Mucosal immunity, in the form of mucosally produced IgA, is an important contributor to protection against a large number of pathogens. Usually, application of an immunogen to a mucosal site is required for induction of a mucosal response. However, VRP possess the ability to induce mucosal responses after parenteral inoculation [23]. Although the mechanism underlying this phenotype remains unclear, cells secreting IgA specific for the vectored gene have been detected in mucosal tissues of animals immunized subcutaneously with VRP (E. Richmond et al., University of North Carolina, unpublished). This suggests that a true mucosal response was elicited by VRP rather than simply the transudation of IgA antibodies from the plasma to a mucosal surface.

Induction of anti-vector (anti-VEEV) immune responses: Anti-VEEV antibodies can be detected in both rodent and primate models following immunization with VRP. However, in vaccination regimens involving multiple inoculations, boosted responses were observed with second and third immunizations. The effect of pre-existing anti-VEEV responses was tested in two experiments. In the first, groups of mice were immunized with diluent or V3014, a molecularly cloned virus containing an attenuating mutation in each of the glycoprotein genes. The 80% plaque reduction neutralization titers ($PRNT_{80}$) averaged 10,000 in the vaccinated group. Both control and vaccinated animals were subsequently immunized with priming and boosting inoculations of VRP expressing the influenza HA gene. The resulting ELISA titers against influenza were approximately 10,000 in both control and V3014 immunized groups.

In a second experiment, one group of mice received two consecutive inoculations of VRP expressing SIVgp140 followed by two consecutive inoculations of VRP expressing a portion of the HIV Gag gene. A control group received four inoculations of diluent. Then, both groups of animals received two consecutive inoculations of VRP expressing influenza HA. The resulting anti-influenza ELISA titers were 3200 and 1800 for the control and pre-immunized groups, respectively. At least in mice therefore, neither sequential vaccination with VRP nor the presence of pre-existing neutralizing antibody to VEEV affected the induction of immune responses by the VEEV replicon vectors.

Targeting of VRP to dendritic cells: Replication of VEEV virus in the mouse is first detected in the draining lymph node, and this feature of VEEV biology strongly suggested that vaccine vectors based on VEEV would be especially effective. Using VRP expressing the green fluorescent protein (GFP) and taking advantage of the fact that VRP express only in the first cell they infect, we demonstrated the initial infection of Langerhans cells at the site of inoculation, the rapid migration of these cells to the draining lymph node, and the continued expression of GFP in dendritic cells (DC) for 4–5 days thereafter [35]. The efficiency of DC infection *in vivo* is strongly influenced by the VEEV glycoprotein envelope of the VRP tested, with different glycoprotein mutants varying with respect to this parameter (A. West et al., University of North Carolina, unpublished). DC targeting also was influenced by the site and route of inoculation. However, with each of these variables, the ability to target DC was positively correlated with the induction of both cellular and humoral responses. Given that DC are among the most potent antigen presenting cells in the body, we feel that the innate ability of VEEV to target these cells is largely responsible for their ability to successfully immunize animals. Analogous studies in non-human primates are in progress.

Vaccine efficacy: VEEV replicon vectors have been used in successful immunization or challenge experiments for a wide range of pathogens and animal model systems. Especially in acute infection models, the VEEV vectors demonstrate the capacity for strong and protective immunization. Examples include influenza viruses in rodents [40] and chickens (H5N1, 42), bacteria and bacterial toxins in rodents [21, 32–34, 43], Ebola virus in rodents [39], Marburg virus in rodents and primates [27, 28], equine arteritis virus in horses [2, 3], Norwalk virus in rodents [23], feline immunodeficiency virus in cats [7], a human papilloma virus tumor model in mice [20, 44], as well as HIV and SIV in rodents [19], in primates [13], and as a contemplated human vaccine for HIV [17, 45].

Conclusions

Study of the molecular genetics of VEEV and its pathogenesis in animal models has provided a number of insights applicable to the design of live attenuated vaccines for VEEV and other emerging zoonotic viruses. Pathogenesis of these viruses may be considered in terms of a pathogenesis pathway providing a working framework for the genetic dissection of pathogenesis using wild-type virus and individual, defined, avirulent mutants, each of which interdicts the pathway at a specific stage(s). Although all avirulent mutants could be considered candidate live attenuated vaccines, pathogenesis studies suggest that: 1) Each mutation will block progression through the pathway at a specific step. 2) There is no selective pressure for reversion to wild-type virulence *per se*, only for reversion in the sense of bypassing the particular genetic blockade imposed by that specific mutation. 3) Preferred attenuating vaccine mutations are more likely found, therefore, among those that block early stages of the pathway. For mutants blocked at a late stage, reversion to bypass the genetic blockade may be tantamount to reversion to wild-type virulence. 4) Mutations which block too early in the pathway may be overattenuated. And 5) Multiple attenuating mutations, each blocking a different step in the pathway, will provide genetic stability to a live attenuated vaccine.

Deletion mutations provide very stable phenotypes, but in the context of a viral genome are often lethal for viral growth even in cell culture. However, the plasticity of viral genomes, especially RNA virus genomes, can give rise to second-site point mutations that partially compensate for a primary deletion. In VEEV mutants deleted for the furin cleavage site required for maturation of the E2 glycoprotein, selection of such second-site resuscitating mutations in cell culture resulted in viruses that were able to grow *in vitro* but were highly attenuated in rodent and primate models. This principle, creating a lethal deletion and then selecting for resuscitating point mutations, seems to apply also to flaviviruses, in which lethal deletions in the capsid gene can be compensated by the selection of point mutations elsewhere to generate a virus which is avirulent in animals [31]. These candidate vaccine strains display remarkable genetic stability, because the deletion is unlikely to be repaired and same-site reversion of the resuscitating point mutation will result in a dead virus rather than a virulent one.

Finally, the VEEV system illustrates the advantages of using a zoonotic virus as a vaccine vector system. Live, attenuated VEEV strains carrying a second 26S promoter and VEEV replicon particles, expressing a variety of immunizing genes from both bacterial and viral pathogens, have proven to be safe and effective vaccines in animal models.

Vaccine strategies do not yet exist for a number of significant human pathogens. The emergence of entirely new human pathogens such as HIV and severe acute respiratory syndrome (SARS) has been both dramatic and frightening, taking an enormous toll on human health, social interactions and economic activity. Because many of these consequences are most evident in poorer developing countries, western governments and business interests have not fully appreciated the urgency of these infectious disease emergencies, nor have they fully accepted the concept of a global imperative for the discovery and application of new vaccine strategies to combat them. We hope that the VEEV model can and will be helpful in the scientific development of additional safe, effective and affordable vaccine strategies in support of global public health needs.

Acknowledgments

The authors acknowledge with great appreciation the contributions of all the persons directly and indirectly associated with the Johnston/Davis laboratory over the more than 20 years encompassed by the research described in this article. Similarly, we would like to acknowledge the generous support our research has received over the years from the North Carolina Biotechnology Center, National Institutes of Health, the U. S. Army Medical Research and Development Command, and the International AIDS Vaccine Initiative.

References

1. Aronson JF, Davis NL, Grieder FB, Charles PC, Knott TA, Brown K, Johnston RE (2000) A single-site mutant and revertants arising *in vivo* define early steps in the pathogenesis of Venezuelan equine encephalitis virus. Virology 270: 111–123
2. Balasuriya UBR, Heidner HW, Davis NL, Wagner HM, Hullinger PJ, Hedges JF, Williams JC, Johnston RE, Wilson WD, Liu IK, MacLachlan NJ (2002) Alphavirus replicon particles expressing the two major envelope proteins of equine arteritis virus induce high level protection against challenge with virulent virus in vaccinated horses. Vaccine 20: 1609–1617
3. Balasuriya UBR, Heidner HW, Hedges JF, Williams JC, Davis NL, Johnston RE, MacLachlan NJ (2000) Expression of the two major envelope proteins of equine arteritis virus as a heterodimer is necessary for induction of neutralizing antibodies in mice immunized with recombinant Venezuelan equine encephalitis virus replicon particles. J Virol 74: 10623–10630
4. Baric RS, Moore DB, Johnston RE (1980) *In vitro* selection of an attenuated variant of Sindbis virus. ICN-UCLA Symp Mol Cell Biol 18: 685–694
5. Baric RS, Trent DW, Johnston RE (1981) A Sindbis virus variant with a cell determined latent period. Virology 110: 237–242
6. Boehme KW, Williams JC, Johnston RE, Heidner HW (2000) Linkage of an alphavirus host-range restriction to the carbohydrate processing phenotypes of the host cells. J Gen Virol 81: 161–170
7. Burkhard MJ, Valenski L, Leavell S, Dean GA, Tompkins WA (2002) Evaluation of FIV protein-expressing VEEV-replicon vaccine vectors in cats. Vaccine 21: 258–268
8. Caley IJ, Betts MR, Davis NL, Swanstrom R, Frelinger JA, Johnston RE (1999) Venezuelan equine encephalitis virus vectors expressing HIV-1 proteins: vector design strategies for improved vaccine efficacy. Vaccine 17: 3124–3135
9. Caley IJ, Betts MR, Irlbeck DM, Davis NL, Swanstrom R, Frelinger JA, Johnston RE (1997) Humoral, mucosal and cellular immunity in response to a human

immunodeficiency virus type 1 immunogen expressed by a Venezuelan equine encephalitis virus vaccine vector. J Virol 71: 3031–3038
10. Charles PC, Walters E, Margolis F, Johnston RE (1995) Mechanism of neuroinvasion of Venezuelan equine encephalitis virus in the mouse. Virology 208: 662–671
11. Davis NL, Brown K, Johnston RE (1996) A viral vaccine vector that expresses foreign genes in lymph nodes and protects against mucosal challenge. J Virol 70: 3781–3787
12. Davis NL, Brown KW, Greenwald GF, Zajac AJ, Zacny VL, Smith JF, Johnston RE (1995) Attenuated mutants of Venezuelan equine encephalitis virus containing lethal mutations in the PE2 cleavage signal combined with a second-site suppressor mutation in E1. Virology 212: 102–110
13. Davis NL, Caley IJ, Brown KW, Betts MR, Irlbeck DM, McGrath KM, Connell MJ, Montefiori DC, Frelinger JA, Swanstrom R, Johnson PR, Johnston RE (2000) Vaccination of macaques against pathogenic simian immunodeficiency virus with Venezuelan equine encephalitis virus replicon particles. J Virol 74: 371–378
14. Davis NL, Fuller FJ, Dougherty WG, Olmsted RA, Johnston RE (1986) A single nucleotide change in the E2 glycoprotein gene of Sindbis virus affects penetration rate in cell culture and virulence in neonatal mice. Proc Natl Acad Sci USA 83: 6771–6775
15. Davis NL, Grieder FB, Smith JF, Greenwald GF, Valenski ML, Sellon DC, Charles PC, Johnston RE (1994) A molecular genetic approach to the study of Venezuelan equine encephalitis virus pathogenesis. Arch Virol Suppl 9: 99–109
16. Davis NL, Powell N, Greenwald GF, Willis LV, Johnson BJB, Smith JF, Johnston RE (1991) Attenuating mutations in the E2 glycoprotein gene of Venezuelan equine encephalitis virus: Construction of single and multiple mutants in a full-length cDNA clone. Virology 183: 20–31
17. Davis NL, West A, Reap E, MacDonald G, Collier M, Dryga S, Maughan M, Connell M, Walker C, McGrath K, Cecil C, Ping L-H, Frelinger J, Olmsted R, Keith P, Swanstrom R, Williamson C, Johnson P, Montefiori D, Johnston RE (2002) Alphavirus replicon particles as candidate HIV vaccines. IUBMB Life 53: 209–211
18. Davis NL, Willis LV, Smith JF, Johnston RE (1989) *In vitro* synthesis of infectious Venezuelan equine encephalitis virus RNA from a cDNA clone: Analysis of a viable deletion mutant. Virology 171: 189–204
19. Dong M, Zhang PF, Grieder F, Lee J, Krishnamurthy G, VanCott T, Broder C, Polonis VR, Yu XF, Shao Y, Faix D, Valente P, Quinnan GV Jr (2003) Induction of primary virus-cross-reactive human immunodeficiency virus type 1-neutralizing antibodies in small animals by using an alphavirus-derived *in vivo* expression system. J Virol 77: 3119–3130
20. Eiben GL, Velders MP, Schreiber H, Cassetti MC, Pullen JK, Smith LR, Kast WM (2002) Establishment of an HLA-A*0201 human papillomavirus type 16 tumor model to determine the efficacy of vaccination strategies in HLA-A*0201 transgenic mice. Cancer Res 62: 5792–5799
21. Gipson CL, Davis NL, Johnston RE, deSilva AM (2003) Evaluation of Venezuelan equine encephalitis (VEEV) replicon-based outer surface protein A (OspA) vaccines in a tick challenge mouse model of Lyme disease. Vaccine 21: 3875–3884
22. Grieder FB, Davis NL, Aronson JF, Charles PC, Sellon DC, Suzuki K, Johnston RE (1995) Specific restrictions in the progression of Venezuelan equine encephalitis virus induced disease resulting from single amino acid changes in the glycoproteins. Virology 206: 994–1006
23. Harrington PR, Yount B, Johnston RE, Davis NL, Moe C, Baric RS (2002) Systemic, mucosal and heterotypic immune induction in mice inoculated with Venezuelan

equine encephalitis replicons expressing Norwalk virus-like particles. J Virol 76: 730–742
24. Heidner HW, Johnston RE (1994) The amino-terminal residue of Sindbis virus glycoprotein E2 influences virus maturation, specific infectivity for BHK cells, and virulence in mice. J Virol 68: 8064–8070
25. Heidner HW, Knott TA, Johnston RE (1996) Differential processing of Sindbis virus glycoprotein PE2 in cultured vertebrate and arthropod cells. J Virol 70: 2069–2073
26. Heidner HW, McKnight KL, Davis NL, Johnston RE (1994) Lethality of PE2 incorporation into Sindbis virus can be suppressed by second-site mutations in E3 and E2. J Virol 68: 2683–2692
27. Hevey M, Negley D, Pushko P, Smith J, Schmaljohn A (1998) Marburg virus vaccines based upon alphavirus replicons protect guinea pigs and nonhuman primates. Virology 251: 28–37
28. Hevey M, Negley D, VanderZanden L, Tammariello RF, Geisbert J, Schmaljohn C, Smith JF, Jahrling PB, Schmaljohn AL (2001) Marburg virus vaccines: comparing classical and new approaches. Vaccine 20: 586–593
29. Johnston RE, Smith JF (1988) Selection for accelerated penetration in cell culture co-selects for attenuated mutants of Venezuelan equine encephalitis virus. Virology 162: 437–443
30. Kofler RM, Heinz FX, Mandl CW (2002) Capsid protein C of tick-borne encephalitis virus tolerates large internal deletions and is a favorable target for attenuation of virulence. J Virol 76: 3534–3543
31. Kofler RM, Leitner A, O'Riordain G, Heinz FX, Mandl CW (2003) Spontaneous mutations restore the viability of tick-borne encephalitis virus mutants with large deletions in protein C. J Virol 77: 443–451
32. Lee JS, Dyas BK, Nystrom SS, Lind CM, Smith JF, Ulrich RG (2002) Immune protection against staphylococcal enterotoxin-induced toxic shock by vaccination with a Venezuelan equine encephalitis virus replicon. J Infect Dis 185: 1192–1196
33. Lee JS, Hadjipanayis AG, Welkos SL (2003) Venezuelan equine encephalitis virus-vectored vaccines protect mice against anthrax spore challenge. Infect Immun 71: 1491–1496
34. Lee JS, Pushko P, Parker MD, Dertzbaugh MT, Smith LA, Smith JF (2001) Candidate vaccine against botulinum neurotoxin serotype A derived from a Venezuelan equine encephalitis virus vector system. Infect Immun 69: 5709–5715
35. MacDonald GH, Johnston RE (2000) The role of dendritic cell targeting in Venezuelan equine encephalitis virus pathogenesis. J Virol 74: 914–922
36. Olmsted RA, Baric RS, Sawyer BA, Johnston RE (1984) Sindbis virus mutants selected for rapid growth in cell culture display attenuated virulence in animals. Science 225: 424–427
37. Paredes AM, Heidner H, Thuman-Commike P, Prasad BVV, Johnston RE, Chiu W (1998) Structural localization of the E3 glycoprotein in attenuated Sindbis virus mutants. J Virol 72: 1534–1541
38. Pratt WD, Davis NL, Johnston RE, Smith JF (2003) Genetically engineered live attenuated vaccines for Venezuelan equine encephalitis: testing in animal models. Vaccine 21: 3854–3862
39. Pushko P, Bray M, Ludwig GV, Parker M, Schmaljohn A, Sanchez A, Jahrling PB, Smith JF (2000) Recombinant RNA replicons derived from attenuated Venezuelan equine encephalitis virus protect guinea pigs and mice from Ebola hemorrhagic fever virus. Vaccine 19: 142–153

40. Pushko P, Parker M, Ludwig GV, Davis NL, Johnston RE, Smith JF (1997) Replicon-helper systems from attenuated Venezuelan equine encephalitis virus: expression of heterologous genes *in vitro* and immunization against heterologous pathogens *in vivo*. Virology 239: 389–401
41. Russell DL, Dalrymple JM, Johnston RE (1989) Sindbis virus mutations which coordinately affect glycoprotein processing, penetration and virulence in mice. J Virol 63: 1619–1629
42. Schultz-Cherry S, Dybing JK, Davis NL, Williamson C, Suarez D, Johnston R, Perdue ML (2000) Influenza virus (A/HK/156/97) hemagglutinin expressed by an alphavirus replicon system protects chickens against lethal infection with Hong Kong-origin H5N1 viruses. Virology 278: 55–59
43. Thomas CE, Van Dam CN, Zhu W, Davis NL, Johnston RE, Sparling PF (2004) Vaccination of mice with gonococcal antigens expressed from Venezuelan equine encephalitis viral replicon particles. Inf Immun (submitted)
44. Velders MP, McElhiney S, Cassetti MC, Eiben GL, Higgins T, Kovacs GR, Elmishad AG, Kast WM, Smith LR (2001) Eradication of established tumors by vaccination with Venezuelan equine encephalitis virus replicon particles delivering human papillomavirus 16 E7 RNA. Cancer Res 61: 7861–7867
45. Williamson C, Johnston RE, Williamson AL (2002) HIV-1 vaccine candidates for South Africa. South African J HIV Med (March): 10–14

Author's address: Robert E. Johnston, Carolina Vaccine Institute, University of North Carolina, Chapel Hill, North Carolina, 27599-7292, U.S.A.; e-mail: robert_johnston@med.unc.edu

Arch Virol (2004) [Suppl] 18: 221–230

Archives of Virology

Recombinant canarypoxvirus vaccine carrying the prM/E genes of West Nile virus protects horses against a West Nile virus-mosquito challenge

J. M. Minke[1], **L. Siger**[2], **K. Karaca**[2], **L. Austgen**[3], **P. Gordy**[3], **R. Bowen**[3], **R. W. Renshaw**[4], **S. Loosmore**[5], **J. C. Audonnet**[1], and **B. Nordgren**[6]

[1]Merial SAS, Lyon, France
[2]Merial Inc., Athens, Georgia, U.S.A.
[3]College of Veterinary Medicine and Biomedical Sciences, Colorado State University, Fort Collins, Colorado, U.S.A.
[4]College of Veterinary Medicine, Cornell University, Ithaca, New York, U.S.A.
[5]Aventis Pasteur, Toronto, Canada
[6]Merial, Duluth, Georgia, U.S.A.

Summary. An ALVAC (canarypoxvirus)-based recombinant (vCP2017) expressing the prM and E genes derived from a 1999 New York isolate of West Nile virus (WNV) was constructed and assessed for its protective efficacy in horses in two different experiments. In the first trial, a dose titration study was conducted to evaluate both serum neutralising antibody responses to WNV and duration of immunity. In the second trial the onset of protection was determined. Twenty-eight adult horses received two doses of vCP2017 administered intramuscularly at 5-week intervals and sixteen horses comprised age-matched non-vaccinated controls. Individual sera were taken periodically and tested for neutralising antibodies against WNV. Horses were challenged by allowing WNV-infected *Aedes albopictus* mosquitoes to feed on them two weeks (second trial) or one year (first trial) after the second vaccination. After challenge, horses were monitored for clinical signs of disease, and blood samples were collected for detection of WNV viremia and antibody.

In both trials, all vaccinated horses developed neutralising antibodies against WNV. None of the vaccinated or control horses developed clinical signs of WNV disease upon challenge. None of the nine horses challenged 2 weeks after primary vaccination and only one of the ten vaccinated horses challenged 1 year after vaccination developed detectable viremia after challenge, whereas more than 80% of the controls became infected. Results from these studies demonstrated that a primary course of two doses of vCP2017 provides both antibody response and an early immunity in horses against WNV viremia.

Introduction

West Nile virus (WNV) is a positive-sense single stranded RNA virus belonging to the Japanese encephalitis virus (JEV) complex of the genus *Flavivirus*, family *Flaviviridae* [6]. West Nile virus infection is well known in Africa, Asia and the Mediterranean basin, where sporadic cases and epidemics have been reported [14]. The virus was detected for the first time in North America in 1999 during an outbreak involving birds, horses and humans in New York City and its environs. Since then the virus has spread rapidly south and west, appearing in almost every state in the continental United States by 2002. The persistence of WNV in mosquitoes throughout the winter makes it difficult or impossible to eradicate the virus and therefore it is no longer considered exotic to the United States [3].

The primary transmission cycle of WNV involves birds and mosquitoes, although direct bird-to-bird transmission [12] and oral infection have been suggested [4] and demonstrated experimentally [9]. Horses become infected through the bite of infected mosquitoes and develop a WNV viremia of low magnitude and short duration. Therefore, infected horses are unlikely to serve as amplifying hosts for WNV and are considered incidental, dead-end hosts [1]. During 2002, more than 14,000 confirmed equine cases of WNV infection were reported in North America [22], with about a 30% mortality rate.

Vaccination with inactivated or modified live vaccines has been an effective mechanism for prevention of flavivirus infection in humans [7, 23] and domestic animals [19]. Recently, an inactivated vaccine against WNV was licensed in the United States for use in horses. There are several disadvantages commonly associated with inactivated vaccines, such as the requirement for the large scale production of an infectious pathogen, the risk of incomplete inactivation, the slow onset of immunity and the need for multiple inoculations. Because of these concerns, second generation vaccines have been developed using recombinant technology. Recently, Davis et al. [2] reported the evaluation of a DNA vaccine against WNV that protected mice and horses against experimental infection. The use of recombinant poxviruses is another potentially effective means of generating protective immunity against flavivirus infection [8, 10, 11]. Canarypoxvirus recombinants have been shown to authentically express the inserted foreign genes in the absence of productive viral replication [18, 21]. This provides a built-in safety feature for vaccination, since there is no potential for transmission of the vaccine vector or the target immunogenic virus. Several canarypoxvirus recombinants have been developed for veterinary use [13, 15, 20, 21] and are now commercially available in North America and Europe.

We have developed a recombinant canarypoxvirus (vCP2017) expressing the premembrane (prM) and envelope (E) proteins of a 1999 New York isolate of WNV. Following safety studies in horses, mice, chickens and crows, this recombinant virus was assessed for its ability to induce immunity in horses and also to determine its potential as a vaccine candidate to prevent WNV infection in them.

Materials and methods

Cell lines and virus strains

The canarypoxvirus-based vector ALVAC [21] was used as the parental strain to generate a recombinant containing the appropriate WNV coding sequences. The virus was grown and titrated in primary specific pathogen free chick embryo fibroblast (CEF) cell monolayers.

The WNV that served as donor RNA for the cDNA synthesis and subsequent construction of vCP2017 was isolated in 1999 at Cornell University from pooled liver and kidney tissue samples of a Mallard duck from Queens, New York.

cDNA cloning of the WNV structural genes

Standard molecular biology techniques were used for the derivation of WNV cDNA sequences and their cloning. Cloning was performed into the pTriEx-1 plasmid vector (Novagen, Madison, WI 53719, USA). Standard coupled reverse transcriptase polymerase chain reaction (RT-PCR) techniques were used to amplify the prM, M and E coding sequences from WNV RNA, using primers ME #2 (forward) (5′-AGCT**GATATC**AACCGGAATTGCAGTCATG-3′) and prM/E (reverse) (5′-CACAGT**CTCGAG**AGCGTGCACGTTCACGGAG-3′). PCR amplified DNA fragment was digested with *Eco*RV and *Xho*I and cloned into the pTriEx vector. The resulting plasmid was designated as pTriEX-WNVME.

Construction of the WNV expression cassette ALVAC donor plasmid

The 0.7 kb *Eco*RV – *Cla*I 5′-end of the WNV gene cassette in pTriEx was amplified by PCR using primers 7600.SL (5′-ATCGC***GATATC***CGTTAAGTTTGTATCGTAATGACCGGAATTG CAGTCATGATTGGCCTG-3′) and 7616.SL (5′-CATCTTCAC***ATCGAT***GGTAGGCTT-3′) and the fragment inserted into pCR2.1 to generate plasmid pDS-2905-2-1. The sequence of the insert was confirmed. Primer 7600.SL contains the 3′-end of the vaccinia H6 promoter. The 1.4 kb *Cla*I – *Xho*I 3′-fragment from pTriEx-WNV was inserted into pUC-4K, generating clone pDS-2889-1. In order to remove one internal TTTTTNT site in the E sequence, site-directed mutagenesis was performed using the Stratagene QuickChange (La Jolla, CA 92037, U.S.A.) kit and primers 7598.SL (5′-GAAGTGGCCA**<u>TCTTCGT</u>**CCATGGACCAACT-3′) and 7599.SL (5′-AGTTGGTCC***GTGCAC***GAAGATGGCCACTTC-3′). Clone pDS-2897-5-1 containing the mutated E sequence was confirmed as correct by sequence analysis. In order to remove the poly-His tag and introduce a translation stop and terminal TTTTTNT, the 1.4 kb 3′ fragment from pDS-2897-5-1 was PCR amplified using primers 7617.SL (5′-AAGCCTACC***ATCGAT***GTGAAGATG-3′) and 7601.SL (5′-GGGCCC***TCTAGA***TAAAAA TTAAGCGTGCACGTTCACGGAGAGGAAGAGCAG-3′). The resultant fragment was inserted into pCR2.1 in order to generate plasmid pDS-2918-1. Plasmid pDS-2918-1 was also confirmed as correct by sequence analysis. The *Eco*RV – *Cla*I 5′-fragment from pDS-2905-2-1 and the *Cla*I – *Xba*I 3′ fragment from pDS-2918-1 were isolated and ligated together with the plasmid pNVQH6C5LSP-18 (plasmid containing the flanking arms for the ALVAC C5 locus), previously digested with *Eco*RV and *Xba*I, for generating pDS-2946-1-1. Clone pDS-2946-1-1 was confirmed by restriction enzyme analysis and the insert was sequenced. The final donor plasmid pDS-2946-1-1 consisted of the vaccinia virus H6 promoter linked to the ATG initiation codon added precisely to the 5′ end of the prM coding sequence and inserted between canarypox flanking arms at the ALVAC C5 locus.

Generation of recombinant ALVAC virus (vCP2017)

Recombination experiments were performed as previously described [16], using pDS-2946-1-1 as donor plasmid and ALVAC as the rescue virus. The pDS-2946-1-1 insertion plasmid was linearized and transfected into primary CEF cells which were subsequently infected with ALVAC to allow *in vitro* recombination to occur between the C5 ORF flanking regions of the viral cDNA and the homologous ALVAC flanking regions in the insertion plasmid. Progeny containing the WNV genes were identified by *in situ* plaque hybridization using a labeled WNV gene probe. A number of WNV-probe positive viral plaques were subjected to six sequential rounds of plaque purification. One representative plaque, designated vCP2017, was then amplified through five passages in primary CEF cells, identified as passage 10, and used to prepare vCP2017 stocks.

Expression of prM and E

Appropriate expression of WNV prM and E genes was confirmed by immunoblot and indirect fluorescence analysis of vCP2017. For the immunoblot assay, CEF cells were infected with ALVAC (parental virus) or vCP2017. After incubation at 37 °C in 5% CO_2 for 24–30 hours, the infected cells were harvested and submitted to electrophoresis on a 12% SDS PAGE gel, which then was transferred to a Millipore P nylon membrane. The membrane was probed with chicken anti-WNV as primary antibody (APHIS, NVSL, Ames, Iowa), and horseradish peroxidase-conjugated donkey anti-chicken serum (Jackson Immunoresearch, West Grove, Pennsylvania) as secondary antibody. The Amersham Enhanced Luminol detection kit (Amersham-Pharmacia, Piscataway, New Jersey) was used for visualization.

Indirect immunofluorescence (IF) tests were performed on acetone fixed infected CEF cells. Monoclonal antibody 538 reacting with prM of WNV [5] and monoclonal F7/101 reacting with E of WNV [16] were used as primary antibodies and a Cy3 coupled goat anti-mouse serum (Jackson Immunoresearch, West Grove, Pennsylvania) as secondary antibody.

Horses

Forty-four adult horses (gelding and mares), weighing between 337 and 600 kilograms, were used in this study. The horses were used under a protocol approved by the Institutional Animal Care and Use Committees at Merial and Colorado State University. None of the horses had detectable antibody to WNV or St. Louis encephalitis virus by a plaque reduction neutralisation test (PRNT) (1) at the commencement of the trial. The horses were fed mixed grains and hay or extruded complete feed in amounts necessary to keep the horses in good body condition. Water was available *ad libitum.*

During the vaccination phase, horses were maintained on pasture or in barns. Horses were regularly treated topically with fly repellent (Tri-Tec®, Farnan Companies Inc., Osborne, Phoenix, Arizona) as per label indications.

Animal vaccination and challenge

The protection provided by vCP2017 was assessed both at 2 weeks and 1 year after the second dose of the primary vaccination course. The vaccines were reconstituted with Carbopol adjuvant (BF Goodrich Chemicals Europe NV, Brussels, Belgium) at a concentration of 4 mg per dose (1 ml) before immunisation.

In the first trial, a dose titration study was conducted in horses to evaluate both serum neutralising antibody titres against WNV as well as the duration of immunity. Three groups of six horses were vaccinated twice 5 weeks apart by the intra-muscular route with graded doses (low-medium-high) of vCP2017 vaccine. One group of six horses was not vaccinated and served as controls.

In the second trial the onset of protection was determined. Ten horses were vaccinated with vCP2017 at low dose according to the same vaccination schedule (see above). An additional 10 horses were not vaccinated and served as controls.

All challenge studies were conducted at the biosafety level 3 containment facilities at Colorado State University by using a previously described mosquito challenge method (1). Each horse was challenged by the bite of 12–15 *Aedes albopictus* mosquitoes that had been infected with the NY99-4132 WNV isolate. Mosquitoes were allowed to feed on horses for 5–8 minutes. After feeding, 3–5 engorged mosquitoes from each horse were macerated and tested for virus by plaque assay. Horses vaccinated with the low and medium doses of vCP2017 and the non-vaccinated control horses from trial #1 and all horses from trial #2 were challenged at 1 year (trial #1) and 2 weeks (trial #2) after the second vaccination.

After challenge, the horses were examined twice daily from days 0 (day of infection) to 13 and once on day 14 for signs of clinical disease; rectal body temperatures also were taken daily. Whole blood samples were collected twice daily for 7 (trial #1) and 13 days (trial #2) and then once daily until day 14 after challenge and tested by plaque assay for the presence of WNV. At the time of euthanasia, a sample of cerebrospinal fluid (CSF) was collected from the atlanto-occipital area from horses from trial #1 and tested for the presence of WNV. Horses were euthanized by an overdose of pentobarbital between days 14 and 24 after challenge, necropsied for gross pathological and histopathological examination (trial #2), and their carcasses were incinerated in the containment facilities.

Virus isolation by plaque assay

Virus in serum samples, mosquito homogenates and CSF was measured by titration in a plaque assay, as previously described [1]. Briefly, 0.1 ml of sample was added to a monolayer of Vero cells in a six-well culture plate and incubated at 37 °C in an atmosphere containing 5% CO_2. Cells were overlaid with 3 ml of 0.5% agarose in M-199 medium without phenol red supplemented with 5% fetal bovine serum and antibiotics. After 48 h, a second 3 ml overlay containing 0.004% neutral red was added. Plaques were scored on days 3, 4 and 5 of incubation and expressed as plaque forming units per ml.

Serology

Bloods were taken from the jugular vein periodically after vaccination and challenge, and the WNV antibody response was evaluated by PRNT. Titration of sera was done in blinded fashion at Colorado State University. Briefly, sera were heat-inactivated at 56 °C for 30 min. and serial two-fold dilutions of those sera were mixed with an equal volume (0.1 ml) containing 200–300 pfu of the NY99-4132 strain of WNV. After an overnight incubation at 4 °C, 0.1 ml of each mixture was then injected onto Vero cell monolayers in 6-well plates and processed as for the plaque assay. Neutralisation endpoints were recorded as the highest dilution of serum giving 80% reduction of 100–150 plaques. Titres ≥ 5 were considered positive.

Results

Expression of prM and E proteins

Expression of the prM and E proteins was detected both by IF (not shown) and Western blotting (Fig. 1). Protein bands of the expected size for WNV E and prM proteins were present in vCP2017 samples, but absent in the ALVAC control samples. The expression of the proteins was found to be phenotypically stable after 10 *in vitro* passages in CEF cells.

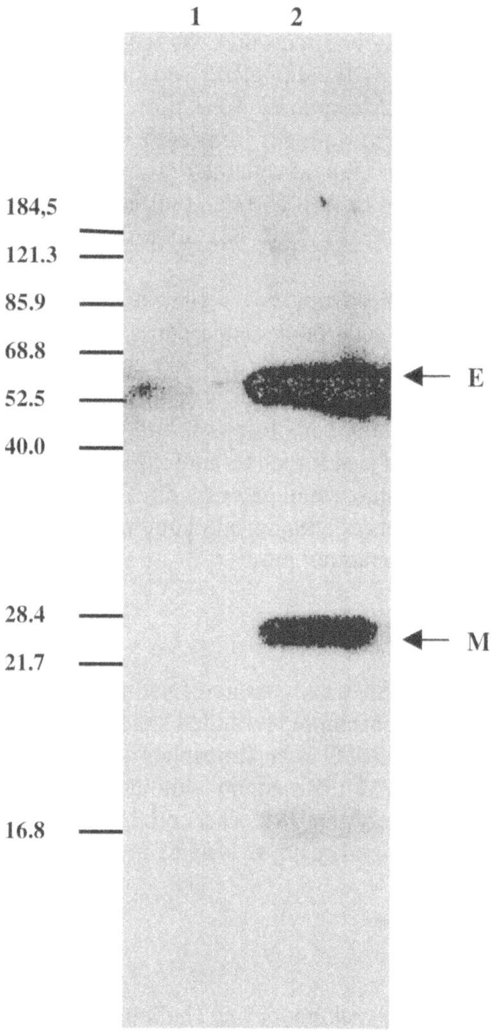

Fig. 1. Western immunoblot analysis of ALVAC (parental virus) and vCP2017 (recombinant ALVAC expressing WNV proteins) virus. Cell pellets from ALVAC (*1*) or CP2017 (*2*) infected CEFs were electrophoresed on a 12% SDS-PAGE gel, then transferred to nylon membrane. The membrane was probed with chicken-anti-WNV (1:1000) as primary antibody, and HRPO labeled donkey anti-chicken antiserum as secondary antibody. Luminol detection kit (Amersham) was used for visualization. vCP2017 infected cells (*2*), but not ALVAC infected cells (*1*), expressed WNV-E and WNV-M proteins (arrows). Sizes (left) are shown in kilodaltons

Antibody responses to vaccination

Antibody to WNV was not detected in any pre-immunization serum. None of the non-vaccinated horses seroconverted to WNV up to the time of challenge. One horse belonging to the high dose group (trial #1) showed an anamnestic antibody response (titre: 80) at 7 days following first vaccination. As we could not rule out the possibility of previous exposure to WNV or antigenically related flaviviruses, this horse was excluded from further analysis. One vaccinated horse from trial #2 died 28 days after first vaccination because of a pre-existing enteric problem unrelated to vaccination and was eliminated from the study.

Table 1 shows the antibody responses after vaccination. None of the vaccinated horses developed detectable antibody after the first vaccination. Two weeks after the second vaccination 15/17 horses from trial #1 and 8/9 horses from trial #2 had detectable antibody (range: 5 to 1280), but no clear dose/related response was seen. The three horses that did not seroconvert at the 80% plaque reduction

Table 1. Serological responses following immunization with vCP2017

Trial	Vaccine dose	Seroconversion*	GMT $PRNT_{80\%}$ (log_{10})	
			Two weeks after 1st vaccination	Two weeks after 2nd vaccination
#1	low	6/6	0	1.1
	medium	5/6	0	0.9
	high	4/5	0	0.8
#2	low	8/9	0	1.3

*Neutralising titres ≥ 5

level all had low titre antibodies at the 50% level. To assess the persistence of the neutralising antibody responses in the horses from trial #1, blood samples were collected 36 weeks after the second vaccination and again assayed by $PRNT_{80}$. Only 2 horses still had low levels of circulating antibody to WNV (titre: 5). Both had the highest PRNT antibody levels (titre: 40) at two weeks after the second vaccination.

Protection against challenge

The quantity of WNV present in the engorged mosquitoes assayed after feeding on the horses ranged from $10^{7.0}$ to $10^{8.0}$ pfu per mosquito in the first trial and between $10^{7.0}$ and $10^{9.1}$ in the second trial.

The results from the two experiments are shown in Table 2. Two vaccinated horses from the duration of immunity study (trial #1) were excluded from the challenge as they became exposed to a field strain of WNV while being kept on pasture during the pre-challenge period as evidenced by seroconversion. No other horses seroconverted prior to challenge. Therefore, 5 vaccinated horses from the medium and low dose groups and 6 control horses were challenged one year after the second vaccination. Five of six control horses exposed to WNV by

Table 2. Antibody response and protection against challenge with WNV

Experiment	Group	N	% Protection against viremia	Seroconversion* (no. positive/no. tested)
#1	low dose	5	100	5/5
	medium dose	5	80	5/5**
	controls	6	16	6/6
#2	low dose	9	100	8/9
	controls	10	20	10/10

*Neutralising titres ≥ 5
**The non protected horse showed a slow antibody response after challenge with WNV

feeding infected mosquitoes on them developed detectable viremias 24 to 72 h after challenge, with peak virus titres between $10^{0.7}$ and $10^{1.8}$ pfu. Virus was not isolated after day 6. Only one vaccinated horse belonging to the medium dose group developed viremia, which lasted 2 days and with a peak viremia of $10^{1.3}$ pfu. This horse had failed to mount a neutralising antibody response after vaccination and had demonstrated a slow (day 14) antibody response after challenge. All other vaccinated horses had antibody as early as day 7 after challenge. Five of 6 control horses developed antibody titres ≥ 5 by day 14 after infection, and one control horse by day 7 after challenge. Clinical signs, including pyrexia were not observed in any of the infected horses. None of the horses had detectable virus in the CSF.

In the second trial, nine vaccinated and ten control horses were challenged two weeks after the second vaccination. Eight of ten control horses became viremic 18 to 72 h after being fed upon by infected mosquitoes. Peak viremia titres ranged from $10^{0.7}$ pfu to $10^{2.4}$ pfu, and virus was not isolated after day 6. None of the nine vaccinated horses developed viremia. All control horses and 8 of 9 vaccinated horses had increased antibody titres after WNV exposure. The horse that did not seroconvert to WNV had a titre of 1280 at the time of challenge. Clinical signs of disease were not observed in any of these horses during the study, with the exception of one control horse which had a transient temperature response (102 °F) on day 6 after mosquito feeding. At necropsy, gross abnormalities were not observed and histopathological examination of representative sections of cortex, brainstem and cerebellum revealed no lesions compatible with WNV infection.

Discussion

We have constructed a recombinant canarypox virus expressing the prM/E proteins of a 1999 New York isolate of WNV. Data from immunoblot experiments demonstrated that cells infected with vCP2017 properly processed the prM and E proteins. Expression was confirmed by IF using monoclonal antibodies specific for E and prM proteins.

The vaccines were well tolerated by the horses and vCP2017 could not be isolated from blood, faeces or nasal swabs taken at regular intervals after vaccination (data not shown). This is consistent with results of previous inoculation of ALVAC in mammals, including humans [8, 18].

The protective capacity of vCP2017 was tested by challenging the vaccinated horses via the bite of WNV infected mosquitoes, a method that approximates natural exposure. The low levels of viremia and the apparent absence of clinical signs in the challenged control horses reported here are similar to those described by Bunning et al. [1]. The vaccine induced neutralising antibodies to WNV and provided early protection against WNV viremia. Moreover, the protective immunity lasted for at least one year after vaccination. Interestingly, protection occurred in the absence of detectable antibody. The requirement for significant levels of neutralising antibody at the time of exposure may not be critical as long as vaccination properly primes the immune system and response is rapid. This has previously been described for other canarypoxvirus constructs containing the protective genes of feline leukaemia virus [20] and canine distemper virus (J. Minke, personal observation). The vaccine may also have induced CTL responses, as was demonstrated for ALVAC-JEV, a canarypoxvirus recombinant against JEV [8]. This construct induced JEV-specific T cell responses in human recipients, despite undetectable neutralising antibody [8].

In our experiments, vCP2017 was tested in the presence of Carbopol adjuvant. Previous studies [13] on an influenza-canarypoxvirus recombinant (vCP1533) for horses, have shown that, although the construct provided protection in the absence of adjuvant, its efficacy was enhanced when combined with Carbopol. The immunological mechanisms underlying Carbopol activity are poorly understood and are currently being investigated.

In summary, a primary vaccination course of two injections with vCP2017 conferred significant protection to horses against WNV viremia and provides a potent and safe alternative to the currently available inactivated WNV vaccine.

Acknowledgments

We would thank Dr. E. Gould for providing anti-WNV monoclonal antibodies and H. Coupier, L. Motes-Kreimeyer, C. Leung and G. Royer and staff for expert technical assistance. We are grateful to H. Hughes, M. Murray and A.L. Glaser for critical review of the manuscript.

References

1. Bunning ML, Bowen RA, Cropp CB, Sullivan KG, Davis BS, Komar N, Godsey MS, Baker D, Hettler DL, Holmes DA, Biggerstaff BJ, Mitchell CJ (2002) Experimental infection of horses with West Nile virus. Emerg Infect Dis 8(4): 380–386
2. Davis BS, Chang GJ, Cropp B, Roehrig JT, Martin DA, Mitchell CJ, Bowen R, Bunning ML (2001) West Nile virus recombinant DNA vaccine protects mouse and horse from virus challenge and expresses in vitro a non-infectious recombinant antigen that can be used in enzyme-linked immunosorbant assays. J Virol 75(9): 4040–4047
3. DeHaven WR (2002) Memorandum NVSL. ProMed Announcements 2002 (06) available at: http://www.promed@promed.isid.harvard.edu
4. Garmendia AE, Van Kruiningen HJ, French RA, Anderson JF, Andreadis TG, Kumar A (2000) Recovery and identification of West Nile virus from a hawk in winter. J Clin Microbiol 3110–3111
5. Gould EA, Buckley A, Higgs S, Gaidamovich S (1990) Antigenicity of flaviviruses. Arch Virol [Suppl 1]: 137–152
6. Heinz FX, Collett MS, Purcell RH, Gould EA, Howard CR, Houghton M (2000) Family *Flaviviridae*. In: van Regenmortel MHV, Fauquet CM, Bishop DHL, Carsten EB, Estes MK, Lemon SM (eds). Virus taxonomy: classification and nomenclature of viruses. Academic Press, San Diego, pp 859–878
7. Hoke CH, Nisalak A, Sangawhipa N, Jatanasen S, Laorakapongse T, Innis BL, Kotchasenee S, Gingrich JB, Latendresse J, Fukai K, Burke DS (1988) Protection against Japanese encephalitis by inactivated vaccines. N Eng J Med 319: 608–614
8. Kanesa-Thasan N, Smucny JJ, Hoke CH, Marks DH, Konishi E, Kurane I, Tang DB, Vaughn DW, Mason PW, Shope RE (2001) Safety and immunogenicity of NYCAC-JEV

and ALVAC-JEV attenuated recombinant Japanese encephalitis virus – poxvirus vaccines in vaccinia-nonimmune and vaccinia-immune humans. Vaccine 19: 483–491
9. Komar N, Langevin S, Hinten S, Nemeth N, Edwards E, Hettler D, Davis B, Bowen R, Bunning M (2003) Experimental infection of North American birds with the New York 1999 strain of West Nile virus. Emerg Infect Dis 9(3): 311–322
10. Konishi E, Kurane I, Mason PW, Shope RE, Ennis FA (1997) Poxvirus-based Japanese encephalitis vaccine candidates induced JE virus-specific CD8+ cytotoxic lymphocytes in mice. Virology 227: 353–360
11. Konishi E, Pincus S, Paoletti E, Laegreid WW, Shope RE, Mason PW (1992) A highly attenuated host range-restricted vaccinia virus strain, NYVAC, encoding the prM, E, and NS1 genes of Japanese encephalitis virus prevents JEV viremia in swine. Virology 190: 454–458
12. McLean RG, Ubico SR, Docherty DE (2001) West Nile virus transmission and ecology in birds. Ann NY Acad Sci 951: 54–57
13. Minke JM, Audonnet JC, Jessett DM, Fischer L, Guigal PM, Coupier H, Pardo MC, Taylor J, Tartaglia J, Mumford JA (2000) Canarypox as a vector for influenza and EHV-1 genes: challenges and rewards. In: Proc Second International Veterinary Vaccines and Diagnostic Conference, Oxford, p 36
14. Murgue B, Murri S, Triki H, Deubel V, Zeller HG (2001) West Nile in the Mediterranean basin: 1950–2000. Ann NY Acad Sci 951: 117–126
15. Pardo MC, Bauman JE, Mackowiak M (1997) Protection of dogs against distemper by vaccination with a canarypox recombinant expressing the canine distemper virus fusion and hemagglutinin glycoproteins. AJVR 58(8): 833–836
16. Peiris JSM, Porterfield JS, Roehrig JT (1982) Monoclonal antibodies against the flavivirus West Nile. J Gen Virol 58: 283–289
17. Perkus ME, Kauffman EB, Taylor J (1993) Methodology of using vaccinia virus to express foreign genes in tissue culture. J Tissue Cult Methods 15: 72–81
18. Plotkin SA, Cadoz M, Meignier B, Meric C, Leroy O, Excler JL, Tartaglia J, Paoleti E, Gonczol E, Chappuis G (1995) The safety and use of canarypox vectored vaccines. Dev Biol Stand 84: 165–170
19. Research Committee for prevention of stillbirth in sows due to Japanese encephalitis (1968) Bull Natl Inst Anim Health 57: 1–8
20. Tartaglia J, Jarrett O, Neil JC, Desmettre Ph, Paoletti E (1993) Protection of cats against feline leukaemia virus by vaccination with a canarypox virus recombinant, ALVAC-FL. J Virol 67(4): 2370–2375
21. Taylor J, Trimarchi C, Weinberg R, Languet B, Guillemin F, Desmettre Ph, Paoletti E (1991) Efficacy studies on a canarypox-rabies recombinant virus. Vaccine 9(3): 190–193
22. Update on current status of West Nile virus. Available at: http://www.aphis.usda.gov
23. Xin YY, Ming ZG, Peng GY, Jian A, Min LH (1988) Safety of a live-attenuated Japanese encephalitis virus vaccine (SA14-14-2) for children. Am J Trop Med Hyg 39: 214–217

Author's address: J. M. Minke, Merial, SAS, 254 rue Marcel Mérieux, 69007 Lyon, France; e-mail: julius.minke@merial.com

Arch Virol (2004) [Suppl] 18: 231–244

Archives of Virology

Diagnosis of zoonotic viral encephalitis

G. Vernet

Molecular Diagnostics, Advanced Technology Unit,
BioMérieux, Marcy-l'Etoile, France

Summary. Human infections by zoonotic encephalitis viruses are usually asymptomatic or symptoms are not specific to these viruses. Some of them have high mortality and morbidity rates and most often no specific treatment exist. This emphasizes the need for a precise identification of arboviruses in clinical specimens from humans and animals. Because these diseases are frequent in developing countries and tend to emerge or re-emerge in others, diagnostic tools must detect the broadest possible range of viruses with a high sensitivity and this is a key factor for surveillance, control of transmission and prevention through vaccination. In countries with limited diagnostic infrastructures, low-cost and easy-to-use tests are required. The diagnosis of arboviral encephalitis has been significantly improved in the recent years. Sensitive ELISA assay to detect antibodies against many arboviruses in serum or CSF are commercially available and can be used to detect early infections. Immunochromatographic rapid tests for the detection of specific IgM that could be used on fingertip blood would be valuable tools in developing countries. A limitation of these serologic assays is their lack of specificity as many arboviruses are antigenically related. Virus isolation or molecular assays from different human or animal tissues are also important diagnostic tools. Molecular assays have been extensively described in the recent years. They are very sensitive and have the advantage over cell culture that specimen transportation is less critical. Real-time detection has even improved sensitivity and reduced time to result. Although the utility of molecular assays for the detection of arboviruses in mosquito pools has been demonstrated, an extensive validation of their pertinence in clinical settings is still required. The use of DNA-microarrays may further extend the range of viruses that can be detected in a single test and allow isolates typing for epidemiology purposes.

Specific constraints

Zoonotic viral encephalitides have several characteristics that may make their diagnosis complex and difficult. Major encephalitis viruses that involve animals in the transmission to humans can be found within the virus families *Flaviviridae*,

Bunyaviridae, *Alphaviridae*, *Paramyxoviridae*, *Rhabdoviridae* and *Arenaviridae*. These virus families are very diverse and, within a family, the viruses are genetically and antigenically distinct and, they may have different epidemiologic, pathologic and transmission characteristics.

In industrialised countries, viral encephalitides are relatively rare and most often due to herpes simplex virus or to enteroviruses. Clinicians and biologists are not often confronted by diseases due to zoonotic infections and may not consider a zoonotic viral aetiology. For example, on average, there are no more than 300 cases of viral encephalitis due to St Louis encephalitis virus (SLEV), La Crosse virus, Western equine encephalitis virus (WEEV) and Eastern equine encephalitis virus (EEEV) each year in the USA. Similarly, fewer than 350 cases of encephalitis due to Tick-borne encephalitis virus (TBEV) have been diagnosed each year between 1990 and 1999 in Western Europe (Germany, France, Switzerland, Austria and Italy). However, these viruses may be responsible for outbreaks and their occurrence has fluctuated widely in recent years. For example, West Nile virus (WNV) has recently emerged in the U.S.A., with more than 3800 cases since 1999. WNV epidemics also have been described in France (1962) and Romania (1996). The international scientific working group on tick-borne encephalitis has reported that TBEV infections have increased in Eastern European countries, including Poland (data available on http://www.tbe-info.com/epidemiology/index.html). In industrialised countries, therefore it is very important to have efficient tools to make a correct diagnosis of the etiologic agent of infected patients and to monitor and prevent the spread of these viruses. The spectrum of viruses covered by these tests must be as broad as possible and a "diagnostic panel" approach would be convenient.

The situation is very different in less advanced countries: between 30,000 and 50,000 Japanese encephalitis virus (JEV) infections are reported each year in South-East Asia, 100,000 to 300,000 Lassa virus infections occur each year in Western Africa and the number of TBEV infections in Russia has ranged from 5225–9427 between 1990 and 1999, although these numbers are surely underestimated. In developing countries, due to very limited financial resources for the diagnosis of infectious diseases, there is a need for low cost diagnostic tests that do not require expensive equipment or highly skilled technicians.

Viral encephalitides may be life threatening and therefore require a rapid and accurate diagnosis. Case-fatality ratios can reach 25% for Russian TBEV infections and 30% for JEV infections. WNV kills about 6.4% of infected people and 30% of infected horses in North America. For most zoonotic viral encephalitides there is no specific antiviral treatment. Still, it is important to establish the aetiology in order to rapidly implement the most appropriate supportive therapies to reduce mortality and minimize neurological sequelae and to provide adequate epidemiologic warning to the community at large.

The diagnosis of zoonotic infections is also important to improve the prevention of disease transmission. Vaccines against JEV and TBEV are commercially available. A better understanding of the epidemiology of those infections at the

Table 1. Biosafety level requirements for working with zoonotic encephalitis viruses

BSL-2	– Dengue viruses, Western equine encephalitis virus, Eastern equine encephalitis virus, lyssaviruses, California encephalitis virus, La Crosse virus
BSL-3	– Venezuelan equine encephalitis virus, tick-borne encephalitis viruses, Japanese encephalitis virus, West Nile virus, Saint Louis encephalitis virus, Murray Valley encephalitis virus
BSL-4	– Russian spring-summer tick-borne encephalitis virus, Hendra and Nipah viruses

population level is important so that we can identify high risk groups that require vaccination. Other active prevention methods include the control of animal reservoirs from which the viruses are transmitted to humans. Diagnostic tools to detect viruses in mosquitoes, ticks or birds must be available to health authorities to implement relevant strategies.

Finally, a very important constraint for the diagnosis of these viruses resides in the fact that at least some of them must be handled in confined laboratories because they are classified as BSL-3 (biosafety level 3) or BSL-4. Table 1 shows the biosafety level required for working with the major zoonotic encephalitis viruses. TBEV, Nipah virus and Hendra virus are BSL-4 viruses. Therefore, only a few laboratories have the equipment and expertise to manipulate them. Even though other viruses are less dangerous, manipulation of the large volumes needed to develop and run diagnostic tests raises important safety issues.

Laboratory diagnosis of zoonotic viral encephalitis

Laboratory diagnosis of zoonotic encephalitides has improved significantly over the last years and relies on three major types of techniques: tissue culture assays and molecular tests for the direct detection of the pathogen, which can be used for human and arthropod diagnosis, and identification of antibodies as an indicator of recent or past exposure to the virus in humans or in other vertebrates. Isolation of the virus by cell culture is still considered as the gold standard but, as for many other viral infections, there is a trend to replace it with nucleic acid (NA) amplification. However, compared to serology the real utility of direct identification remains to be established because, for many viruses (EEEV, WEEV, SLEV, TBEV, JEV, etc.) the viremic period is short and usually precedes the onset of symptoms. Thus, antibody techniques are central to the diagnosis of these and other viral infections.

Antibody detection

Classical serologic techniques, including hemagglutination-inhibition (HI), complement fixation, plaque-reduction neutralisation test (PRNT) and indirect fluorescent antibody assays (IFA) have been often replaced by enzyme linked immunosorbent assay (ELISA) but IFA and PRNT are still used as reference techniques, even though they are technically demanding.

IgM usually is the first class of antibodies appearing after an infection and is usually no longer detected 6 or 8 weeks after infection. Therefore, detection

Table 2. POC-ELISA: virus panels adapted to different areas of the world (from [26])

Test panel	Viruses included
Western United States	WEEV, SLEV, LACV, CEV, VEEV, WNV
Eastern United States	EEEV, SLEV, LACV, POWV, WNV
Europe	TBEV, WNV
Central America and Caribbean	WEEV, EEEV, VEEV, SLEV, DENV
South America	WEEV, EEEV, VEEV, SLEV, DENV
Africa	DENV, WNV
Australia and Oceania	DENV, MVEV
Asia and Middle East	JEV, TBEV, DENV, WNV

WEEV: western equine encephalitis virus; *SLEV*: St. Louis encephalitis virus; *LACV*: La Crosse encephalitis virus; *CEV*: California encephalitis virus; *VEEV*: Venezuelan equine encephalitis virus; *DENV*: dengue viruses; *WNV*: West Nile virus; *EEEV*: eastern equine encephalitis virus; *POWV*: Powassan virus; *TBEV*: tick-borne encephalitis virus; *MVEV*: Murray Valley encephalitis virus; *JEV*: Japanese encephalitis virus

of IgM antibody can be used as a reliable assay to diagnose early infection. However, because IgM antibody sometimes persists for an extended period, it is not a wholly satisfactory technique. A significant improvement in IgM detection has been made through the introduction of IgM antibody capture assays (MAC-ELISA) which reduce the number of false-positives due to presence of rheumatoid factor and the number of false-negatives due to competition with IgG antibody for antigen. Taking advantage of the ELISA format and of the antigenic similarity between viruses of the same genera, Martin et al. [26] have recently developed 7 geographically-based testing panels that allow the simultaneous screening of all recognized viruses in the corresponding area (Table 2). In this test format, broadly group-reactive monoclonal antibodies (one for alphaviruses, one for California group viruses and one for flaviviruses) have been used in combination with virus-specific antigens, providing sensitivity and specificity similar to those of PRNT.

IgG is produced later during infection (usually after 12 days) and may persist for the life of the individual. Quantitative IgG assays are used in conjunction with IgM assays for diagnosis of an infection. In this configuration, it is necessary to test acute and convalescent-phase serum or cerebrospinal fluid (CSF) specimens and to measure the increase in IgG titres (at least fourfold) to distinguish a recent infection from the immunity provided by a remote infection or by prior vaccination. Several companies have commercialised IgG assays for diagnosis of zoonotic encephalitis viruses. However, the performance of these assays is variable. Niedrig et al. [29] compared six different commercial IgG ELISA kits for TBEV antibodies and found sensitivities between 73 and 99%. In that study, which included specimens from patients infected with dengue and yellow fever viruses, the specificity of the assays also widely varied from 14 to 81%, illustrating the need to carefully select the system to reduce cross-reactions between antibodies to related viruses.

Cross-reactions are especially obvious between flaviviruses (SLEV, WNV, dengue viruses and Murray Valley encephalitis virus) and between alphaviruses (including Venezuelan equine encephalitis virus). Distinction between antigenically related viruses usually is made by detecting neutralising antibodies in CSF or serum. However, this test requires up to a week before a result is obtained. As an alternative to neutralisation tests, Blitvich et al. [3] described a sensitive epitope-blocking ELISA using monoclonal antibodies against WNV and SLEV to detect antibodies to these viruses in bird sera. One of the monoclonal antibodies (3.1112G), is able to perfectly distinguish antibodies against the two viruses: the mean inhibition is 66.5% (38.5%–88.6%) with 13 sera from various bird species infected with WNV and 6.5% (0–17.7%) with 7 sera from birds infected with SLEV. On the contrary, binding of monoclonal antibody 6B5A-2 is preferentially inhibited by sera from birds infected with SLEV. The competition format allows one to detect both IgG and IgM and may be a valuable tool to discriminate between infections by antigenically related viruses. Competition assays also have been commercialised for the detection of antibodies against hepatitis B virus core protein and toxoplasma. The performance of these assays for the diagnosis of encephalitis in clinical settings remains to be established.

Field tests

Several assay formats can be adapted to the difficult conditions of diagnosis in less advanced countries. Immunochromatographic methods are currently commercialised for the direct or indirect detection of infectious agents. Although some of them have poor performance in term of sensitivity, specificity and robustness, efforts have been made by commercial companies and several immunochromatographic tests now have characteristics comparable to those of standard ELISA. For detection of antibody to Human Immunodeficiency virus (HIV) for which very high performances are required, an assay based on this format, Determine HIV-1/2 (Abbott Laboratories) has recently been shown to be highly sensitive and specific [1]. These tests are very simple, easy-to-read and rapid. They can be used on anticoagulated whole blood, serum and plasma samples and have been validated for use with fingertip blood samples. This format has been used recently for the detection of IgM against Puumala virus, the causative agent of Nephropathia epidemic, a mild form of hemorrhagic fever with renal syndrome, using serum, plasma or fingertip blood (Fig. 1A). Its sensitivity and specificity were higher than 97.1 and 96.2%, respectively, as compared with classical ELISA and IFA, on a panel including frozen samples and samples from patients infected with viruses inducing antibodies potentially cross-reacting with Puumala antigens [13].

Solomon et al. [35] have described a capture dot immunoassay (MAC-Dot) for the detection of JEV IgM on the model of the assay used for the diagnosis of infections with dengue viruses [18]. Although dot-blot assays are usually less easy-to-use than are immunochromatography assays, the dengue assay has been validated in multicenter field studies. The JEV MAC-Dot has proved to have high positive (0.98) and negative (0.99) predictive values for the diagnosis of JEV infections on a panel of 60 positive and 119 negative patients.

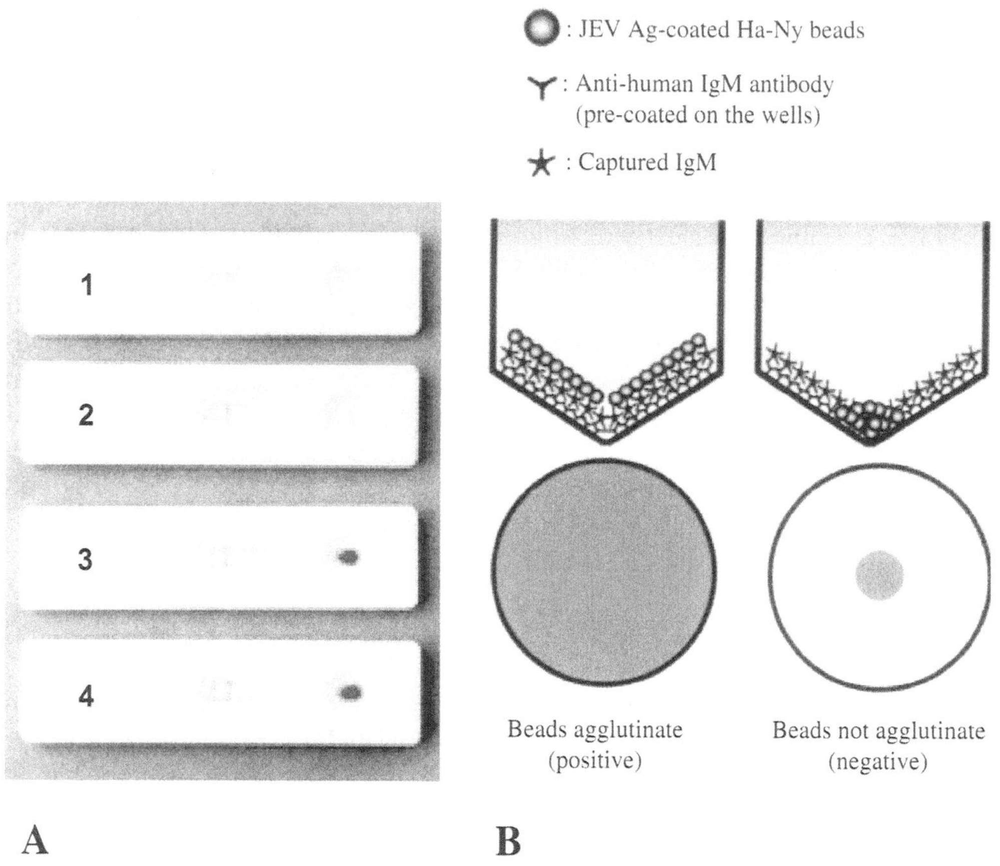

Fig. 1. Two rapid tests for the detection of IgM antibody. **A** POC-PUUMALA rapid test: IgM-positive serum (*1*), IgM-negative serum (*2*), a fingertip blood sample spiked with IgM-positive serum (*3*), and an IgM-negative fingertip blood sample (*4*). **B** Schematic presentation of Particle Agglutinin reactions of Japanese encephalitis virus antigen-coated, hydroxyapatite-coated nylon beads

Particle agglutination assays also may be valuable diagnostic tools in laboratories with insufficient resources. Yamamoto et al. [40] have described hydroxyapatite-coated nylon beads with a very high protein adsorption capability useful in agglutination tests for the detection of JEV IgG in human sera (Fig. 1B). Although tested on a limited panel of confirmed specimen, the results obtained with this assay were consistent with those from HI, plaque-reduction neutralisation or ELISA.

Once such assays are developed by commercial organizations and validated as clinical diagnostics, the cost will continue to be an important constraint for use in developing countries.

Cell culture

Direct virus detection is the method of choice for zoonotic encephalitis viruses in mosquitoes and vertebrate tissues used for virus surveillance in endemic areas.

Direct detection is also useful for diagnosis of human infections by some but not all encephalitis viruses. Viruses can be directly detected by cell culture, passage in suckling mice, antigen ELISA or NA amplification. Vero cells are most often used for this but pig kidney cell lines have been used for isolation of TBEV and mouse neuroblastoma cells have been used to detect rabies virus. For detection following culture, the reference test is still plaque assay or CPE. Immunofluorescence detection assays have been improved by the use of monoclonal antibodies.

Antigen detection

Antigen assays are based on the use of capture and detection antibodies and monoclonal antibodies are the source of major improvements for these techniques. However, the sensitivity of these assays is not sufficient in those infections in which low viremias are frequent. Kramer et al. [15] have compared plaque assay, antigen detection assay and reverse-trancription-polymerase chain reaction (RT-PCR) for the detection of WEEV and SLEV in mosquitoes and in avian tissues. They found that the antigen detection assay is the least sensitive, detecting only 38% of positive mosquito pools. In their study, they found that RT-PCR is the most sensitive assay for detecting the virus in avian tissues.

Nucleic acids amplification

As for other viral infections, for a number of reasons NA testing to detect zoonotic viruses is replacing cell culture methods. One is the fact that NA amplification has the advantage of being less sensitive to the transport conditions of the specimen than are culture assays. Two recent studies [14, 38] have shown that mosquito pools do not require cold chain handling to be efficiently processed by RT-PCR.

However, even though NA amplification is a valuable tool for the detection of these infections in arthropods or animals, it is not yet certain that it can be used to diagnose infections in humans hospitalised with encephalitis. Briese et al. [4] using a blinded panel of samples from patients infected with WNV during the 1999 New York epidemic, showed that viral RNA could be found in CSF from most of the patients. Nevertheless, in that study, 3 of 10 patients with CSF IgM had no WNV RNA. They also found that the detection of WNV RNA in CSF may provide a negative prognostic. On the contrary, TBEV detection by NA amplification may be less useful because the viremia in TBEV infections usually precedes the onset of neurological symptoms (Fig. 2). As stated by Holzmann [12], very few samples taken from TBEV-infected patients before seroconversion or with IgM have been found positive only by RT-PCR. Similarly, it is difficult to isolate JEV from clinical specimens, probably because the levels of circulating viruses are low and neutralizing antibodies appear rapidly and quench the viremia [35]. Alphaviruses may also be difficult to detect by NA amplification, according to Cinque et al. [8].

However, new NA amplification technologies or protocols (see below), with higher analytical sensitivity, may improve the diagnosis utility of viral RNA

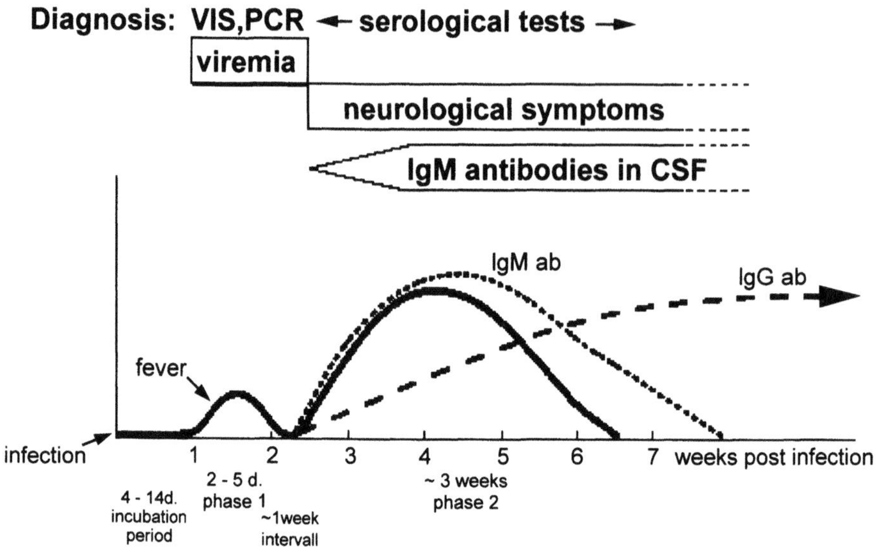

Fig. 2. Clinical and diagnostic course of human tick-borne encephalitis virus infection. *VIS*: virus isolation

Table 3. Clinical sensitivity of RT-PCR assays for TBEV, JEV and EEEV

Virus	Format	Analytical sensitivity	Clinical sensitivity	Authors
TBEV	TaqMan/ABI PRISM 7700	10 copies	14 sera + 21 CSF with patients with suspected TBEV infection: 1 positive CSF. Sensitivity compared to serology not studied	Schwaiger and Cassinotti [34]
JEV	RT-PCR	ND	25 sera + 13 CSF from 16 patients with suspected JEV infection. 14 patients were detected using RT-PCR + hemagglutination-inhibition. One infection was missed by RT-PCR	Sun et al. [36]
	Nested RT-PCR Flaviviruses universal primers	ND	45/52 patients admitted with an established JEV infections were diagnosed correctly	Meiyu et al. [28]
EEEV	Nested RT-PCR	30 copies	8 sera from patients infected with VEE. All are diagnosed correctly	Linssen et al. [23]

TBEV: tick-borne encephalitis virus; *JEV*: Japanese encephalitis virus; *EEEV*: eastern equine encephalitis virus

detection in clinical samples. Table 3 lists studies showing the clinical sensitivity of such assays for patients with encephalitis.

End-point detection techniques

NA amplification is based either on RT-PCR or on transcription methods such as NASBA (Nucleic Acid Sequence-Based Amplification) or TMA (Transcription

Mediated Amplification). For JEV, highly sensitive RT-PCR assays have been used to identify the majority of infected patients. Similarly, all 8 patients with established Venezuelan equine encephalitis virus infections were correctly identified by RT-PCR on sera (Linssen et al.) [23]. However, PCR for TBEV seems to identify only a limited number of infected patients. No commercially available kit exists for the detection of zoonotic encephalitis viruses and, for now, this diagnosis relies on home brew assays only. RT-PCR protocols have been described for flaviviruses [21, 30, 27], for alphaviruses [19, 23], for bunyaviruses [17] including Crimean-Congo haemorrhagic fever virus [5] and for lyssaviruses [11]. NASBA assays have been described for rabies virus [39], flaviviruses [21] and alphaviruses [19].

Real-time amplification

Recently, several real-time assays have been developed for detecting encephalitis viral RNA. These assays offer the advantage of increased sensitivity, higher throughput, shorter time-to-result and better reproducibility and quantification. The TaqMan and LightCycler technologies (Roche Molecular Systems) use RT-PCR as an amplification method. The TaqMan assay is based on virus-specific 5′ nuclease oligoprobes labelled at their 5′ extremity with a fluorescent dye (for example FAM) and at their 3′ extremity with a quencher (TAMRA) (Fig. 3A). The LightCycler assays can use either the SYBR Green dye, 5′ nuclease oligoprobes or fluorescent adjacent oligoprobes (the FRET technology, Fig. 3C) and has the advantage of being performed in closed capillaries that limit cross-contaminations. TaqMan is generally considered better adapted to the needs of laboratory diagnosis than is LightCycler, which is better suited for research laboratories. The NASBA

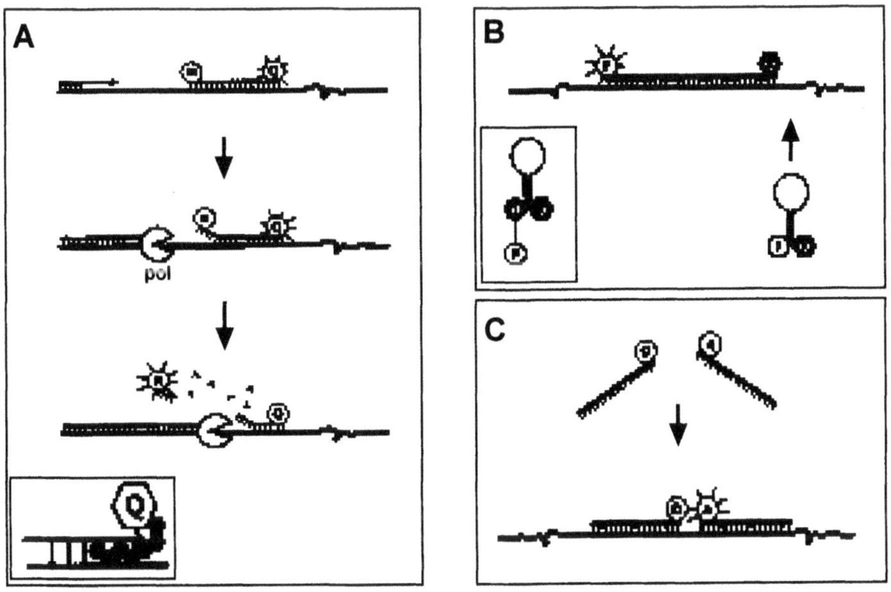

Fig. 3. Probe technologies used with the different real-time platforms. **A** 5′-nuclease oligoprobes. **B** Molecular beacons. **C** FRET probes (from [25])

technology can also be adapted to real-time detection through the use of molecular beacons (Fig. 3B) [21]. TaqMan assays have been described for the detection of flaviviruses [20, 34], alphaviruses [19] and rabies virus [2]. LightCycler assays are less frequently described in the literature but have been developed for Lassa and Crimean-Congo hemorrhagic fever viruses [9]. Real-time NASBA (bioMerieux) which only recently has been introduced, has been described for WNV and SLEV only [21]. Other manufacturers also propose real-time platforms, iCycler from BIO-RAD, SmartCycler from Cepheyd, which have not yet been used for virus NA amplification. Interestingly, some of these have transportable versions that may be used for field studies.

Consensus and multiplex amplification

Because the pattern of zoonotic viral encephalitis is changing and international travelling is increasing sharply, it is very important not to limit the diagnosis to viruses known to circulate in a given area but to have tools that exhaustively detect the important viruses found in various parts of the world. Consensus primers have been optimised to detect such viruses within certain genera. This is the case for bunyaviruses and alphaviruses [16] and flaviviruses [6, 7, 10, 16, 28, 33]. The critical question with regard to these universal assays is the sensitivity, because many primers must be degenerate to be able to detect all viruses of interest. However, using consensus primers for flaviviruses, Scaramozzino et al. [33] could identify WNV directly from CSF of a patient suffering from severe encephalitis. A method to detect any of several viruses in CSF in the same reaction is multiplex amplification, which has been used by different authors to detect herpesviruses, enteroviruses, and JC/BK virus (see for example, Read and Kurtz [31]). Information regarding zoonotic encephalitis viruses is limited [22]. As for consensus primers, the sensitivity of such assays is usually lower than that of assays detecting individual viruses because the unique thermocycling conditions must be optimised to accommodate the detection of all viruses in a single tube. Theoretically, real-time PCR platforms can perform multiplex amplification by the use of either multiple light-emitting diodes or of a light source with a broad range of wavelengths, allowing several oligoprobes labelled with different reporter dyes to be used. However, at this time, one of the best viral, multiplex real-time PCR protocols can detect (and discriminate) four retroviral sequences, whereas conventional end-point PCR usually can detect more than five target viruses [25]. Another approach for multiplex real-time assays using LightCycler and SYBR green takes advantage of differences in melting temperatures of the amplicons generated from different viruses. Read et al. [32] have described a real-time assay based on this principle for herpes simplex virus, varicella zoster virus and enteroviruses in CSF. These technologies should be evaluated for their capacity to detect zoonotic encephalitis viruses.

Microarrays

Any panel of viruses to be detected in order to make a comprehensive diagnosis of a zoonotic encephalitis must contain a large number of viruses. In addition, the genomic regions targeted by the assays can have natural polymorphisms that prevent the use of single probes for their detection following amplification. Medium- or high-density DNA-microarrays, containing hundreds or thousands of different oligonucleotides can be used as a detection technology following amplification with NASBA or (RT)-PCR for the detection of virus NA. Photolithography chemistry, a technology developed by Affymetrix, a California based company, allows robust and reproducible manufacturing of high-density DNA-microarrays that meet the quality criteria for *in vitro* diagnostics [24]. These DNA-microarrays have been used for the identification, strain typing and subtyping and for detection of antibiotic resistant mutants of mycobacteria [37], staphylococci and antiviral resistant mutants of HIV. An assay based on RT-PCR and a DNA-microarray, useful in detecting principal zoonotic encephalitis viruses and other encephalitis viruses, including herpesviruses, enteroviruses, polyomaviruses and paramyxoviruses, is currently under development in our laboratory. It will contain sets of 20-mer probes covering sequence signatures specific for each virus and will be associated with several multiplex amplification protocols corresponding to different geographical areas or different profiles of patients (immunocompetent adults, children, immunosuppressed patients) and will be validated for use as a differential diagnostic tool in large clinical studies.

Conclusion

Currently, laboratory diagnosis of zoonotic viral encephalitis most often relies on serologic assays. These assays will remain a major component of diagnosis because their cost is low and, especially for immunochromatography tests, they can be done in areas where health infrastructures are not well developed. Recent improvements in sensitivity and specificity have made serologic and immunochromatographic assays more efficient for the detection of antibody to a wide range of encephalitis viruses. Nucleic acid detection tests will gradually replace reference cell culture assays. However, there is still need for improvement of all these assays and, more important, a need for extensive clinical validation studies of human or animal specimens. Like other infections, those causing viral encephalitis will benefit from the development of new and promising diagnostic technologies, including field tests, real-time NA amplification and DNA microarrays.

References

1. Arai H, Petchclai B, Khupulsup K, Kurimura T, Takeda K (1999) Evaluation of a rapid immunochromatographic test for the detection of antibodies to human immunodeficiency virus. J Clin Microbiol 37: 367–370
2. Black EM, lowings JP, Smith J, Heaton PR, McElhinney LM (2002) A rapid RT-PCR method to differentiate six established genotypes of rabies and rabies-related viruses using TaqMan technology. J Virol Methods 105: 25–35
3. Blitvich BJ, Marlenee NL, Hall RA, Calisher CH, Bowen RA, Roehrig JT, Komar N, Langevin SA, Beaty BJ (2003) Epitope-blocking enzyme-linked immunosorbent assays for the detection of serum antibodies to West Nile virus in multiple avian species. J Clin Microbiol 41: 1041–1047

4. Briese T, Glass WG, Lipkin WI (2000) Detection of West Nile virus sequences in cerebrospinal fluid. Lancet 355: 1614–1615
5. Burt FJ, Leman PA, Smith JF, Swanepoel R (1998) The use of reverse transcription-polymerase chain reaction for the detection of viral nucleic acid in the diagnosis of Crimean-Congo haemorrhagic fever. J Virol Methods 70: 129–137
6. Chang GJ, Trent DW, Vorndam AV, Vergne E, Kinney RM, Mitchell CJ (1994) An integrated target amplification assay, reverse transcriptase-PCR-enzyme-linked immunosorbent assay, to detect and characterize flaviviruses. J Clin Microbiol 32: 477–483
7. Chow VTK, Seah CLK, Chan YC (1993) Use of NS3 consensus primers for the polymerase chain reaction amplification and sequencing of dengue viruses and other flaviviruses. Arch Virol 133: 157–170
8. Cinque P, Bossolasco S, Lundkvist A (2003) Molecular analysis of cerebrospinal fluid in viral diseases of the central nervous system. J Clin Virol 26: 1–28
9. Drosten C, Kümmerrer BM, Schmitz H, Günther S (2003) Molecular diagnostics of viral hemorrhagic fevers. Antiviral Res 57: 61–87
10. Fulop L, Barrett ADT, Phillpotts R, Martin K, Leslie D, Titball RW (1993) Rapid identification of flaviviruses based on conserved NS5 gene sequences. J Virol Methods 44: 179–188
11. Heaton PR, McElhinney LM, Lowings JP (1999) Detection and identification of rabies and rabies-related viruses using rapid-cycle PCR. J Virol Methods 81: 63–69
12. Holtzmann H (2003) Diagnosis of tick-borne encephalitis. Vaccine 21: SI/36–SI/40
13. Hujakka H, Koistinen V, Eerikaïnen P, Kuronen I, Mononen I, Parviainen M, Lundkvist A, Vaheri A, Nârvänen A, Vapalahti O (2001) New immunochromatographic rapid test for diagnosis of acute Puumala virus infection. J Clin Microbiol 39: 2146–2150
14. Kramer LD, Chiles RE, Do TD, Fallah HM (2001) Detection of St. Louis encephalitis and western equine encephalomyelitis RNA in mosquitoes tested without maintenance of a cold chain. J Am Mosq Control Assoc 17: 213–215
15. Kramer LD, Wolfe TM, Green EN, Chiles RE, Fallah H, Fang Y, Reisen WK (2002) Detection of encephalitis viruses in mosquitoes (Diptera: Culicidae) and avian tissues. J Med Entomol 39: 312–323
16. Kuno G (1998) Universal diagnostic RT-PCR protocol for arboviruses. J Virol Methods 72: 27–41
17. Kuno G, Mitchell CJ, Chang GJJ, Smith GC (1996) Detecting bunyaviruses of the Bunyamwera and California serogroups by a PCR technique. J Clin Microbiol 34: 1184–1188
18. Lam SK, Fong MY, Chungue E, Doraisingham S, Igarashi A, Khin MA, Kyaw ZT, Nisalak A, Roche C, Vaughn DW, Vordnam V (1996) Multicentre evaluation of dengue IgM dot enzyme immunoassay. Clin Diagn Virol 7: 93–98
19. Lambert AJ, Martin D, Lanciotti R (2003) Detection of north American eastern and western equine encephalitis viruses by nucleic acid amplification assays. J Clin Microbiol 41: 379–385
20. Lanciotti RS, Kerst AJ, Nasci RS, Godsey MS, Mitchell CJ, Savage HM, Komar N, Panella NA, Allen BC, Volpe KE, Davis BS, Roehrig JT (2000) Rapid detection of West Nile virus from human clinical specimens, field-collected mosquitoes, and avian samples by a TaqMan reverse transcriptase-PCR assay. J Clin Microbiol 38: 4066–4071
21. Lanciotti RS, Kerst AJ (2001) Nucleic acid sequence-based amplification assays for rapid detection of West Nile and St. Louis encephalitis viruses. J Clin Microbiol 39: 4506–4513

22. Lee JH, Tennessen K, Lilley BG, Unnasch TR (2002) Simultaneous detection of three mosquito-borne encephalitis viruses (eastern equine, La Crosse, and St. Louis) with a single-tube multiplex reverse transcriptase polymerase chain reaction assay. J Am Mosq Control Assoc 18: 26–31
23. Linssen B, Kinney RM, Aguilar P, Russell KL, Watts DM, Kaaden OR, Pfeffer M (2000) Development of reverse transcription-PCR assays specific for detection of equine encephalitis viruses. J Clin Microbiol 38: 1527–1535
24. Lipshutz RJ, Morris D, Chee M, Hubbell E, Kozal MJ, Shah N, Shen N, Yang R, Fodor SPA (1995) Using oligonucleotide probe arrays to access genetic diversity. BioTechniques 19: 442–447
25. Mackay IM, Arden KE, Nitsche A (2002) Real-time PCR in virology. Nucleic Acids Res 30: 1292–1305
26. Martin DA, Muth DA, Brown T, Johnson AJ, Karabatsos N, Roehrig JT (2000) Standardization of immunoglobulin M capture enzyme-linked immunosorbent assays for routine diagnosis of arboviral infections. J Clin Microbiol 38: 1823–1826
27. McMinn PC, Carman PG, Smith DW (2000) Early diagnosis of Murray Valley encephalitis by reverse transcriptase-polmerase chain reaction. Pathology 32: 49–51
28. Meiyu F, Huosheng C, Cuiha C, Xiaodong T, Lianhua J, Yifei P, Weijun C, Huiyu G (1997) Detection of flaviviruses by reverse transcriptase-polymerase chain reaction with the universal primer set. Microbiol Immunol 41: 209–213
29. Niedrig M, Vaisviliene D, Teichmann A, Klockmann U, Niel SS (2001) Comparison of six different commercial IgG-ELISA kits for the detection of TBEV-antibodies. J Clin Virol 20: 179–182
30. Paranjpe S, Banerjee K (1998) Detection of Japanese encephalitis virus by reverse transcription/polymerase chain reaction. Acta Virol 42: 5–11
31. Read SJ, Kurtz JB (1999) Laboratory diagnosis of common viral infections of the central nervous system by using a single multiplex PCR screening assay. J Clin Microbiol 37: 1352–1355
32. Read SJ, Mitchell JL, Fink CG (2001) LightCycler multiplex PCR for the laboratory diagnosis of common viral infections of the central nervous system. J Clin Microbiol 39: 3056–3059
33. Scaramozzino N, Crance JM, Jouan A, DeBriel DA, Stoll F, Garin D (2001) Comparison of Flavivirus universal primer pairs and development of a rapid, highly sensitive heminested reverse transcription-PCR assay for detection of flaviviruses targeted to a conserved region of the NS5 gene sequence. J Clin Microbiol 39: 1922–1927
34. Schwaiger M, Cassinotti P (2003) Development of a quantitative real-time RT-PCR assay with internal control for the laboratory detection of tick borne encephalitis virus (TBEV) RNA. J Clin Virol 27: 136–145
35. Solomon T, Thao LTT, Dung NM, Kneen R, Hung NT, Nisalak A, Vaughn DW, Farrar J, Hien TT, White NJ, Cardosa MJ (1998) Rapid diagnosis of Japanese encephalitis by using immunoglobulin M dot enzyme immunoassay. J Clin Microbiol 36: 2030–2034
36. Sun J, Tso SJ, Chen BQ, Liu QZ (2000) Detection of Japanese encephalitis virus in samples of JE patients by RT-PCR. Zhonghua Shi Yan He Lin Chuang Bing Du Xue Za Zhi 14: 184–187
37. Troesch A, Nguyen H, Miyada CG, Desvarennes S, Gingeras TR, Kaplan TM, Cros P, Mabilat C (1999) *Mycobacterium* species identification and rifampin resistance testing with high density DNA probe arrays. J Clin Microbiol 37: 49–55
38. Turell MJ, Spring AR, Miller MK, Cannon CE (2001) Effect of conditions on the detection of West Nile viral RNA by reverse transcriptase-polymerase chain reaction from mosquito (Diptera: Culicidae) pools. J Med Entomol 39: 1–3

39. Wacharapluesadee S, Hemachudha T (2001) Nucleic-acid sequence based amplification in the rapid diagnosis of rabies. Lancet 358: 892–893
40. Yamamoto A, Nakayama M, Tashiro M, Ogawa T, Kurane I (2000) Hydroxyapatite-coated nylon beads as a new reagent to develop a particle agglutination assay system for detecting Japanese encephalitis virus-specific human antibodies. J Clin Virol 19: 195–204

Author's address: Guy Vernet, Molecular Diagnostics, Advanced Technology Unit, BioMérieux, 69280 Marcy-l'Etoile, France; e-mail: guy.vernet@eu.biomerieux.com

Springer-Verlag
and the Environment

WE AT SPRINGER-VERLAG FIRMLY BELIEVE THAT AN international science publisher has a special obligation to the environment, and our corporate policies consistently reflect this conviction.

WE ALSO EXPECT OUR BUSINESS PARTNERS – PRINTERS, paper mills, packaging manufacturers, etc. – to commit themselves to using environmentally friendly materials and production processes.

THE PAPER IN THIS BOOK IS MADE FROM NO-CHLORINE pulp and is acid free, in conformance with international standards for paper permanency.

GPSR Compliance

The European Union's (EU) General Product Safety Regulation (GPSR) is a set of rules that requires consumer products to be safe and our obligations to ensure this.

If you have any concerns about our products, you can contact us on ProductSafety@springernature.com

In case Publisher is established outside the EU, the EU authorized representative is:

Springer Nature Customer Service Center GmbH
Europaplatz 3
69115 Heidelberg, Germany

Batch number: 10732720

Printed by Printforce, the Netherlands